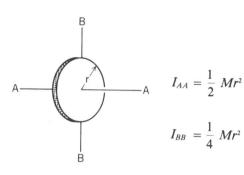

$$I_{AA} = \frac{1}{2}\, Mr^2$$

$$I_{BB} = \frac{1}{4}\, Mr^2$$

Thin disc

$$I_{AA} = Mr^2$$

$$I_{BB} = \frac{M}{2}\left(r^2 + \frac{h^2}{6}\right)$$

Thin-walled cylinder

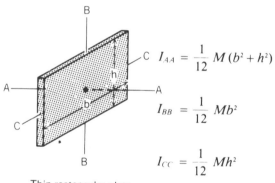

$$I_{AA} = \frac{1}{12}\, M(b^2 + h^2)$$

$$I_{BB} = \frac{1}{12}\, Mb^2$$

$$I_{CC} = \frac{1}{12}\, Mh^2$$

Thin rectangular plate

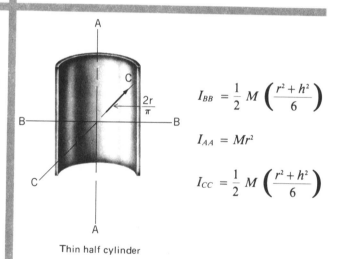

$$I_{BB} = \frac{1}{2}\, M\left(\frac{r^2 + h^2}{6}\right)$$

$$I_{AA} = Mr^2$$

$$I_{CC} = \frac{1}{2}\, M\left(\frac{r^2 + h^2}{6}\right)$$

Thin half cylinder

$$V = \frac{2}{3}\, \pi r^3$$

$$I_{AA} = \frac{2}{5}\, Mr^2$$

$$I_{BB} = \frac{2}{5}\, Mr^2$$

Hemisphere

D1303095

# SELECTED DIMENSIONAL EQUIVALENTS

| | |
|---|---|
| LENGTH | $1 \text{ m} \equiv 3.281 \text{ ft} \equiv 39.37 \text{ in.}$ <br> $1 \text{ mi} \equiv 5280 \text{ ft} \equiv 1.609 \text{ km}$ <br> $1 \text{ km} \equiv .6214 \text{ mi}$ |
| TIME | $1 \text{ hr} \equiv 60 \text{ min} \equiv 3600 \text{ sec}$ |
| MASS | $1 \text{ kg} \equiv 2.2046 \text{ lbm} \equiv .068521 \text{ slug}$ |
| FORCE | $1 \text{ N} \equiv .2248 \text{ lbf}$ <br> $1 \text{ dyne} \equiv 10 \ \mu\text{N}$ |
| SPEED | $1 \text{ mi/hr} \equiv 1.609 \text{ km/hr} \equiv 1.467 \text{ ft/sec}$ <br> $1 \text{ km/hr} = .6214 \text{ mi/hr}$ <br> $1 \text{ knot} = 1.152 \text{ mi/hr} \equiv 1.853 \text{ km/hr}$ <br> $\equiv 1.689 \text{ ft/sec}$ |
| ENERGY | $1 \text{ J} \equiv 1 \text{ N-m}$ <br> $1 \text{ Btu} \equiv 778.16 \text{ ft-lbf} \equiv 1.055 \text{ kJ}$ <br> $1 \text{ watt-hour} \equiv 2.778 \times 10^{-4} \text{ J}$ |
| VOLUME | $1 \text{ gal} \equiv .16054 \text{ ft}^3 \equiv .0045461 \text{ m}^3$ <br> $1 \text{ liter} \equiv .03531 \text{ ft}^3 = .2642 \text{ gal}$ |
| POWER | $1 \text{ w} \equiv 1 \text{ J/S}$ <br> $1 \text{ hp} \equiv 550 \text{ ft-lb/sec} \equiv .7068 \text{ Btu/sec}$ <br> $\equiv 746 \text{ w}$ |

# Preface

With the publication of the fourth edition, this text moves into the fourth decade of its existence. In the spirit of the times, the first edition introduced a number of "firsts" in an introductory engineering mechanics textbook. These "firsts" included

**a)** the first treatment of space mechanics
**b)** the first use of the control volume for linear momentum considerations of fluids
**c)** the first introduction to the concept of the tensor

Users of the earlier editions will be glad to know that the 4th edition continues with the same approach to engineering mechanics. The goal has always been to aim toward working problems as soon as possible from first principles. Thus, examples are carefully chosen during the development of a series of related areas to instill continuity in the evolving theory and then, after these areas have been carefully discussed with rigor, come the problems. Furthermore at the ends of each chapter, there are many problems that have not been arranged by text section. The instructor is encouraged as soon as he/she is well along in the chapter to use these problems. The instructors manual will indicate the nature of each of these problems as well as the degree of difficulty. The text is not chopped up into many methodologies each with an abbreviated discussion followed by many examples for using the specific methodology and finally a set of problems carefully tailored for the methodology. The nature of the format in this and preceding editions is more than ever first to discourage excessive mapping of homework problems from the examples. And second, it is to lessen the memorization of specific, specialized methodologies in lieu of absorbing basic principles.

A new feature in the fourth edition is a series of starred sections called "Looking Ahead . . . ." These are simplified discussions of topics that appear in later engineering courses and tie in directly or indirectly to the topic under study. For instance, after discussing free body diagrams, there is a short "Looking Ahead" section in which the concept and use of the control volume is presented as well as the system concepts that appear in fluid mechanics and thermodynamics. In the chapter on virtual work for particles and rigid bodies, there is a simplified discussion of the displacement methods and force methods for deformable bodies that will show up later in solids courses. After finding the forces for simple trusses, there is a "Looking Ahead" section discussing briefly what has to be done to get displacements. There are quite a few others in the text. It has been found that many students find these interesting and later when they come across these topics in other courses or work, they report that the connections so formed coming out of their sophomore mechanics courses have been most valuable.

Over 400 new problems have been added to the fourth edition equally divided between the statics and dynamics books. A complete word-processed solutions manual accompanies the text. The illustrations needed for problem statement and solution are taken as enlargements from the text. Generally, each problem is on a separate page. The instructor will be able conveniently to select problems in order to post solutions or to form transparencies as desired. Also, there are 30 computer projects in which, for a number of cases, the student prepares his/her own software or engages in design. As an added bonus, the student will be able to maintain his/her proficiency in programming. Carefully prepared computer programs as well as computer outputs will be included in the manual. I normally assign one or two such projects during a semester over and above the usual course material. Also included in the manual is a disk that has the aforementioned programs for each of the computer projects. The computer programs for these projects generally run about 30 lines of FORTRAN and run on a personal computer. The programming required involves skills developed in the freshman course on FORTRAN.

Another important new feature of the fourth edition is an organization that allows one to go directly to the three dimensional chapter on dynamics of rigid bodies (Chapter 13) and then to easily return to plane motion (Chapter 6). Or one can go the opposite way. Footnotes indicate how this can be done, and complimentary problems are noted in the Solutions Manual.

FOURTH
EDITION

# Engineering Mechanics

Statics

## Irving H. Shames

*Professor*
*Dept. of Civil, Mechanical and Environmental Engineering*
*The George Washington University*

Prentice Hall, Upper Saddle River, New Jersey 07458

Mohawk Valley Community College Library

Shames, Irving Herman.
    Engineering mechanics. Statics / Irving H. Shames.— 4th ed.
        p.   cm.
    Includes bibliographical references and index.
    ISBN0–13–356908–X
    1.  Statics   2.  Mechanics, Applied.   I.  Title
TA351.S48   1996
620.1′03—dc20                                96–629
                                             CIP

Acquisitions Editor: William Stenquist
Editor in Chief: Marcia Horton
Production Editor: Rose Kernan
Text Designer: Christine Wolf
Cover Designer: Amy Rosen
Editorial Assistant: Meg Weist
Manufacturing Buyer: Donna Sullivan

© 1997, 1980, 1966, 1959, 1958 by Prentice-Hall, Inc.
Simon & Schuster/A Viacom Company
Upper Saddle River, New Jersey 07458

The author and publisher of this book have used their best efforts in preparing this book.
These efforts include the development, research, and testing of the theories and programs
to determine their effectiveness. The author and publisher shall not be liable in any event for
incidental or consequential damages with, or arising out of, the furnishing, performance, or
use of these programs.

All rights reserved. No part of this book may be reproduced or
transmitted in any form or by any means, without written permission
in writing from the publisher.

Printed in the United States of America

10  9  8  7  6  5  4  3  2


ISBN 0-13-356908-X


Prentice-Hall International (UK) Limited, London
Prentice-Hall of Australia Pty. Limited, Sydney
Prentice-Hall Canada Inc., Toronto
Prentice-Hall Hispanoamericana, S.A., Mexico
Prentice-Hall of India Private Limited, New Delhi
Prentice-Hall of Japan, Inc., Tokyo
Simon & Schuster Asia Pte. Ltd., Singapore
Editora Prentice Hall do Brasil, Ltda., Rio de Janeiro

TA
351
.S48
1997

*To my Dear, Wonderful Wife Sheila*

# Contents

Preface    ix

## 1 Fundamentals of Mechanics Review I    3

†**1.1**  Introduction   3
†**1.2**  Basic Dimensions and Units of Mechanics,   4
†**1.3**  Secondary Dimensional Quantities   7
†**1.4**  Law of Dimensional Homogeneity   8
†**1.5**  Dimensional Relation between Force and Mass   9
**1.6**  Units of Mass   10
**1.7**  Idealizations of Mechanics   12
†**1.8**  Vector and Scalar Quantities   14
**1.9**  Equality and Equivalence of Vectors   17
†**1.10** Laws of Mechanics   19
**1.11** Closure   22

## 2 Elements of Vecor Algebra Review II    23

†**2.1**  Introduction   23
†**2.2**  Magnitude and Multiplication of a Vector by a Scalar   23
†**2.3**  Addition and Subtraction of Vectors   24

**2.4**  Resolution of Vectors; Scalar Components   30
†**2.5**  Unit Vectors   33
**2.6**  Useful Ways of Representing Vectors   35
**2.7**  Scalar or Dot Product of Two Vectors   41
**2.8**  Cross Product of Two Vectors   47
**2.9**  Scalar Triple Product   51
**2.10** A Note on Vector Notation   54
**2.11** Closure   56

## 3 Important Vector Quantities    61

**3.1**  Position Vector   61
**3.2**  Moment of a Force about a Point   62
**3.3**  Moment of a Force about an Axis   69
**3.4**  The Couple and Couple Moment   77
**3.5**  The Couple Moment as a Free Vector   79
**3.6**  Addition and Subtraction of Couples   80
**3.7**  Moment of a Couple About a Line   82
**3.8**  Closure   89

# 4 Equivalent Force Systems 93

**4.1** Introduction 93
**4.2** Translation of a Force to a Parallel Position 94
**4.3** Resultant of a Force System 102
**4.4** Simplest Resultants of Special Force Systems 106
**4.5** Distributed Force Systems 107
**4.6** Closure 143

# 5 Equations of Equilibrium 151

**5.1** Introduction 151
**5.2** The Free-body Diagram 152
**5.3** Free Bodies Involving Interior Sections 154
***5.4** Looking Ahead— Control Volumes 158
**5.5** General Equations of Equilibrium 162
**5.6** Problems of Equilibrium I 164
**5.7** Problems of Equilibrium II 183
**5.8** Two Point Equivalent Loading 199
**5.9** Problems Arising from Structures 200
**5.10** Static Indeterminacy 204
**5.11** Closure 210

# 6 Introduction to Structural Mechanics 221

## Part A: Trusses 221

**6.1** The Structural Model 221
**6.2** The Simple Truss 224
**6.3** Solution of Simple Trusses 225
**6.4** Method of Joints 225
**6.5** Method of Sections 238
***6.6** Looking Ahead—Deflection of a Simple, Linearly Elastic Truss 242

## Part B: Section Forces in Beams 247

**6.7** Introduction 247
**6.8** Shear Force, Axial Force, and Bending Moment 247
**6.9** Differential Relations for Equilibrium 259

## Part C: Chains and Cables 266

**6.10** Introduction 266
**6.11** Coplanar Cables; Loading is a Function of $x$ 266
**6.12** Coplanar Cables: Loading is the Weight of the Cable Itself 270
**6.13** Closure 277

# 7 Friction Forces 281

**7.1** Introduction 281
**7.2** Laws of Coulomb Friction 282
***7.3** A Comment Concerning the Use of Coulomb's Law 284
**7.4** Simple Contact Friction Problems 284
**7.5** Complex Surface Contact Friction Problems 299
**7.6** Belt Friction 301
**7.7** The Square Screw Thread 317
***7.8** Rolling Resistance 319
**7.9** Closure 323

# 8 Properties of Surfaces 331

**8.1** Introduction 331
**8.2** First Moment of an Area and the Centroid 331
**8.3** Other Centers 342
**8.4** Theorems of Pappus-Guldinus 347
**8.5** Second Moments and the Product of Area of a Plane Area 355
**8.6** Tranfer Theorems 356

**8.7**   Computations Involving Second
          Moments and Products of Area   357
**8.8**   Relation Between Second Moments
          and Products of Area   366
**8.9**   Polar Moment of Area   369
**8.10**  Principal Axes   370
**8.11**  Closure   375

# 9 Moments and Products of Inertia   379

**9.1**   Introduction   379
**9.2**   Formal Definition of
          Inertia Quantities   379
**9.3**   Relation Between Mass-Inertia Terms
          and Area-Inertia Terms   386
**9.4**   Translation of Coordinate Axes   392
***9.5**  Transformation Properties
          of the Inertia Terms   395
***9.6**  Looking Ahead—Tensors   400
***9.7**  The Inertia Ellipsoid and Principal
          Moments of Inertia   407
**9.8**   Closure   410

# 10 *Methods of Virtual Work and Stationary Potential Energy   413

**10.1**  Introduction   413

**Part A: Method of Virtual Work   414**

**10.2**  Principle of Virtual Work
          for a Particle   414
**10.3**  Principle of Virtual Work
          for Rigid Bodies   415
**10.4**  Degrees of Freedom and the
          Solution of Problems   418
***10.5** Looking Ahead—
          Deformable Solids   424

**Part B: Method of Total Potential
Energy   432**

**10.6**  Conservative Systems   432
**10.7**  Condition of Equilibrium
          for a Conservative System   434
**10.8**  Stability   441
***10.9** Looking Ahead—More on Total
          Potential Energy   443
**10.10** Closure   446

# APPENDIX I
## Integration Formulas   xvii

# APPENDIX II
## Computation of Principal Moments of Inertia   xix

# APPENDIX III
## Computer Projects: Statics   xxi

## Projects   xxiii

## Answers   xxxviii

## Index   xlv

Another change is Chapter 16 on plane motion. It has been reworked with the aim of attaining greater rigor and clarity particularly in the solving of problems.

There has also been an increase in the coverage and problems for hydrostatics as well as examples and problems that will preview problems coming in the solids course that utilize principles from statics.

It should also be noted that the notation used has been chosen to correspond to that which will be used in more advanced courses in order to improve continuity with upper division courses. Thus, for moments and products of inertia I use $I_{xx}$, $I_{yy}$, $I_{xz}$ *etc. rather than* $I_x$, $I_y$, $P_{xy}$ etc. The same notation is used for second moments and products of area to emphasize the direct relation between these and the preceding quantities. Experience indicates that there need be no difficulty on the student's part in distinguishing between these quantities; the context of the discussion suffices for this purpose. The concept of the tensor is presented in a way that for years we have found to be readily understood by sophomores even when presented in large classes. This saves time and makes for continuity in all mechanics courses, particularly in the solid mechanics course. For bending moment, shear force, and stress use is made of a common convention for the sign—namely the convention involving the normal to the area element and the direction of the quantity involved be it bending moment, shear force or stress component. All this and indeed other steps taken in the book will make for smooth transition to upper division course work.

In overall summary, two main goals have been pursued in this edition. They are

1. To encourage working problems from first principles and thus to mini-mize excessive mapping from examples and to discourage rote learning of specific methodologies for solving various and sundry kinds of specific problems.
2. To "open-end" the material to later course work in other engineering sciences with the view toward making smoother transitions and to provide for greater continuity. Also, the purpose is to engage the interest and curiosity of students for further study of mechanics.

During the 13 years after the third edition, I have been teaching sophomore mechanics to large classes at SUNY, Buffalo, and, after that, to regular sections of students at The George Washington University, the latter involving an international student body with very diverse backgrounds. During this time, I have been working on improving the clarity and strength of this book under classroom conditions giving it the most severe test as a text. I believe the fourth edition as a result will be a distinct improvement over the previous editions and will offer a real choice for schools desiring a more mature treatment of engineering mechanics.

I believe sophomore mechanics is probably the most important course taken by engineers in that much of the later curriculum depends heavily on this course. And for **all** engineering programs, this is usually the first real engineering course where students can and must be creative and inventive in solving problems. Their old habits of mapping and rote learning of specific problem methodologies will not suffice and they must learn to see mechanics as an integral science. The student must "bite the bullet" and work in the way he/she will have to work later in the curriculum and even later when getting out of school altogether. No other subject so richly involves mathematics, physics, computers, and down to earth common sense simultaneously in such an interesting and challenging way. We should take maximum advantage of the students exposure to this beautiful subject to get him/her on the right track now so as to be ready for upper division work.

At this stage of my career, I will risk impropriety by presenting now an extended section of acknowledgments. I want to give thanks to SUNY at Buffalo where I spent 31 happy years and where I wrote many of my books. And I want to salute the thousands (about 5000) of fine students who took my courses during this long stretch. I wish to thank my eminent friend and colleague Professor Shahid Ahmad who among other things taught the sophomore mechanics sequence with me and who continues to teach it. He gave me a very thorough review of the fourth edition with many valuable suggestions. I thank him profusely. I want particularly to thank Professor Michael Symans, from Washington State University, Pullman for his superb contributions to the entire manuscript. I came to The George Washington University at the invitation of my longtime friend and former Buffalo colleague Dean Gideon Frieder and the faculty in the Civil, Mechanical and Environmental Engineering Department. Here, I came back into contact with two well-known scholars that I knew from the early days of my career, namely Professor Hal Liebowitz (president-elect of the National Academy of Engineering) and Professor Ali Cambel (author of recent well-received book on chaos). I must give profound thanks to the chairman of my new department at G.W., Professor Sharam Sarkani. He has allowed me to play a vital role in the academic program of the department. I will be able to continue my writing at full speed as a result. I shall always be grateful to him. Let me not forget the two dear ladies in the front office of the department. Mrs. Zephra Coles in her decisive efficient way took care of all my needs even before I was aware of them. And Ms. Joyce Jeffress was no less helpful and always had a humorous comment to make.

I was extremely fortunate in having the following professors as reviewers.

Professor Shahid Ahmad, SUNY at Buffalo
Professor Ravinder Chona, Texas A&M University
Professor Bruce H. Karnopp. University of Michigan
Professor Richard F. Keltie, North Carolina State University
Professor Stephen Malkin, University of Massachusetts
Professor Sudhakar Nair, Illinois Institute of Technology
Professor Jonathan Wickert, Carnegie Mellon University

I wish to thank these gentlemen for their valuable assistance and encouragement.

I have two people left. One is my good friend Professor Bob Jones from V.P.I. who assisted me in the third edition with several hundred excellent statics problems and who went over the entire manuscript with me with able assistance and advice. I continue to benefit in the new edition from his input of the third edition. And now, finally, the most important person of all, my dear wife Sheila. She has put up all these years with the author of this book, an absent-minded, hopeless workaholic. Whatever I have accomplished of any value in a long and ongoing career, I owe to her.

# About the Author

Irving Shames presently serves as a Professor in the Department of Civil, Mechanical, and Environmental Engineering at The George Washington University. Prior to this appointment Professor Shames was a Distinguished Teaching Professor and Faculty Professor at The State University of New York—Buffalo, where he spent 31 years.

Professor Shames has written up to this point in time 10 textbooks. His first book Engineering Mechanics, Statics and Dynamics was originally published in 1958, and it is going into its fourth edition in 1996. All of the books written by Professor Shames have been characterized by innovations that have become mainstays of how engineering principles are taught to students. Engineering Mechanics, Statics and Dynamics was the first widely used Mechanics book based on vector principles. It ushered in the almost universal use of vector principles in teaching engineering mechanics courses today.

Other textbooks written by Professor Shames include:

- *Mechanics of Deformable Solids*, Prentice-Hall, Inc.
- *Mechanics of Fluids*, McGraw-Hill
- *Introduction to Solid Mechanics*, Prentice-Hall, Inc.
- *Introduction to Statics*, Prentice-Hall, Inc.
- *Solid Mechanics—A Variational Approach* (with C.L. Dym), McGraw-Hill
- *Energy and Finite Element Methods in Structural Mechanics*, (with C.L. Dym), Hemisphere Corp., of Taylor and Francis
- *Elastic and Inelastic Stress Analysis* (with F. Cozzarelli), Prentice-Hall, Inc.

In recent years, Professor Shames has expanded his teaching activities and has held two summer faculty workshops in mechanics sponsored by the State of New York, and one national workshop sponsored by the National Science Foundation. The programs involved the integration both conceptually and pedagogically of mechanics from the sophomore year on through graduate school.

# Statics

REVIEW I*

# Fundamentals of Mechanics

## †1.1 Introduction

*Mechanics* is the physical science concerned with the dynamical behavior (as opposed to chemical and thermal behavior) of bodies that are acted on by mechanical disturbances. Since such behavior is involved in virtually all the situations that confront an engineer, mechanics lies at the core of much engineering analysis. In fact, no physical science plays a greater role in engineering than does mechanics, and it is the oldest of all the physical sciences. The writings of Archimedes covering buoyancy and the lever were recorded before 200 B.C. Our modern knowledge of gravity and motion was established by Isaac Newton (1642–1727), whose laws founded Newtonian mechanics, the subject matter of this text.

In 1905, Einstein placed limitations on Newton's formulations with his theory of relativity and thus set the stage for the development of relativistic mechanics. The newer theories, however, give results that depart from those of Newton's formulations only when the speed of a body approaches the speed of light (186,000 miles/sec). These speeds are encountered in the large-scale phenomena of dynamical astronomy. Furthermore for small-scale phenomena involving subatomic particles, quantum mechanics must be used rather than Newtonian mechanics. Despite these limitations, it remains nevertheless true that, in the great bulk of engineering problems, Newtonian mechanics still applies.

---

*The reader is urged to be sure that Section 1.9 is thoroughly understood since this section is vital for a good understanding of statics in particular and mechanics in general.

Also, the notation † before the titles of certain sections indicates that specific questions concerning the contents of these sections requiring verbal answers are presented at the end of the chapter. The instructor may wish to assign these sections as a reading assignment along with the requirement to answer the aforestated associated questions as the author routinely does himself.

## †1.2   Basic Dimensions and Units of Mechanics

To study mechanics, we must establish abstractions to describe those characteristics of a body that interest us. These abstractions are called *dimensions*. The dimensions that we pick, which are independent of all other dimensions, are termed *primary* or *basic dimensions*, and the ones that are then developed in terms of the basic dimensions we call *secondary dimensions*. Of the many possible sets of basic dimensions that we could use, we will confine ourselves at present to the set that includes the dimensions of length, time, and mass. Another convenient set will be examined later.

**Length—A Concept for Describing Size Quantitatively.**   In order to determine the size of an object, we must place a second object of known size next to it. Thus, in pictures of machinery, a man often appears standing disinterestedly beside the apparatus. Without him, it would be difficult to gage the size of the unfamiliar machine. Although the man has served as some sort of standard measure, we can, of course, only get an approximate idea of the machine's size. Men's heights vary, and, what is even worse, the shape of a man is too complicated to be of much help in acquiring a precise measurement of the machine's size. What we need, obviously, is an object that is constant in shape and, moreover, simple in concept. Thus, instead of a three-dimensional object, we choose a one-dimensional object.[1] Then, we can use the known mathematical concepts of geometry to extend the measure of size in one dimension to the three dimensions necessary to characterize a general body. A straight line scratched on a metal bar that is kept at uniform thermal and physical conditions (as, e.g., the meter bar kept at Sèvres, France) serves as this simple invariant standard in one dimension. We can now readily calculate and communicate the distance along a certain direction of an object by counting the number of standards and fractions thereof that can be marked off along this direction. We commonly refer to this distance as length, although the term "length" could also apply to the more general concept of size. Other aspects of size, such as volume and area, can then be formulated in terms of the standard by the methods of plane, spherical, and solid geometry.

A *unit* is the name we give an accepted measure of a dimension. Many systems of units are actually employed around the world, but we shall only use the two major systems, the American system and the SI system. The basic unit of length in the American system is the foot, whereas the basic unit of length in the SI system is the meter.

**Time—A Concept for Ordering the Flow of Events.**   In observing the picture of the machine with the man standing close by, we can sometimes tell approximately when the picture was taken by the style of clothes the man is

---

[1] We are using the word "dimensional" here in its everyday sense and not as defined above.

wearing. But how do we determine this? We may say to ourselves: "During the thirties, people wore the type of straw hat that the fellow in the picture is wearing." In other words, the "when" is tied to certain events that are experienced by, or otherwise known to, the observer. For a more accurate description of "when," we must find an action that appears to be completely repeatable. Then, we can order the events under study by counting the number of these repeatable actions and fractions thereof that occur while the events transpire. The rotation of the earth gives rise to an event that serves as a good measure of time—the day. But we need smaller units in most of our work in engineering, and thus, generally, we tie events to the second, which is an interval repeatable 86,400 times a day.

**Mass—A Property of Matter.** The student ordinarily has no trouble understanding the concepts of length and time because he/she is constantly aware of the size of things through his/her senses of sight and touch, and is always conscious of time by observing the flow of events in his/her daily life. The concept of mass, however, is not as easily grasped since it does not impinge as directly on our daily experience.

Mass is a property of matter that can be determined from *two* different actions on bodies. To study the first action, suppose that we consider two hard bodies of entirely different composition, size, shape, color, and so on. If we attach the bodies to identical springs, as shown in Fig. 1.1, each spring will extend some distance as a result of the attraction of gravity for the bodies. By grinding off some of the material on the body that causes the greater extension, we can make the deflections that are induced on both springs equal. Even if we raise the springs to a new height above the earth's surface, thus lessening the deformation of the springs, the extensions induced by the pull of gravity will be the same for both bodies. And since they are, we can conclude that the bodies have an equivalent innate property. This property of each body that manifests itself in the *amount of gravitational attraction* we call *mass*.

The equivalence of these bodies, after the aforementioned grinding operation, can be indicated in yet a second action. If we move both bodies an equal distance downward, by stretching each spring, and then release them at the same time, they will begin to move in an identical manner (except for small variations due to differences in wind friction and local deformations of the bodies). We have imposed, in effect, the same mechanical disturbance on each body and we have elicited the same dynamical response. Hence, despite many obvious differences, the two bodies again show an equivalence.

*The property of mass, then, characterizes a body both in the action of gravitational attraction and in the response to a mechanical disturbance.*

To communicate this property quantitatively, we may choose some convenient body and compare other bodies to it in either of the two above-

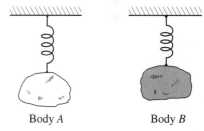

Body *A*                    Body *B*

**Figure 1.1.** Bodies restrained by identical springs.

mentioned actions. The two basic units commonly used in much American engineering practice to measure mass are the *pound mass*, which is defined in terms of the attraction of gravity for a standard body at a standard location, and the *slug*, which is defined in terms of the dynamical response of a standard body to a standard mechanical disturbance. A similar duality of mass units does not exist in the SI system. There only the *kilogram* is used as the basic measure of mass. The kilogram is measured in terms of response of a body to a mechanical disturbance. Both systems of units will be discussed further in a subsequent section.

We have now established three basic independent dimensions to describe certain physical phenomena. It is convenient to identify these dimensions in the following manner:

$$\text{length} \quad [L]$$
$$\text{time} \quad [t]$$
$$\text{mass} \quad [M]$$

These formal expressions of identification for basic dimensions and the more complicated groupings to be presented in Section 1.3 for secondary dimensions are called "dimensional representations."

Often, there are occasions when we want to change units during computations. For instance, we may wish to change feet into inches or millimeters. In such a case, we must replace the unit in question by a *physically equivalent* number of new units. Thus, a foot is replaced by 12 inches or 305 millimeters. A listing of common systems of units is given in Table 1.1, and a table of equivalences between these and other units is given on the inside covers. Such relations between units will be expressed in this way:

$$1 \text{ ft} \equiv 12 \text{ in.} \equiv 305 \text{ mm}$$

The three horizontal bars are not used to denote *algebraic* equivalence; instead, they are used to indicate physical equivalence. Here is another way of expressing the relations above:

**Table 1.1   Common Systems of Units**

| *cgs* | | *SI* | |
|---|---|---|---|
| Mass | Gram | Mass | Kilogram |
| Length | Centimeter | Length | Meter |
| Time | Second | Time | Second |
| Force | Dyne | Force | Newton |
| *English* | | *American Practice* | |
| Mass | Pound mass | Mass | Slug or pound mass |
| Length | Foot | Length | Foot |
| Time | Second | Time | Second |
| Force | Poundal | Force | Pound force |

$$\left(\frac{1 \text{ ft}}{12 \text{ in.}}\right) \equiv 1, \quad \left(\frac{1 \text{ ft}}{305 \text{ mm}}\right) \equiv 1$$

$$\left(\frac{12 \text{ in.}}{1 \text{ ft}}\right) \equiv 1, \quad \left(\frac{305 \text{ mm}}{1 \text{ ft}}\right) \equiv 1 \tag{1.1}$$

The unity on the right side of these relations indicates that the numerator and denominator on the left side are physically equivalent, and thus have a 1:1 relation. This notation will prove convenient when we consider the change of units for secondary dimensions in the next section.

# †1.3  Secondary Dimensional Quantities

When physical characteristics are described in terms of basic dimensions by the use of suitable definitions (e.g., velocity is defined[2] as a distance divided by a time interval), such quantities are called *secondary dimensional quantities*. In Section 1.4, we will see that these quantities may also be established as a consequence of natural laws. The dimensional representation of secondary quantities is given in terms of the basic dimensions that enter into the formulation of the concept. For example, the dimensional representation of velocity is

$$[\text{velocity}] \equiv \frac{[L]}{[t]}$$

That is, the dimensional representation of velocity is the dimension length divided by the dimension time. The units for a secondary quantity are then given in terms of the units of the constituent basic dimensions. Thus,

$$[\text{velocity units}] \equiv \frac{[\text{ft}]}{[\text{sec}]}$$

A *change* of units from one system into another usually involves a change in the scale of measure of the secondary quantities involved in the problem. Thus, one scale unit of velocity in the American system is 1 foot per second, while in the SI system it is 1 meter per second. How may these scale units be correctly related for complicated secondary quantities? That is, for our simple case, how many meters per second are equivalent to 1 foot per second? The formal expressions of dimensional representation may be put to good use for such an evaluation. The procedure is as follows. Express the dependent quantity dimensionally; substitute existing units for the basic dimensions; and finally, change these units to the equivalent numbers of units in the new system. The result gives the number of scale units of the quantity in the new system of units that is equivalent to 1 scale unit of the quantity in the old system. Performing these operations for velocity, we would thus have

$$1\left(\frac{\text{ft}}{\text{sec}}\right) \equiv 1\left(\frac{.305 \text{ m}}{\text{sec}}\right) \equiv .305\left(\frac{\text{m}}{\text{sec}}\right)$$

---

[2]A more precise definition will be given in the chapters on dynamics.

which means that .305 scale unit of velocity in the SI system is equivalent to 1 scale unit in the American system.

Another way of changing units when secondary dimensions are present is to make use of the formalism illustrated in relations 1.1. To change a unit in an expression, multiply this unit by a ratio physically equivalent to unity, as we discussed earlier, so that the old unit is canceled out, leaving the desired unit with the proper numerical coefficient. In the example of velocity used above, we may replace ft/sec by m/sec in the following manner:

$$1\left(\frac{ft}{sec}\right) \equiv \left(\frac{1\,ft}{sec}\right) \cdot \left(\frac{.305m}{1\,ft}\right) \equiv .305\left(\frac{m}{sec}\right)$$

It should be clear that, when we multiply by such ratios to accomplish a change of units as shown above, we do not alter the magnitude of the *actual physical quantity* represented by the expression. Students are strongly urged to employ the above technique in their work, for the use of less formal methods is generally an invitation to error.

# †1.4  Law of Dimensional Homogeneity

Now that we can describe certain aspects of nature in a quantitative manner through basic and secondary dimensions, we can by careful observation and experimentation learn to relate certain of the quantities in the form of equations. In this regard, there is an important law, the law of *dimensional homogeneity*, which imposes a restriction on the formulation of such equations. This law states that, because natural phenomena proceed with no regard for man-made units, *basic equations representing physical phenomena must be valid for all systems of units*. Thus, the equation for the period of a pendulum, $t = 2\pi\sqrt{L/g}$, must be valid for all systems of units, and is accordingly said to be *dimensionally homogeneous*. It then follows that the fundamental equations of physics are dimensionally homogeneous; and all equations derived analytically from these fundamental laws must also be dimensionally homogeneous.

What restriction does this condition place on an equation? To answer this, let us examine the following arbitrary equation:

$$x = ygd + k$$

For this equation to be dimensionally homogeneous, the numerical equality between both sides of the equation must be maintained for all systems of units. To accomplish this, the change in the scale of measure of each group of terms must be the same when there is a change of units. That is, if the numerical measure of one group such as $ygd$ is doubled for a new system of units, so must that of the quantities $x$ and $k$. *For this to occur under all systems of units, it is necessary that every grouping in the equation have the same dimensional representation.*

In this regard, consider the dimensional representation of the above equation expressed in the following manner:

$$[x] = [ygd] + [k]$$

From the previous conclusion for dimensional homogeneity, we require that

$$[x] \equiv [ygd] \equiv [k]$$

As a further illustration, consider the dimensional representation of an equation that is *not* dimensionally homogeneous:

$$[L] = [t]^2 + [t]$$

When we change units from the American to the SI system, the units of feet give way to units of meters, but there is no change in the unit of time, and it becomes clear that the numerical value of the left side of the equation changes while that of the right side does not. The equation, then, becomes invalid in the new system of units and hence is not derived from the basic laws of physics. Throughout this book, we shall invariably be concerned with dimensionally homogeneous equations. Therefore, we should dimensionally analyze our equations to help spot errors.

## †1.5 Dimensional Relation Between Force and Mass

We shall now employ the law of dimensional homogeneity to establish a new secondary dimension—namely *force*. A superficial use of Newton's law will be employed for this purpose. In a later section, this law will be presented in greater detail, but it will suffice at this time to state that the acceleration of a particle[3] is inversely proportional to its mass for a given disturbance. Mathematically, this becomes

$$a \propto \frac{1}{m} \qquad (1.2)$$

where $\propto$ is the proportionality symbol. Inserting the constant of proportionality, $F$, we have, on rearranging the equation,

$$F = ma \qquad (1.3)$$

The mechanical disturbance, represented by $F$ and called *force*, must have the following dimensional representation, according to the law of dimensional homogeneity:

$$[F] \equiv [M]\frac{[L]}{[t]^2} \qquad (1.4)$$

The type of disturbance for which relation 1.2 is valid is usually the action of one body on another by direct contact. However, other actions, such as magnetic, electrostatic, and gravitational actions of one body on another involving no contact, also create mechanical effects that are valid in Newton's equation.

---

[3]We shall define particles in Section 1.7.

We could have initiated the study of mechanics by considering *force* as a basic dimension, the manifestation of which can be measured by the elongation of a standard spring at a prescribed temperature. Experiment would then indicate that for a given body the acceleration is directly proportional to the applied force. Mathematically,

$$F \propto a; \text{ therefore, } F = ma$$

from which we see that the proportionality constant now represents the property of mass. Here, mass is now a secondary quantity whose dimensional representation is determined from Newton's law:

$$M \equiv [F]\frac{[t]^2}{[L]} \tag{1.5}$$

As was mentioned earlier, we now have a choice between two systems of basic dimensions—the *MLt* or the *FLt* system of basic dimensions. Physicists prefer the former, whereas engineers usually prefer the latter.

## 1.6    Units of Mass

As we have already seen, the concept of mass arose from two types of actions —those of motion and gravitational attraction. In American engineering practice, units of mass are based on both actions, and this sometimes leads to confusion. Let us consider the *FLt* system of basic dimensions for the following discussion. The unit of force may be taken to be the pound-force (lbf), which is defined as a force that extends a standard spring a certain distance at a given temperature. Using Newton's law, we then define the *slug* as the amount of mass that a 1-pound force will cause to accelerate at the rate of 1 foot per second per second.

On the other hand, another unit of mass can be stipulated if we use the gravitational effect as a criterion. Here, the *pound mass* (lbm) is defined as the amount of matter that is drawn by gravity toward the earth by a force of 1 pound-force (lbf) at a specified position on the earth's surface.

We have formulated two units of mass by two different actions, and to relate these units we must subject them to the *same* action. Thus, we can take 1 pound mass and see what fraction or multiple of it will be accelerated 1 ft/sec$^2$ under the action of 1 pound of force. This fraction or multiple will then represent the number of units of pound mass that are equivalent to 1 slug. It turns out that this coefficient is $g_0$, where $g_0$ has the value corresponding to the acceleration of gravity at a position on the earth's surface where the pound mass was standardized. To three significant figures, the value of $g_0$ is 32.2. We may then make the statement of equivalence that

$$1 \text{ slug} \equiv 32.2 \text{ pounds mass}$$

To use the pound-mass unit in Newton's law, it is necessary to divide by $g_0$ to form units of mass, that have been derived from Newton's law. Thus,

$$F = \frac{m}{g_0} a \qquad (1.6)$$

where $m$ has the units of pound mass and $m/g_0$ has units of slugs. Having properly introduced into Newton's law the pound-mass unit from the viewpoint of physical equivalence, let us now consider the dimensional homogeneity of the resulting equation. The right side of Eq. 1.6 must have the dimensional representation of $F$ and, since the unit here for $F$ is the pound force, the right side must then have this unit. Examination of the units on the right side of the equation then indicates that the units of $g_0$ must be

$$[g_0] \equiv \frac{[\mathrm{lbm}][\mathrm{ft}]}{[\mathrm{lbf}][\mathrm{sec}]^2} \qquad (1.7)$$

How does *weight* fit into this picture? Weight is defined as *the force of gravity on a body*. Its value will depend on the position of the body relative to the earth's surface. At a location on the earth's surface where the pound mass is standardized, a mass of 1 pound (lbm) has the weight of 1 pound (lbf), but with increasing altitude the weight will become smaller than 1 pound (lbf). The mass, however, remains at all times a 1-pound mass (lbm). If the altitude is not exceedingly large, the measure of weight, in lbf, will practically equal the measure of mass, in lbm. Therefore, it is unfortunately the practice in engineering to think erroneously of weight at positions other than on the earth's surface as the measure of mass, and consequently to use the symbol $W$ to represent either lbm or lbf. In this age of rockets and missiles, it behooves us to be careful about the proper usage of units of mass and weight throughout the entire text.

If we know the weight of a body at some point, we can determine its mass in slugs very easily, provided that we know the acceleration of gravity, $g$, at that point. Thus, according to Newton's law,

$$W(\mathrm{lbf}) = m(\mathrm{slugs}) \times g(\mathrm{ft/sec}^2)$$

Therefore,

$$m(\mathrm{slugs}) = \frac{W(\mathrm{lbf})}{g(\mathrm{ft/sec}^2)} \qquad (1.8)$$

Up to this point, we have only considered the American system of units. In the SI system of units, a *kilogram* is the amount of mass that will accelerate 1 m/sec$^2$ under the action of a force of 1 newton. Here we do not have the problem of 2 units of mass; the kilogram is the basic unit of mass. However, we do have another kind of problem—that the kilogram is unfortunately also used as a measure of force, as is the newton. One kilogram of force is the weight of 1 kilogram of mass at the earth's surface, where the acceleration of gravity (i.e., the acceleration due to the force of gravity) is

9.81 m/sec². A newton, on the other hand, is the force that causes 1 kilogram of mass to have an acceleration of 1 m/sec². Hence, 9.81 newtons are equivalent to 1 kilogram of force. That is,

$$9.81 \text{ newtons} \equiv 1 \text{ kilogram(force)} \equiv 2.205 \text{ lbf}$$

Note from the above that the newton is a comparatively small force, equaling approximately one-fifth of a pound. A kilonewton (1000 newtons), which will be used often, is about 200 lb. In this text, we shall *not* use the kilogram as a unit of force. However, you should be aware that many people do.[4]

Note that at the earth's surface the weight $W$ of a mass $M$ is:

$$W(\text{newtons}) = [M(\text{kilograms})](9.81)(\text{m/s}^2) \qquad (1.9)$$

Hence:

$$M(\text{kilograms}) = \frac{W(\text{newtons})}{9.81 \ (\text{m/s}^2)} \qquad (1.10)$$

Away from the earth's surface, use the acceleration of gravity $g$ rather than 9.81 in the above equations.

## 1.7    Idealizations of Mechanics

As we have pointed out, basic and secondary dimensions may sometimes be related in equations to represent a physical action that we are interested in. We want to represent an action using the known laws of physics, and also to be able to form equations simple enough to be susceptible to mathematical computational techniques. Invariably in our deliberations, we must replace the actual physical action and the participating bodies with hypothetical, highly simplified substitutes. We must be sure, of course, that the results of our substitutions have some reasonable correlation with reality. All analytical physical sciences must resort to this technique, and, consequently, their computations are not cut-and-dried but involve a considerable amount of imagination, ingenuity, and insight into physical behavior. We shall, at this time, set forth the most fundamental idealizations of mechanics and a bit of the philosophy involved in scientific analysis.

**Continuum.** Even the simplification of matter into molecules, atoms, electrons, and so on, is too complex a picture for many problems of engineering mechanics. In most problems, we are interested only in the average measurable manifestations of these elementary bodies. Pressure, density, and temperature are actually the gross effects of the actions of the many molecules and atoms, and they can be conveniently assumed to arise from a hypothetically continuous distribution of matter, which we shall call the *continuum*, instead of from a conglomeration of discrete, tiny bodies. Without such an

---

[4]This is particularly true in the marketplace where the word "kilos" is often heard.

artifice, we would have to consider the action of each of these elementary bodies—a virtual impossibility for most problems.

**Rigid Body.** In many cases involving the action on a body by a force, we simplify the continuum concept even further. The most elemental case is that of a rigid body, which is a continuum that undergoes theoretically no deformation whatever. Actually, every body must deform to a certain degree under the actions of forces, but in many cases the deformation is too small to affect the desired analysis. It is then preferable to consider the body as rigid, and proceed with simplified computations. For example, assume that we are to determine the forces transmitted by a beam to the earth as the result of a load $P$ (Fig. 1.2). If $P$ is small enough, the beam will undergo little deflection, and we can carry out a straightforward simple analysis using the *undeformed geometry* as if the body were indeed rigid. If we were to attempt a more accurate analysis—even though a slight increase in accuracy is not required—we would then need to know the exact position that the load assumes relative to the beam *after* the beam has ceased to deform, as shown in an exaggerated manner in Fig. 1.3. To do this accurately is a hopelessly difficult task, especially when we consider that the support must also "give" in a certain way. Although the alternative to a rigid-body analysis here leads us to a virtually impossible calculation, situations do arise in which more realistic models must be employed to yield the required accuracy. For example, when determining the internal force distribution in a body, we must often take the deformation into account, however small it might be. Other cases will be presented later. *The guiding principle is to make such simplifications as are consistent with the required accuracy of the results.*

**Point Force.** A finite force exerted on one body by another must cause a finite amount of local deformation, and always creates a finite area of contact between the bodies through which the force is transmitted. However, since we have formulated the concept of the rigid body, we should also be able to imagine a finite force to be transmitted through an infinitesimal area or point. This simplification of a force distribution is called a *point force*. In many cases where the actual area of contact in a problem is very small but is not known exactly, the use of the concept of the point force results in little sacrifice in accuracy. In Figs. 1.2 and 1.3, we actually employed the graphical representation of the point force.

**Particle.** The *particle* is defined as an object that has no size but that has a mass. Perhaps this does not sound like a very helpful definition for engineers to employ, but it is actually one of the most useful in mechanics. For the trajectory of a planet, for example, it is the mass of the planet and not its size that is significant. Hence, we can consider planets as particles for such computations. On the other hand, take a figure skater spinning on the ice. Her revolutions are controlled beautifully by the orientation of the body. In this motion, the size and distribution of the body are significant, and since a

**Figure 1.2.** Rigid-body assumption—use original geometry.

**Figure 1.3.** Deformable body.

particle, by definition, can have no distribution, it is patently clear that a particle cannot represent the skater in this case. If, however, the skater should be billed as the "human cannonball on skates" and be shot out of a large air gun, it would be possible to consider her as a single particle in ascertaining her trajectory, since arm and leg movements that were significant while she was spinning on the ice would have little effect on the arc traversed by the main portion of her body.

You will learn later that the *center of mass* or *mass center* is a hypothetical point at which one can concentrate the mass of the body for certain dynamics calculations. Actually in the previous examples of the planet and the "human cannonball on skates," the particle we refer to is actually the mass center whose motion is sufficient for the desired information. Thus, when the motion of the mass center of a body suffices for the information desired, we can replace the body by a particle, namely the mass center.

Many other simplifications pervade mechanics. The perfectly elastic body, the frictionless fluid, and so on, will become familiar as you study various phases of mechanics.

## †1.8    Vector and Scalar Quantities

We have now proposed sets of basic dimensions and secondary dimensions to describe certain aspects of nature. However, more than just the dimensional identification and the number of units are often needed to convey adequately the desired information. For instance, to specify fully the motion of a car, which we may represent as a particle at this time, we must answer the following questions:

1. How fast?
2. Which way?

The concept of velocity entails the information desired in questions 1 and 2. The first question, "How fast?", is answered by the speedometer reading, which gives the value of the velocity in miles per hour or kilometers per hour. The second question, "Which way?", is more complicated, because two separate factors are involved. First, we must specify the angular orientation of the velocity relative to a reference frame. Second, we must specify the sense of the velocity, which tells us whether we are moving *toward or away from* a given point. The concepts of angular orientation of the velocity and sense of the velocity are often collectively denoted as the *direction* of the velocity. Graphically, we may use a *directed line segment* (an arrow) to describe the velocity of the car. The *length* of the directed line segment gives information as to "how fast" and is the *magnitude* of the velocity. The angular orientation of the directed line segment and the position of the arrowhead give information as to "which way"—that is, as to the *direction* of the velocity. The directed line segment itself is called the *velocity*, whereas the length of the directed line segment—that is, the magnitude—is called the *speed*.

There are many physical quantities that are represented by a directed line segment and thus are describable by specifying a magnitude and a direction. The most common example is force, where the magnitude is a measure of the intensity of the force and the direction is evident from how the force is applied. Another example is the *displacement vector* between two points on the path of a particle. The magnitude of the displacement vector corresponds to the distance moved along a *straight line* between two points, and the direction is defined by the orientation of this line relative to a reference, with the sense corresponding to which point is being approached. Thus, $\boldsymbol{\rho}_{AB}$ (see Fig. 1.4) is the displacement vector from $A$ to $B$ (while $\boldsymbol{\rho}_{BA}$ goes from $B$ to $A$).

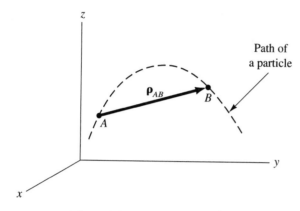

**Figure 1.4.** Displacement vector $\boldsymbol{\rho}_{AB}$.

Certain quantities having magnitude and direction combine their effects in a special way. Thus, the combined effect of two forces acting on a particle, as shown in Fig. 1.5, corresponds to a single force that may be shown by experiment to be equal to the diagonal of a parallelogram formed by the graphical representation of the forces. That is, the quantities add according to the *parallelogram law*. All quantities that have magnitude and direction and that add according to the parallelogram law are called *vector quantities*. Other quantities that have only magnitude, such as temperature and work, are called *scalar quantities*. A vector quantity will be denoted with a boldface italic letter, which in the case of force becomes $F$.[5]

The reader may ask: Don't all quantities having magnitude and direction combine according to the parallelogram law and, therefore, become

**Figure 1.5.** Parallelogram law.

----

[5]Your instructor on the blackboard and you in your homework will not be able to use boldface notation for vectors. Accordingly, you may choose to use a superscript arrow or bar, e.g., $\vec{F}$ or $\bar{F}$ ($\underline{F}$ or $\underset{\sim}{F}$ are other possibilities).

vector quantities? No, not all of them do. One very important example will be pointed out after we reconsider Fig. 1.5. In the construction of the parallelogram it matters not which force is laid out first. In other words, "$F_1$ combined with $F_2$" gives the same result as "$F_2$ combined with $F_1$." In short, the combination is *commutative*. If a combination is not commutative, it cannot in general be represented by a parallelogram operation and is thus not a vector. With this in mind, consider the *finite* angle of rotation of a body about an axis. We can associate a magnitude (degrees or radians) and a direction (the axis and a stipulation of clockwise or counterclockwise) with this quantity. However, the finite angle of rotation cannot be considered a vector because in general two finite rotations about different axes cannot be replaced by a single

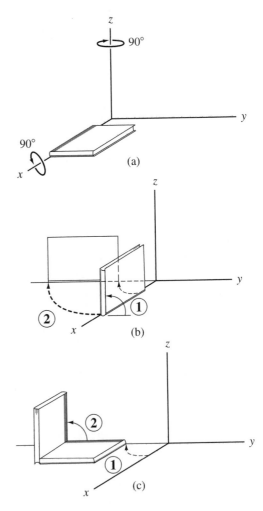

**Figure 1.6.** Successive rotations are not commutative.

finite rotation consistent with the parallelogram law. The easiest way to show this is to demonstrate that the combination of such rotations is not commutative. In Fig. 1.6(a) a book is to be given two rotations—a 90° counterclockwise rotation about the $x$ axis and a 90° clockwise rotation about the $z$ axis, both looking in toward the origin. This is carried out in Figs. 1.6(b) and (c). In Fig. 1.6(c), the sequence of combination is reversed from that in Fig. 1.6(b), and you can see how it alters the final orientation of the book. Finite angular rotation, therefore, is not a vector quantity, since the parallelogram law is not valid for such a combination.[6]

You may now wonder why we tacked on the parallelogram law for the definition of a vector and thereby excluded finite rotations from this category. The answer to this query is as follows. In the next chapter, we will present certain sets of very useful operations termed *vector algebra*. These operations are valid in general *only* if the parallelogram law is satisfied as you will see when we get to Chapter 2. Therefore, we had to restrict the definition of a vector in order to be able to use this kind of algebra for these quantities. Also, it is to be pointed out that later in the text we will present yet a third definition consistent with our latest definition. This next definition will have certain advantages as we will see later.

Before closing the section, we will set forth one more definition. The *line of action* of a vector is a hypothetical infinite straight line collinear with the vector (see Fig. 1.7). Thus, the velocities of two cars moving on different lanes of a straight highway have different lines of action. Keep in mind that the line of action involves no connotation as to sense. Thus, a vector $V'$ collinear with $V$ in Fig. 1.7 and with opposite sense would nevertheless have the same line of action.

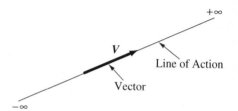

**Figure 1.7.** Line of action of a vector.

# 1.9   Equality and Equivalence of Vectors

We shall avoid many pitfalls in the study of mechanics if we clearly make a distinction between the equality and the equivalence of vectors.

*Two vectors are equal if they have the same dimensions, magnitude, and direction.* In Fig. 1.8, the velocity vectors of three particles have equal length, are identically inclined toward the reference $xyz$, and have the same sense. Although they have different lines of action, they are nevertheless equal according to the definition.

*Two vectors are equivalent in a certain capacity if each produces the very same effect in this capacity.* If the criterion in Fig. 1.8 is change of elevation of the particles or total distance traveled by the particles, all three vectors give the same result. They are, in addition to being equal, also

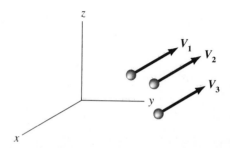

**Figure 1.8.** Equal-velocity vectors.

---

[6]However, *vanishingly small* rotations can be considered as vectors since the commutative law applies for the combination of such rotations. A proof of this assertion is presented in Appendix IV. The fact that infinitesimal rotations are vectors in accordance with our definition will be an important consideration when we discuss angular velocity in Chapter 15.

Mohawk Valley Community College Library

equivalent for these capacities. If the absolute height of the particles above the $xy$ plane is the question in point, these vectors will not be equivalent despite their equality. Thus, it must be emphasized that *equal vectors need not always be equivalent; it depends entirely on the situation at hand.* Furthermore, vectors that are not equal may still be equivalent in some capacity. Thus, in the beam in Fig. 1.9, forces $F_1$ and $F_2$ are unequal, since their magnitudes are 10 lb and 20 lb, respectively. However, it is clear from elementary physics that their moments about the base of the beam are equal, and so the forces have the same "turning" action at the fixed end of the beam. In that capacity, the forces are equivalent. If, however, we are interested in the deflection of the free end of the beam resulting from each force, there is no longer an equivalence between the forces, since each will give a different deflection.

**Figure 1.9.** $F_1$ and $F_2$ equivalent for moment about $A$.

To sum up, the *equality* of two vectors is determined by the vectors themselves, and the *equivalence* between two vectors is determined by the task involving the vectors.

In problems of mechanics, we can profitably delineate three classes of situations concerning equivalence of vectors:

1. *Situations in which vectors may be positioned anywhere in space without loss or change of meaning provided that magnitude and direction are kept intact.* Under such circumstances the vectors are called *free vectors.* For example, the velocity vectors in Fig. 1.8 are free vectors as far as total distance traveled is concerned.

2. *Situations in which vectors may be moved along their lines of action without change of meaning.* Under such circumstances the vectors are called *transmissible vectors.* For example, in towing the object in Fig. 1.10, we may apply the force anywhere along the rope $AB$ or may push at point $C$. The resulting motion is the same in all cases, so the force is a transmissible vector for this purpose.

3. *Situations in which the vectors must be applied at definite points.* The point may be represented as the tail or head of the arrow in the graphical representation. For this case, no other position of application leads to

equivalence. Under such circumstances, the vector is called a *bound vector*. For example, if we are interested in the deformation induced by forces in the body in Fig. 1.10, we must be more selective in our actions than we were when all we wanted to know was the motion of the body. Clearly, force $F$ will cause a different deformation when applied at point $C$ than it will when applied at point $A$. The force is thus a bound vector for this problem.

**Figure 1.10.** $F$ is transmissible for towing.

We shall be concerned throughout this text with considerations of equivalence.

# †1.10 Laws of Mechanics

The entire structure of mechanics rests on relatively few basic laws. Nevertheless, for the student to comprehend these laws sufficiently to undertake novel and varied problems, much study will be required.

We shall now discuss briefly the following laws, which are considered to be the foundation of mechanics:

**1.** Newton's first and second laws of motion.
**2.** Newton's third law.
**3.** The gravitational law of attraction.
**4.** The parallelogram law.

**Newton's First and Second Laws of Motion.** These laws were first stated by Newton as

*Every particle continues in a state of rest or uniform motion in a straight line unless it is compelled to change that state by forces imposed on it.*

*The change of motion is proportional to the natural force impressed and is made in a direction of the straight line in which the force is impressed.*

Notice that the words "rest," "uniform motion," and "change of motion" appear in the statements above. For such information to be meaningful, we must have some frame of reference relative to which these states of motion can be described. We may then ask: relative to what reference in space does every particle remain at "rest" or "move uniformly along a straight line" in the absence of any forces? Or, in the case of a force acting on the particle, relative to what reference in space is the "change in motion proportional to the force"? Experiment indicates that the "fixed" stars act as a reference for which the first and second laws of Newton are highly accurate. Later, we will see that any other system that moves uniformly and without rotation relative to the fixed stars may be used as a reference with equal accuracy. All such references are called *inertial references*. The earth's surface is usually employed as a reference in engineering work. Because of the rotation of the earth and the variations in its motion around the sun, it is not, strictly speaking, an inertial reference. However, the departure is so small for most situations (exceptions are the motion of guided missiles and spacecraft) that the error incurred is very slight. We shall, therefore, usually consider the earth's surface as an inertial reference, but will keep in mind the somewhat approximate nature of this step.

As a result of the preceding discussion, we may define *equilibrium* as *that state of a body in which all its constituent particles are at rest or moving uniformly along a straight line relative to an inertial reference.* The converse of Newton's first law, then, stipulates for the equilibrium state that there must be no force (or equivalent action of no force) acting on the body. Many situations fall into this category. The study of bodies in equilibrium is called *statics*, and it will be an important consideration in this text.

In addition to the reference limitations explained above, a serious limitation was brought to light at the turn of this century. As pointed out earlier, the pioneering work of Einstein revealed that the laws of Newton become increasingly more approximate as the speed of a body increases. Near the speed of light, they are untenable. In the vast majority of engineering computations, the speed of a body is so small compared to the speed of light that these departures from Newtonian mechanics, called *relativistic effects*, may be entirely disregarded with little sacrifice in accuracy. In considering the motion of high-energy elementary particles occurring in nuclear phenomena, however, we cannot ignore relativistic effects. Finally, when we get down to very small distances, such as those between the protons and neutrons in the nucleus of an atom, we find that Newtonian mechanics cannot explain many observed phenomena. In this case, we must resort to quantum mechanics, and then Newton's laws give way to the Schrödinger equation as the key equation.

**Newton's Third Law.** Newton stated in his third law:

> *To every action there is always opposed an equal reaction, or the mutual actions of two bodies upon each other are always equal and directed to contrary points.*

This is illustrated graphically in Fig. 1.11, where the action and reaction between two bodies arise from direct contact. Other important actions in which Newton's third law holds are gravitational attractions (to be discussed next) and electrostatic forces between charged particles. It should be pointed out that there are actions that do not follow this law, notably the electromagnetic forces between charged moving bodies.[7]

**Law of Gravitational Attraction.** It has already been pointed out that there is an attraction between the earth and the bodies at its surface, such as *A* and *B*

---

[7]Electromagnetic forces between charged moving particles are equal and opposite but are not collinear and hence are not "directed to contrary points."

in Fig. 1.11. This attraction is mutual and Newton's third law applies. There is also an attraction between the two bodies $A$ and $B$ themselves, but this force because of the small size of both bodies is extremely weak. However, the mechanism for the mutual attraction between the earth and each body is the same as that for the mutual attraction between the bodies. These forces of attraction may be given by the *law of gravitational attraction*:

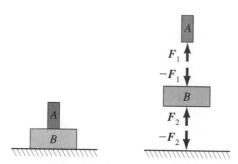

**Figure 1.11.** Newton's third law.

*Two particles will be attracted toward each other along their connecting line with a force whose magnitude is directly proportional to the product of the masses and inversely proportional to the distance squared between the particles.*

Avoiding vector notation for now, we may thus say that

$$F = G \frac{m_1 m_2}{r^2} \qquad (1.11)$$

where $G$ is called the *universal gravitational constant.* In the actions involving the earth and the bodies discussed above, we may consider each body as a particle, with its entire mass concentrated at its center of gravity.[8] Hence, if we know the various constants in formula 1.11, we can compute the weight of a given mass at different altitudes above the earth.

**Parallelogram Law.** Stevinius (1548–1620) was the first to demonstrate that forces could be combined by representing them by arrows to some suitable scale, and then forming a parallelogram in which the diagonal represents the sum of the two forces. As we pointed out, all vectors must combine in this manner.

---

[8]To be studied in detail in Chapter 4.

# 1.11 Closure

In this chapter, we have introduced the basic dimensions by which we can describe in a quantitative manner certain aspects of nature. These basic, and from them secondary, dimensions may be related by dimensionally homogeneous equations which, with suitable idealizations, can represent certain actions in nature. The basic laws of mechanics were thus introduced. Since the equations of these laws relate vector quantities, we shall introduce a useful and highly descriptive set of vector operations in Chapter 2 in order to learn to handle these laws effectively and to gain more insight into mechanics in general. These operations are generally called *vector algebra*.

---

*Check-Out for Sections with* †

**1.1.** What are two kinds of limitations on Newtonian mechanics?

**1.2.** What are the two phenomena wherein mass plays a key role?

**1.3.** If a pound force is defined by the extension of a standard spring, define the pound mass and the slug.

**1.4.** Express mass density dimensionally. How many scale units of mass density (mass per unit volume) in the SI units are equivalent to 1 scale unit in the American system using (a) slugs, ft, sec and (b) lbm, ft, sec?

**1.5.** **(a)** What is a necessary condition for *dimensional homogeneity* in an equation?

   **(b)** In the Newtonian viscosity law, the frictional resistance $\tau$ (force per unit area) in a fluid is proportional to the distance rate of change of velocity $dV/dy$. The proportionality constant $\mu$ is called the *coefficient of viscosity*. What is its dimensional representation?

**1.6.** Define a vector and a scalar.

**1.7.** What is meant by *line of action* of a vector?

**1.8.** What is a *displacement* vector?

**1.9.** What is an *inertial reference*?

REVIEW II*

# Elements of Vector Algebra

## †2.1   Introduction

In Chapter 1, we saw that a scalar quantity is adequately given by a magnitude, while a vector quantity requires the additional specification of a direction. The basic algebraic operations for the handling of scalar quantities are those familiar ones studied in grade school, so familiar that you now wonder even that you had to be "introduced" to them. For vector quantities, these methods may be cumbersome since the directional aspects must be taken into account. Therefore, an algebra has evolved that clearly and concisely allows for certain very useful manipulations of vectors. It is not merely for elegance or sophistication that we employ vector algebra. Indeed, we can achieve greater insight into the subject matter—particularly into dynamics—by employing the more powerful and descriptive methods introduced in this chapter.

## †2.2   Magnitude and Multiplication of a Vector by a Scalar

The magnitude of a quantity, in strict mathematical parlance, is always a *positive* number of units whose value corresponds to the numerical measure of the quantity. Thus, the magnitude of a quantity of measure −50 units is +50 units. Note that the magnitude of a quantity is its absolute value. The mathematical symbol for indicating the magnitude of a quantity is a set of vertical lines enclosing the quantity. That is,

$$|-50 \text{ units}| = \text{absolute value } (-50 \text{ units}) = +50 \text{ units}$$

---

*The reader is urged to pay particular attention to Section 2.4 on **Resolution of Vectors** and Section 2.6 on **Useful Ways of Representing Vectors**.

†Again, as in Chapter 1, we have used the symbol † for certain section headings to indicate that at the end of the chapter there are questions to be answered in writing pertaining to these sections. The instructor may wish to assign the reading of these sections along with the aforementioned questions.

Similarly, the magnitude of a vector quantity is a positive number of units corresponding to the length of the vector in those units. Using our vector symbols, we can say that

$$\text{magnitude of vector } A = |A| \equiv A$$

Thus, $A$ is a positive scalar quantity. We may now discuss the multiplication of a vector by a scalar.

The definition of the product of vector $A$ by scalar $m$, written simply as $mA$, is given in the following manner:

> *$mA$ is a vector having the same direction as $A$ and a magnitude equal to the ordinary scalar product between the magnitudes of $m$ and $A$. If $m$ is negative, it means simply that the vector $mA$ has a direction directly opposite to that of $A$.*

The vector $-A$ may be considered as the product of the scalar $-1$ and the vector $A$. Thus, from the statement above we see that $-A$ differs from $A$ in that it has an opposite sense. Furthermore, these operations have nothing to do with the line of action of a vector, so $A$ and $-A$ may have different lines of action. This will be the case of the couple to be studied in Chapter 3.

## †2.3  Addition and Subtraction of Vectors

In adding a number of vectors, we may repeatedly employ the parallelogram construction. We can do this graphically by scaling the lengths of the arrows according to the magnitudes of the vector quantities they represent. The magnitude of the final arrow can then be interpreted in terms of its length by employing the chosen scale factor. As an example, consider the coplanar[1] vectors $A$, $B$, and $C$ shown in Fig. 2.1(a). The addition of the vectors $A$, $B$, and $C$ has been accomplished in two ways. In Fig. 2.1(b) we first add $B$ and $C$ and then add the resulting vector (shown dashed) to $A$. This combination can be represented by the notation $A + (B + C)$. In Fig. 2.1(c), we add $A$ and $B$, and then add the resulting vector (shown dashed) to $C$. The representation of this combination is given as $(A + B) + C$. Note that the final vector is identical for both procedures. Thus,

$$A + (B + C) = (A + B) + C \tag{2.1}$$

When the quantities involved in an algebraic operation can be grouped without restriction, the operation is said to be *associative*. Thus, the addition of vectors is both commutative, as explained earlier, and associative.

To determine a summation of, let us say, two vectors without recourse to graphics, we need only make a simple sketch of the vectors approximately to scale. By using familiar trigonometric relations, we can get a direct evaluation of the result. This is illustrated in the following examples.

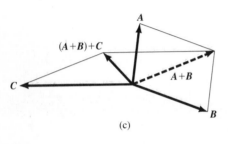

**Figure 2.1.** Addition by parallelogram law.

[1]Coplanar, meaning "same plane," is a word used often in mechanics.

## Example 2.1

Add the forces acting on a particle situated at the origin of a two-dimensional reference frame (Fig. 2.2). One force has a magnitude of 10 lb acting in the positive $x$ direction, whereas the other has a magnitude of 5 lb acting at an angle of 135° with a sense directed away from the origin.

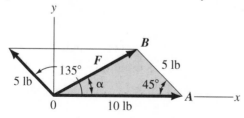

**Figure 2.2.** Find $F$ and $\alpha$ using trigonometry.

To get the sum (shown as $F$), we may use the law of cosines[2] for one of the triangular portions of the sketched parallelogram. Thus, using triangle $OBA$,

$$|F| = \left[10^2 + 5^2 - (2)(10)(5)\cos 45°\right]^{1/2}$$
$$= (100 + 25 - 70.7)^{1/2} = \sqrt{54.3} = 7.37 \text{ lb}$$

The direction of the vector may be described by giving the angle and the sense. The angle is determined by employing the law of sines for triangle $OBA$.[3]

$$\frac{5}{\sin \alpha} = \frac{7.37}{\sin 45°}$$

$$\sin \alpha = \frac{(5)(0.707)}{7.37} = 0.480$$

Therefore,

$$F = 7.37 \text{ lb}$$
$$\alpha = 28.6°$$

The sense is shown using the directed line segment.

---

[2]

You will recall from trigonometry that the *law of cosines* for side $b$ of a triangle is given as
$$b^2 = a^2 + c^2 - 2ac \cos \beta$$

[3]The *law of sines* is given as follows for a triangle:

$$\frac{a}{\sin \alpha} = \frac{b}{\sin \beta} = \frac{c}{\sin \gamma}$$

## ■ Example 2.2

A simple slingshot (see Fig. 2.3) is about to be "fired." If the entire rubber band requires 3 lb per inch of elongation, what force does the band exert on the hand? The total unstretched length of the rubber band is 5 in.

The top view of the slingshot is shown in Fig. 2.4. The change in overall length of the rubber band $\Delta L$ from its unstretched length is

$$\Delta L = 2\left(1.5^2 + 8^2\right)^{1/2} - 5 = 11.28 \text{ in.}$$

The tension in the entire extended rubber band is then (11.28)(3) lb. Consequently, the force $F$ transmitted by *each leg* of the slingshot is

$$F = (11.28)(3) = \boxed{33.84 \text{ lb}}$$

and the value of $\theta$

$$\theta = \tan^{-1}\frac{1.5}{8} = 10.62°$$

In Fig. 2.5, we show a parallelogram involving the forces $F$ and their sum $R$ where $R$ is the force that the band exerts on the hand. We can use the law of cosines on either of the triangles to get $R$. Thus

$$R^2 = 33.84^2 + 33.84^2 - (2)(33.84)(33.84)\cos \alpha$$

Noting that $\alpha = 180° - (2)(10.62°) = 158.8°$ we have

$$R = \left[(2)(33.84)^2(1 - \cos 158.8°)\right]^{1/2} = \boxed{66.52 \text{ lb}}$$

A more direct calculation can be used by considering two right triangles within the chosen triangle. Then using elementary trigonometry we have

$$R = (2)(33.84) \cos (10.62)° = 66.52 \text{ lb}$$

**Figure 2.3.** Simple slingshot.

**Figure 2.4.** Top view of the slingshot.

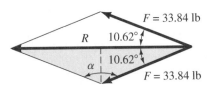

**Figure 2.5.** Parallelogram of forces.

It must be emphatically pointed out that the addition of vectors $A$ and $B$ only involves the vectors themselves and *not* their lines of actions or their positions along their respective lines of action. That is, we can change their lines of action and move them along their respective lines of action so as to form two sides of a parallelogram. For the additional vector algebra that we will develop in this chapter, we can take similar liberties with the vectors involved.

We may also add the vectors by moving them successively to parallel positions so that the head of one vector connects to the tail of the next vector, and so on. The sum of the vectors will then be a vector whose tail connects to the tail of the first vector and whose head connects to the head of the last vector. This last step will form a polygon from the vectors, and we say that the vector sum then "closes the polygon." Thus, adding the 10-lb vector to the

5-lb vector in Fig. 2.2, we would form the sides $OA$ and $AB$ of a triangle. The sum $F$ then closes the triangle and is $OB$. Also, in Fig. 2.6(a), we have shown three coplanar vectors $F_1$, $F_2$, and $F_3$. The vectors are connected in Fig. 2.6(b) as described. The sum of the vectors then is the dashed vector that closes the polygon. In Fig. 2.6(c), we have laid off the vectors $F_1$, $F_2$, and $F_3$ in a different sequence. Nevertheless, it is seen that the sum is the same vector as in Fig. 2.6(b). Clearly, the *order* of laying off the vectors is not significant.

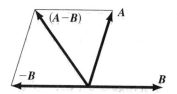

**Figure 2.7.** Subtraction of vectors.

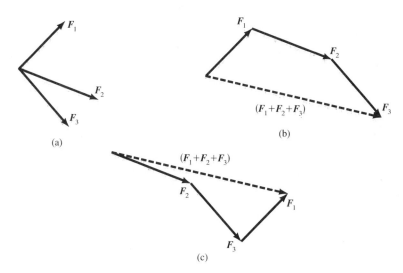

**Figure 2.6.** Addition by "closing the polygon."

A simple physical interpretation of the above vector sum can be formed for vectors each of which represents a movement of a certain distance and direction (i.e., a *displacement* vector). Then, traveling along the system of given vectors you start from one point (the tail of the first vector) and end at another point (the head of the last vector). The vector *sum* that closes the polygon is equivalent to the system of given vectors, in that it takes you from the same initial to the same final point.

The polygon summation process, like the parallelogram of addition, can be used as a graphical process, or, still better, can be used to generate analytical computations with the aid of trigonometry. The extension of this procedure to any number of vectors is obvious.

The process of *subtraction* of vectors is defined in the following manner: to subtract vector $B$ from vector $A$, we reverse the direction of $B$ (i.e., multiply by $-1$) and then add this new vector to $A$ (Fig. 2.7).

This process may also be used in the polygon construction. Thus, consider coplanar vectors $A$, $B$, $C$, and $D$ in Fig. 2.8(a). To form $A + B - C - D$, we proceed as shown in Fig. 2.8(b). Again, the order of the process is not significant, as can be seen in Fig. 2.8(c).

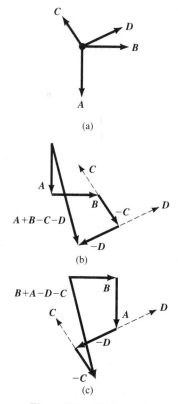

**Figure 2.8.** Addition and subtraction using polygon construction.

**2.1.** Add a 20-N force pointing in the positive $x$ direction to a 50-N force at an angle 45° to the $x$ axis in the first quadrant and directed away from the origin.

**2.2.** Subtract the 20-N force in Problem 2.1 from the 50-N force.

**2.3.** Add the vectors in the $xy$ plane. Do this first graphically, using the force polygon, and then do it analytically.

**Figure P.2.3**

**2.4.** A lightweight homemade plane is being observed as it flies at constant altitude but in a series of separate constant directions. At the outset, it goes due east for 5 km, then due north for 7 km, then southeast for 4 km, and finally, southwest for 8 km. Graphically determine the shortest distance from the starting point to the end point of the previous observations. See Fig. P.2.4.

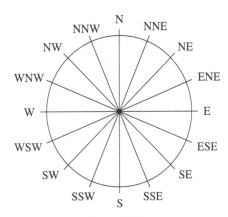

**Figure P.2.4**

**2.5.** A homing pigeon is released at point $A$ and is observed. It flies 10 km due south, then goes due east for 15 km. Next it goes southeast for 10 km and finally goes due south 5 km to reach its destination $B$. Graphically determine the shortest distance between $A$ and $B$. Neglect the earth's curvature.

**2.6.** Force $A$ (given as a horizontal 10-N force) and $B$ (vertical) add up to a force $C$ that has a magnitude of 20 N. What is the

magnitude of force $B$ and the direction of force $C$? (For the simplest results, use the force polygon, which for this case is a right triangle, and perform analytical computations.)

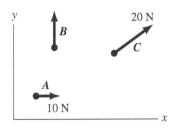

**Figure P.2.6**

**2.7.** A light cable from a Jeep is tied to the peak of an A-frame and exerts a force of 450 N along the cable. A 1,000-kg log is suspended from a second cable, which is fastened to the peak. What is the total force from the cables on the A-frame?

**Figure P.2.7**

**2.8.** Find the total force and its direction from the cable acting on each of the three pulleys, each of which is free to turn. The 100-N weight is stationary.

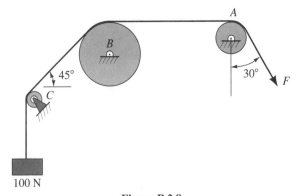

**Figure P.2.8**

28

**2.9.** If the difference between forces $B$ and $A$ in Fig. P.2.6 is a force $D$ having a magnitude of 25 N, what is the magnitude of $B$ and the direction of $D$?

**2.10.** What is the sum of the forces transmitted by the structural rods to the pin at $A$?

**Figure P.2.10**

**2.11.** Suppose in Problem 2.10 we require that the total force transmitted by the members to pin $A$ be inclined 12° to the horizontal. If we do not change the force transmitted by the horizontal member, what must be the new force for the other member whose direction remains at 40°? What is the total force?

**2.12.** Using the parallelogram law, find the tensile force in cable $AC$, $T_{AC}$, and the angle $\alpha$. (We will do this problem differently in Example 5.4.)

**Figure P.2.12**

**2.13.** In the preceding problem, what should the angle $\delta$ be so that the sum of the forces from cable $DE$ and cable $EA$ is colinear with the boom $GE$? Verify that $\delta = 55°$.

**2.14.** Three forces act on the block. The 500-N and the 600-N forces act, respectively, on the upper and lower faces of the block, while the 1,000-N force acts along the edge. Give the magnitude of the sum of these forces using the parallelogram law twice.

**Figure P.2.14**

**2.15.** A man pulls with force $W$ on a rope through a simple frictionless pulley to raise a weight $W$. What total force is exerted on the pulley?

**Figure P.2.15**

**2.16.** Add the three vectors using the parallelogram law twice. The 100-N force is in the $xz$ plane, while the other two forces are parallel to the $yz$ plane and do not intersect. Give the magnitude of the sum and the angle it forms with the $x$ axis.

**Figure P.2.16**

**2.17.** A mass $M$ is supported by cables (1) and (2). The tension in cable (1) is 200 N, whereas the tension in (2) is such as to maintain the configuration shown. What is the mass of $M$ in kilograms? (You will learn very shortly that the weight of $M$ must be equal and opposite to the vector sum of the supporting forces for equilibrium.)

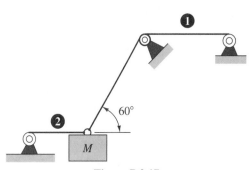

**Figure P.2.17**

29

**2.18.** Two football players are pushing a blocking dummy. Player *A* pushes with 100-lb force while player *B* pushes with 150-lb force toward bow *C* of the dummy. What is the total force exerted on the dummy by the players?

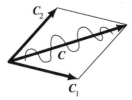

**Figure P.2.18**

**2.19.** What are the forces $F_1$ and $F_2$ and the angle $\beta$ for any given angle $\alpha$ to relieve the force of gravity *W* from the horizontal support of the block at *A*? The rollers on the side of the block do not contribute to the vertical support of the block. The wires connect to the geometric center of the block *C*. The weight *W* is 500 N. Form three independent equations for any given $\alpha$ involving the unknowns $F_1$, $F_2$, and $\beta$.

**Figure P.2.19**

**\*2.20.** Do problem 2.19 and then form an interactive computer program so that the user at a prompt is asked to insert an angle $\alpha$ in radians for which the program will deliver the correct values of $F_1$, $F_2$, and $\beta$.

**2.21.** Two soccer players approach a stationary ball 10 ft away from the goal. Simultaneously, a player on team *O* (offense) kicks the ball with force 100 lb for a split second while a player on team *D* (defense) kicks with force 70 lb during the same time interval. Does the offense score (assuming that the goalie is asleep)?

**Figure P.2.21**

## 2.4 Resolution of Vectors; Scalar Components

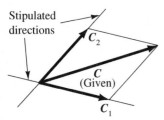

**Figure 2.9.** Two-dimensional resolution of vector *C*.

The opposite action of addition of vectors is called *resolution*. Thus, for a given vector *C*, we may find a pair of vectors in any two stipulated directions coplanar with *C* such that the two vectors, called *components*, sum to the original vector. This is a *two-dimensional* resolution involving two component vectors *coplanar* with the original vector. We shall discuss three-dimensional resolution involving three noncoplanar component vectors later in the section. The two-dimensional resolution can be accomplished by graphical construction of the parallelogram, or by using simple helpful sketches and then employing trigonometric relations. An example of two-dimensional resolution is shown in Fig. 2.9. The two vectors $C_1$ and $C_2$ formed in this way are the *component* vectors. We often replace a vector by its components since the components are always equivalent in rigid-body mechanics to the original vector. When this is done, it is often helpful to indicate that the original vector is no longer operative by drawing a wavy line through the original vector as shown in Fig. 2.10.

**Figure 2.10.** Vector *C* is replaced by its components and is no longer operative.

# Example 2.3

A sailboat cannot go directly into the wind, but must *tack* from side to side as shown in Fig. 2.11 wherein a sailboat is going from marker *A* to marker *B* 5,000 meters apart. What is the additional distance $\Delta L$ beyond 5,000 m that the sailboat must travel to get from *A* to *B*?

Clearly the displacement vector[4] $\boldsymbol{\rho}_{AB}$ is equivalent to the vector sum of displacement vectors $\boldsymbol{\rho}_{AC}$ plus $\boldsymbol{\rho}_{CB}$ in that the same starting points *A*, and the same destination points *B*, are involved in each case. Thus, vectors $\boldsymbol{\rho}_{AC}$ and $\boldsymbol{\rho}_{CB}$ are two-dimensional components of vector $\boldsymbol{\rho}_{AB}$. Accordingly, we can show a parallelogram for those vectors for which triangle *ABC* forms half of the parallelogram (see Fig. 2.12). We leave it for you to justify the various angles indicated in the diagram. Now we first use the *law of sines*.

$$\frac{AC}{\sin \beta} = \frac{5,000}{\sin \alpha}$$

$$\therefore AC = \frac{5,000 \sin 20°}{\sin 135°} = 2,418.4 \text{ m}$$

And

$$\frac{BC}{\sin \gamma} = \frac{5,000}{\sin \alpha}$$

$$\therefore BC = \frac{5,000 \sin 25°}{\sin 135°} = 2,988.4 \text{ m}$$

Hence the increase in distance $\Delta L$ is

$$\Delta L = (2,418.4 + 2,988.4) - 5,000 = \boxed{406.8 \text{ m}}$$

[4]A *displacement vector*, we remind you, connects two points *A* and *B* in space and is often denoted as $\boldsymbol{\rho}_{AB}$. The order of the subscripts gives the sense of the vector—here going from *A* to *B*.

**Figure 2.11.** Sailboat tacking.

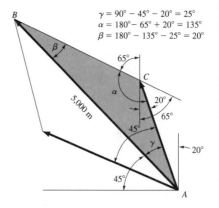

$\gamma = 90° - 45° - 20° = 25°$
$\alpha = 180° - 65° + 20° = 135°$
$\beta = 180° - 135° - 25° = 20°$

**Figure 2.12.** Enlarged parallelogram.

It is also readily possible to find *three* components *not in the same plane as* **C** that add up to **C**. This is the aforementioned three-dimensional resolution. Consider the specification of three *orthogonal* directions[5] for the resolution of **C** positioned in the first quadrant, as is shown in Fig. 2.13. The resolution may be accomplished in two steps. Resolve **C** along the *z* direction, and along the

[5]Although the vector can be resolved along three *skew* directions (hence nonorthogonal), the orthogonal directions are used most often in engineering practice.

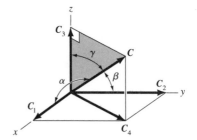

**Figure 2.13.** Orthogonal or rectangular components.

intersection of the $xy$ plane and the plane formed by $C$ and the $z$ axis. This is a two-dimensional resolution with the parallelogram becoming a rectangle because of the normalcy of the $z$ axis to the $xy$ plane. This gives orthogonal vectors $C_3$ and $C_4$ that replace original vector $C$. Next take vector $C_4$ and resolve it along axes $x$ and $y$ by a second two-dimensional resolution involving a rectangle once again thus forming orthogonal vectors $C_1$ and $C_2$ that may replace vector $C_4$. Clearly orthogonal vectors $C_1$, $C_2$, and $C_3$ add up to $C$ and accordingly can replace $C$ under any and all circumstances. Hence $C_1$, $C_2$, and $C_3$ are called *orthogonal* or *rectangular component vectors* of vector $C$.

The direction of a vector $C$ relative to an orthogonal reference is given by the cosines of the angles formed by the vector and the respective coordinate axes. These are called *direction cosines* and are denoted as

$$\cos\ (C,\ x) = \cos\ \alpha \equiv l$$
$$\cos\ (C,\ y) = \cos\ \beta \equiv m \qquad (2.2)$$
$$\cos\ (C,\ z) = \cos\ \gamma \equiv n$$

where $\alpha$, $\beta$, and $\gamma$ are associated with the $x$, $y$, and $z$ axes, respectively. Now let us consider the right triangle, whose sides are $C$ and the component vector, $C_3$, shown shaded in Fig. 2.13. It then becomes clear, from trigonometric considerations of the right triangle, that for the first quadrant

$$|C_3| = |C| \cos\ \gamma = |C|\ n \qquad (2.3)$$

If we had decided to resolve $C$ first in the $y$ direction instead of the $z$ direction, we would have produced a geometry from which we could conclude that $|C_2| = |C|m$. Similarly, we can say that $|C_1| = |C|l$. We can then express $|C|$ in terms of its orthogonal components in the following manner, using the Pythagorean theorem.[6]

$$|C| = \left[\left(|C|l\right)^2 + \left(|C|m\right)^2 + \left(|C|n\right)^2\right]^{1/2} \qquad (2.4)$$

From this equation we can define the *orthogonal* or *rectangular scalar components* of the vector $C$ having *any* orientation as

$$C_x = |C|l, \qquad C_y = |C|m, \qquad C_z = |C|n \qquad (2.5)$$

Note that $C_x$, $C_y$, and $C_z$ may be negative, depending on the sign of the direction cosines. Finally, it must be pointed out that *although $C_x$, $C_y$, and $C_z$ are associated with certain axes and hence certain directions, they have been developed as scalars and must be handled as scalars.* Thus, an equation such as $10V = V_x \cos \beta$ is not correct, because the left side is a vector and the right side is a scalar. This should spur you to observe care in your notation.

Sometimes only *one* of the scalar orthogonal components of a vector (often called a rectangular scalar component) is desired. Then, just one direction is prescribed, as shown in Fig. 2.14. Thus, the scalar rectangular component $C_s$ is $|C| \cos\delta$. Note we have shown a pair of other rectangular

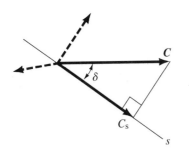

**Figure 2.14.** Rectangular component of $C$.

components as dashed vectors in Fig. 2.14. However, it is only the single component $C_s$ that we often use, disregarding other rectangular components. It is always the case that the triangle formed by the vector and its scalar rectangular component is a right triangle. In establishing $C_s$ we therefore speak of "dropping a perpendicular from $C$ to $s$" or of "projecting along $s$."

The scalar rectangular component $C_s$ could also be the result of a *two-dimensional* orthogonal resolution wherein the other component is in the plane of $C$ and $C_s$ and is *normal* to $C_s$. It is important to remember, however, that a component of a *nonorthogonal* two-dimensional resolution is *not* a rectangular component.

As a final consideration, let us examine vectors $A$ and $B$, which, along with direction $s$, form a plane as is shown in Fig. 2.15. The sum of the vectors $A$ and $B$ is found by the parallelogram law to be $C$. We shall now show that the projection of $C$ along $s$ is the same as the sum of the projections of the two-dimensional components of $A$ and $B$, taken along $s$. That is,

$$C_s = A_s + B_s$$

On the diagram, then, the following relation must be verified:

$$ac = ad + ab \tag{a}$$

But

$$ac = ab + bc \tag{b}$$

Also, it is clear that

$$ad = bc \tag{c}$$

By substituting from Eqs. (b) and (c) into Eq. (a), we reduce Eq. (a) to an identity which shows that the projection of the sum of two vectors is the same as the sum of the projections of the two vectors.

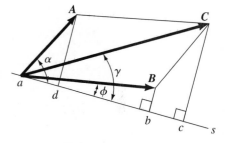

**Figure 2.15.** $C_s = A_s + B_s$.

## 2.5   Unit Vectors

It is sometimes convenient to express a vector $C$ as the product of its magnitude and a vector $a$ of unit magnitude and having direction corresponding to the vector $C$. The vector $a$ is called a *unit vector*. The unit vector is also at times denoted as $\hat{a}$. (You will write it as $\hat{a}$.) It has no dimensions. We formulate this vector as follows:

$$a(\text{unit vector in direction } C) = \frac{C}{|C|} \tag{2.6}$$

Clearly, this development fulfills the requirements that have been set forth for this vector. We can then express the vector $C$ in the form

$$C = |C|a \tag{2.7}$$

The unit vector, once established, does not have, per se, an inherent line of action. This will be determined entirely by its use. In the preceding equation,

**Figure 2.16.** Unit vector *a*.

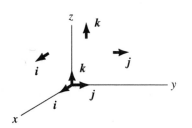

**Figure 2.17.** Unit vectors for *xyz* axes.

the unit vector *a* might be considered collinear with the vector *C*. However, we can represent the vector *D*, shown in Fig. 2.16 parallel to *C*, by using the unit vector *a* as follows:

$$D = |D|a \qquad (2.7a)$$

It thus acts as a free vector. Occasionally, it is useful to label a unit vector meant to have the line of action of a certain vector with the lowercase letter of the capital letter associated with that vector. Thus, in Eqs. 2.7 and 2.7(a) for this purpose we might have employed in the place of *a* the letters *c* and *d* (in your case $\hat{c}$ and $\hat{d}$), respectively. Next, if a given vector is represented using a lowercase letter, such as the vector *r*, then we often make use of the circumflex mark to indicate the associated unit vector. Thus,

$$r = |r|\hat{r} \qquad (2.7b)$$

Unit vectors that are of particular use are those directed along the directions of coordinate axes of a rectangular reference, where *i*, *j*, and *k* (your instructor will probably use the notation $\hat{i}$, $\hat{j}$, and $\hat{k}$) correspond to the *x*, *y*, and *z* directions, as shown in Fig. 2.17.[7]

Since the sum of a set of concurrent vectors is equivalent in all situations to the original vector, we can always replace the vector *C* by its rectangular scalar components in the following manner:

$$C = C_x i + C_y j + C_z k \qquad (2.8)$$

In Chapter 1, we saw that vectors that are equal have the same magnitude and direction. Hence, if $A = B$, we can say that

$$A_x i + A_y j + A_z k = B_x i + B_y j + B_z k \qquad (2.9)$$

Then, since the unit vectors have mutually different directions, we conclude from above that

$$A_x i = B_x i$$
$$A_y j = B_y j$$
$$A_z k = B_z k$$

It then follows that

$$A_x = B_x$$
$$A_y = B_y$$
$$A_z = B_z$$

[7]Curvilinear coordinate systems have associated sets of unit vectors just as do the rectangular coordinate systems. As will be seen later, however, certain of these unit vectors do not all have fixed directions in space for a given reference as do the vectors *i*, *j*, and *k*.

Hence, the vector equation, $A = B$, has resulted in three scalar equations that in totality are equivalent in every way to the vector statement of equality. Thus, in Newton's law we would have

$$F = ma \qquad\qquad (2.10a)$$

as the vector equation, and

$$F_x = ma_x, \quad F_y = ma_y, \quad F_z = ma_z \qquad\qquad (2.10b)$$

as the corresponding scalar equations.

## 2.6   Useful Ways of Representing Vectors

Quite often, we show a rectangular parallelepiped with sides oriented parallel to the coordinate axes and positioned somewhere along the line of action of a vector (see Fig. 2.18) such that this line of action coincides with an inside diagonal of the rectangular parallelepiped. The purpose of this rectangular parallelepiped and diagonal is to allow for the easy determination of the orientation of the line of action and hence the orientation of a vector. $AB$ in the diagram is such a diagonal used for the determination of the line of action of vector $F$. Numbers for this purpose are shown along the sides of the rectangular parallelepiped without units. *Any* set of numbers can be used as long as the *ratios* of these numbers remain the ones required for the proper determination of the orientation of the vector. That determination proceeds by first replacing the *displacement vector* $\rho_{AB}$ from corner $A$ to corner $B$ by a set of three vector displacements going from $A$ to $B$ along the sides of the rectangular parallelepiped. We thereby can replace the vector $\rho_{AB}$ by the sum of its rectangular components. Thus, for the case shown in Fig. 2.18 we can say[8]

$$\rho_{AB} = 10j - 4i + 6k = -4i + 10j + 6k$$

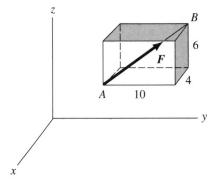

**Figure 2.18.** Rectangular parallelepiped used for specifying the direction of a vector.

---

[8]Imagine you are "walking" from $A$ to $B$ but restricting your movements to be along the coordinate directions. This movement is equivalent to going directly from $A$ to $B$ in that the same endpoints result.

Now using the Pythagorean theorem, divide $\boldsymbol{\rho}_{AB}$ by its magnitude, namely: $\sqrt{4^2 + 10^2 + 6^2}$. We thus form the unit vector $\hat{\boldsymbol{\rho}}_{AB}$. That is,

$$\hat{\boldsymbol{\rho}}_{AB} = \frac{\boldsymbol{\rho}_{AB}}{|\boldsymbol{\rho}_{AB}|} = \frac{-4i + 10j + 6k}{\sqrt{4^2 + 10^2 + 6^2}} = -.3244i + .8111j + .4867k$$

As a final step we can give vector $F$ as follows:

$$F = F(-.3244i + .8111j + .4867k)$$

If $F = 100$ N we then can say:

$$F = -32.44i + 81.11j + 48.67k \text{ N}$$

Note that the rectangular parallelepiped can be anywhere along the line of action of $F$ including cases where $F$ is not inside of the parallelepiped or extends beyond the parallelepiped (see Fig. 2.19). In the two-dimensional case, a right triangle serves the same purpose as the rectangular parallelepiped in three dimensions. This is shown in Fig. 2.20 where vector $V$ is in the $xy$ plane. Here we can say,

$$V = V\left[\frac{9}{\sqrt{2^2 + 9^2}}i + \frac{2}{\sqrt{2^2 + 9^2}}j\right] = V(.9762i + .2169j)$$

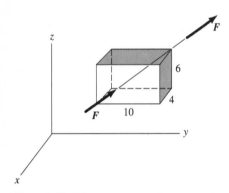

**Figure 2.19.** Other ways to use the rectangular parallelepiped.

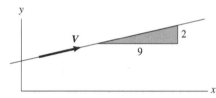

**Figure 2.20.** Right triangle used for specifying the direction of a vector in two dimensions.

There are times when the rectangular parallelpiped is not shown explicitly. However, the length of the sides of one having the proper diagonal may be available so that the replacement of the diagonal displacement vector into rectangular components can be readily achieved. The simplest procedure is to mentally move from the beginning point of the diagonal to the final point always moving along coordinate directions, or, in other words, always moving along the sides of the hypothetical rectangular parallelepiped. Thus, in Fig. 2.21, for the vector $F_1$ we can consider $AB$ to be the diagonal and in going from $A$ to $B$ we could first move in the minus $x$ direction by an amount $-1$, then move in the plus $y$ direction by the amount 1.5, and finally in the $z$ direction by an amount 3. This would take us from initial point $A$ to final point $B$. The corresponding displacement vector would then be

$$\boldsymbol{\rho}_{AB} = -1i + 1.5j + 3j$$

The following example will illustrate the use of orthogonal resolution as well as the use of rectangular components of vectors.

# Example 2.4

A crane (not shown) is supporting a 2,000-N crate (see Fig. 2.21) through three cables: *AB*, *CB*, and *DB*. Note that *D* is at the center of the outer edge of the crate; *C* is 1.6 m from the corner of this edge; and *B* is directly above the center of the crate. What are the forces $F_1$, $F_2$, and $F_3$ transmitted by the cables?

We will soon learn formally what our common sense tells us, namely that the vector sum of force $F_1$, force $F_2$, and force $F_3$ must equal $2,000k$ N. We first express these three forces in terms of rectangular components. Thus,

$$F_{AB} = F_1\left(\frac{\boldsymbol{\rho}_{AB}}{|\boldsymbol{\rho}_{AB}|}\right) = F_1\left[\frac{-1i + 1.5j + 3k}{\sqrt{1^2 + 1.5^2 + 3^2}}\right]$$
$$= F_1(-.2857i + .4286j + .8571k)\text{ N}$$

$$F_{DB} = F_2\left(\frac{\boldsymbol{\rho}_{DB}}{|\boldsymbol{\rho}_{DB}|}\right) = F_2\left[\frac{1i + 0j + 3k}{\sqrt{1^2 + 3^2}}\right] = F_2(.3162i + .9487k)\text{ N}$$

$$F_{CB} = F_3\left(\frac{\boldsymbol{\rho}_{CB}}{|\boldsymbol{\rho}_{CB}|}\right) = F_3\left[\frac{-.6i - 1.5j + 3k}{\sqrt{.6^2 + 1.5^2 + 3^2}}\right]$$
$$= F_3(-.1761i - .4402j + .8805k)\text{ N}$$

We now sum the three forces to equal $2,000k$ N.

$$F_1(-.2857i + .4286j + .8571k) + F_2(.3162i + .9487k)$$
$$+ F_3(-.1761i - .4402j + .8805k) = 2,000k$$

We have three scalar equations from the previous equation.

$$-.2857\,F_1 + .3162\,F_2 - .1761\,F_3 = 0$$
$$.4286\,F_1 + 0 - .4402\,F_3 = 0$$
$$.8571\,F_1 + .9487\,F_2 + .8805\,F_3 = 2,000$$

Solving simultaneously, we get the following results:

$$F_1 = 648.1\text{ N}$$
$$F_2 = 937.1\text{ N}$$
$$F_3 = 631.1\text{ N}$$

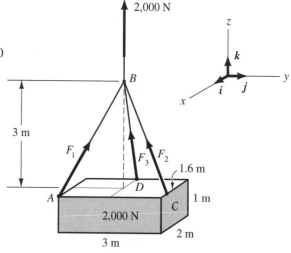

**Figure 2.21.** A crate is supported by three forces.

**2.22.** Resolve the 100-lb force into a set of components along the slot shown and in the vertical direction.

100 lb

20°

20°

**Figure P.2.22**

**2.23.** A farmer needs to build a fence from the corner of his barn to the corner of his chicken house 30 m away in the NE direction. However, he wants to enclose as much of the barnyard as possible. Thus, he runs the fence east, from the corner of his barn to the property line and then NNE to the corner of his chicken house. How long is the fence?

**2.24.** Resolve the force $F$ into a component perpendicular to $AB$ and a component parallel to $BC$.

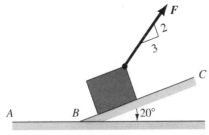

$F$

2

3

C

A

B

20°

**Figure P.2.24**

**2.25.** A simple truss (to be studied later in detail) supports two forces. If the forces in the members are colinear with the members, what are the forces in the members? *Hint:* The forces in the members must have a vector sum equal and opposite to the vector sum of $F_1$ and $F_2$. The entire system is coplanar.

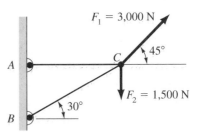

$F_1 = 3,000$ N

45°

A

C

B

30°

$F_2 = 1,500$ N

**Figure P.2.25**

**2.26.** Two tugboats are maneuvering an ocean liner. The desired total force is 3,000 lb at an angle of 15° as shown in the diagram. If the tugboat forces have directions as shown, what must the forces $F_1$ and $F_2$ be?

$F_1$

$\alpha = 30°$

15°

$\beta = 45°$

3,000 lb

$F_2$

**Figure P.2.26**

**2.27.** In the previous problem, if $F_2 = 1,000$ lb and $\beta = 40°$, what should $F_1$ and $\alpha$ be so that $F_1 + F_2$ yields the indicated 3,000-lb force?

**2.28.** A 1,000-N force is resolved into components along $AB$ and $AC$. If the component along $AB$ is 700 N, determine the angle $\alpha$ and the value of the component along $AC$.

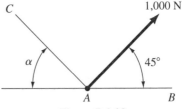

C

1,000 N

$\alpha$

45°

A

B

**Figure P.2.28**

**2.29.** Two men are trying to pull a crate which will not move until a 150-lb total force is applied in any one direction. Man $A$ can pull only at 45° to the desired direction of crate motion, whereas man $B$ can pull only at 60° to the desired motion. What force must each man exert to start the box moving as shown?

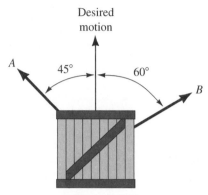

**Figure P.2.29**

**2.30.** What is the sum of the three forces? The 2,000-N force is in the $yz$ plane.

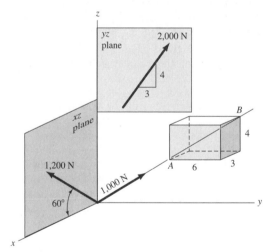

**Figure P.2.30**

**2.31.** The 500-N force is to be resolved into components along the $AC$ and $AB$ directions in the $xy$ plane measured by the angles $\alpha$ and $\beta$. If the component along $AC$ is to be 1,000 N and the component along $AB$ is to be 800 N, compute $\alpha$ and $\beta$.

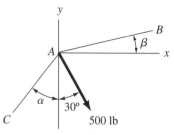

**Figure P.2.31**

**2.32.** The orthogonal components of a force are:
$x$ component 10 lb in positive $x$ direction
$y$ component 20 lb in positive $y$ direction
$z$ component 30 lb in negative $z$ direction
(a) What is the magnitude of the force itself?
(b) What are the direction cosines of the force?

**2.33.** What are the rectangular components of the 100-lb force? What are the direction cosines for this force?

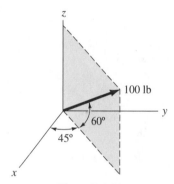

**Figure P.2.33**

**2.34.** The 1,000-N force is parallel to the displacement vector $\vec{OA}$ while the 2,000-N force is parallel to the displacement vector $\vec{CB}$. What is the vector sum of these forces?

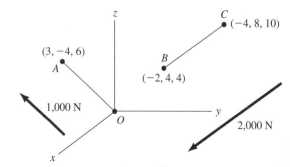

**Figure P.2.34**

**2.35.** A 50-m-long diagonal member $OE$ in a space frame is inclined at $\alpha = 70°$ and $\beta = 30°$ to the $x$ and $y$ axes, respectively. What is $\gamma$? How long must members $OA$, $AC$, $OB$, $BC$, and $CE$ be to support end $E$ of $OE$?

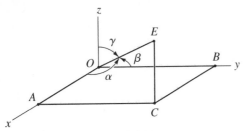

**Figure P.2.35**

**2.36.** What is the orthogonal total force component in the $x$ direction of the force transmitted to pin $A$ of a roof truss by the four members? What is the total component in the $y$ direction?

**Figure P.2.36**

**2.37.** A 30-ton light tank is being lowered from a ship using two cables $AB$ and $BC$. What are the forces in these cables? Use a parallelogram sketch. Also, do this problem using rectangular components.

Cables are in the $yz$ plane

**Figure P.2.37**

**2.38.** Suppose two points $A$ and $B$ in space are designated and a velocity vector $V$ is colinear with the line of action of the displacement vector $\rho_{AB}$ as has been shown. Express $V$ in terms of its rectangular components.

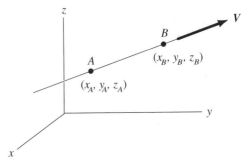

**Figure P.2.38**

**2.39.** Given the following force expressed as a function of position:

$$F = (10x - 6)i + x^2zj + xyk$$

What are the direction cosines of the force at position $(1, 2, 2)$? What is the position along the $x$ coordinate where $F_x = 0$? Plot $F_y$ versus the $x$ coordinate for an elevation $z = 1$.

**2.40.** What is the sum of the following set of three vectors?
$A = 6i + 10j + 16k$ lb
$B = 2i - 3j$ lb
$C$ is a vector in the $xy$ plane at an inclination of $45°$ to the positive $x$ axis and directed away from the origin; it has a magnitude of 25 lb.

**2.41.** What is the unit vector for the displacement vector from point $(2, 1, 9)$ to point $(7, 4, 2)$? Express a 10-m displacement vector in the same direction in terms of $i, j,$ and $k$.

**2.42.** A vector $A$ has a line of action that goes through the coordinates $(0, 2, 3)$ and $(-1, 2, 4)$. If the magnitude of this vector is 10 units, express the vector in terms of the unit vectors $i, j,$ and $k$.

**2.43.** Express the force $F$ in terms of the unit vectors $i, j,$ and $k$.

**Figure P.2.43**

**2.44.** Express the 100-N force in terms of the unit vectors $i, j,$ and $k$. What is the unit vector in the direction of the 100-N force? The force lies along diagonal $AB$.

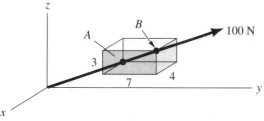

**Figure P.2.44**

**2.45.** Express the unit vectors $i, j,$ and $k$ in terms of unit vectors $\epsilon_{\bar{r}}, \epsilon_\theta,$ and $\epsilon_z$. (These are unit vectors for *cylindrical coordinates*.) Express the 1,000-lb force going through the origin and through point (2, 4, 4) in terms of the unit vectors $i, j, k$ and $\epsilon_{\bar{r}}, \epsilon_\theta, \epsilon_z$ with $\theta = 60°$. (See the footnote on p. 34.)

**Figure P.2.45**

# 2.7 Scalar or Dot Product of Two Vectors

In elementary physics, work was defined as the product of the force component, in the direction of a displacement, times the displacement. In effect, two vectors, force and displacement, are employed to give a scalar, work. In other physical problems, vectors are associated in this same manner so as to result in a scalar quantity. A vector operation that represents such operations concisely is the scalar product (or dot product), which, for the vector $A$ and $B$ in Fig. 2.22, is defined as[9]

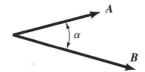

**Figure 2.22.** $\alpha$ is smallest angle between $A$ and $B$.

$$A \cdot B = |A| \, |B| \cos \alpha \qquad (2.11)$$

where $\alpha$ is the smaller angle between the two vectors. Note that the dot product may involve vectors of different dimensional representation, and may be positive or negative, depending on whether the smaller included angle $\alpha$ is less than or greater than 90°. Note also that $A \cdot B$ is equivalent to first projecting vector $A$ onto the line of action of vector $B$ (this gives us $|A| \cos \alpha$), and then multiplying by the magnitude of vector $B$ (or vice versa). The appropriate sign must, of course, be assigned positive if the projected component of vector $A$ and vector $B$ point in the same direction; negative, if not.

The work concept for a force $F$ acting on a particle moving along a path described by $s$ can now be given as

$$W = \int F \cdot ds$$

where $ds$ is a displacement on the path along which the particle is moved.

[9]To ensure that there is no confusion between the dot product of two vectors and the ordinary product of two scalars that you have used up to now, we urge you to read $A \cdot B = C$ as "$A$ dotted into $B$ yields $C$."

41

As with addition and subtraction of vectors, the dot product operation involves only the vectors themselves and not their respective lines of action. Accordingly, for a dot product of two vectors, we can move the vectors so as to intersect at their tails as in Fig. 2.22. Remember in so doing we must not alter the magnitudes and directions of the vectors.

Let us next consider the scalar product of $m\boldsymbol{A}$ and $n\boldsymbol{B}$. If we carry it out according to our definitions:

$$(m\boldsymbol{A}) \bullet (n\boldsymbol{B}) = |m\boldsymbol{A}| \, |n\boldsymbol{B}| \cos (m\boldsymbol{A}, n\boldsymbol{B})$$
$$= (mn) \, |\boldsymbol{A}| \, |\boldsymbol{B}| \cos (\boldsymbol{A}, \boldsymbol{B}) = (mn) \, (\boldsymbol{A} \bullet \boldsymbol{B}) \qquad (2.13)$$

Hence, the scalar coefficients in the dot product of two vectors multiply in the ordinary way, while only the vectors themselves undergo the vectorial operation as we have defined it.

From the definition, clearly the dot product is *commutative*, since the number $|\boldsymbol{A}| \, |\boldsymbol{B}| \cos (\boldsymbol{A}, \boldsymbol{B})$ is independent of the order of multiplication of its terms. Thus,

$$\boldsymbol{A} \bullet \boldsymbol{B} = \boldsymbol{B} \bullet \boldsymbol{A} \qquad (2.14)$$

Let us now consider $\boldsymbol{A} \bullet (\boldsymbol{B} + \boldsymbol{C})$. By definition, we may project the vector $(\boldsymbol{B} + \boldsymbol{C})$ onto the line of action of $\boldsymbol{A}$ and then, assigning the appropriate sign, multiply the magnitude of $\boldsymbol{A}$ times the projection of $\boldsymbol{B} + \boldsymbol{C}$. However, in Section 2.4 we showed that the projection of the sum of two vectors is the same as the sum of the projections of the vectors, which means that

$$\boldsymbol{A} \bullet (\boldsymbol{B} + \boldsymbol{C}) = \boldsymbol{A} \bullet \boldsymbol{B} + \boldsymbol{A} \bullet \boldsymbol{C} \qquad (2.15)$$

An operation on a sum of quantities that is the same as the sum of the operations on the quantities is called a *distributive operation*. Thus, the dot product is distributive.

The scalar product between unit vectors will now be carried out. The product $\boldsymbol{i} \bullet \boldsymbol{j}$ is 0, since the angle $\alpha$ in Eq. 2.11 is 90°, which makes $\cos \alpha = 0$. On the other hand, $\boldsymbol{i} \bullet \boldsymbol{i} = 1$. We can thus conclude that the dot product of equal orthogonal unit vectors for a given reference is unity and that of unequal orthogonal unit vectors is zero.

If we express the vectors $\boldsymbol{A}$ and $\boldsymbol{B}$ in Cartesian components when taking the dot product, we get

$$\boldsymbol{A} \bullet \boldsymbol{B} = (A_x\boldsymbol{i} + A_y\boldsymbol{j} + A_z\boldsymbol{k}) \bullet (B_x\boldsymbol{i} + B_y\boldsymbol{j} + B_z\boldsymbol{k})$$
$$= A_xB_x + A_yB_y + A_zB_z \qquad (2.16)$$

Thus, we see that a scalar product of two vectors is the sum of the ordinary products of the respective components.[10]

---

[10]Thus the ordinary grade school product of two numbers, i.e. $(a)(b)$, is a special case of the dot product $\boldsymbol{a} \bullet \boldsymbol{b}$ where the the vectors have the same direction. Thus

$$a\boldsymbol{i} \bullet b\boldsymbol{i} = (a)(b)$$

If a vector is multiplied by itself as a dot product, the result is the square of the magnitude of the vector. That is,

$$A \cdot A = |A| \, |A| = A^2 \tag{2.17}$$

Conversely, the square of a number may be considered to be the dot product of two equal vectors having a magnitude equal to the number. Note also that

$$A \cdot A = A_x^2 + A_y^2 + A_z^2 = A^2 \tag{2.18}$$

We can conclude from Eq. 2.18 that

$$A = \sqrt{A_x^2 + A_y^2 + A_z^2}$$

which checks with the Pythagorean theorem.

The dot product may be of immediate use in expressing the scalar rectangular component of a vector along a given direction as discussed in Section 2.4. If you refer back to Fig. 2.14, you will recall that the component of $C$ along the direction $s$ is given as

$$C_s = |C| \cos \delta$$

Now let us consider a unit vector $s$ along the direction of the line $s$. If we carry out the dot product of $C$ and $s$ according to our fundamental definition, the result is

$$C \cdot s = |C| \, |s| \cos \delta$$

But since $|s|$ is unity, when we compare the preceding two equations, it is apparent that

$$C_s = C \cdot s$$

Similarly, the following useful relations are valid:

$$C_x = C \cdot i, \quad C_y = C \cdot j, \quad C_z = C \cdot k$$

Finally, express the unit vector $\hat{r}$ directed out from the origin (see Fig. 2.23) in terms of the orthogonal scalar components:

$$\hat{r} = (\hat{r} \cdot i)i + (\hat{r} \cdot j)j + (\hat{r} \cdot k)k$$

But

$$\hat{r} \cdot i = |\hat{r}| \, |i| \cos (\hat{r}, x) = l$$

Similarly, $\hat{r} \cdot j = m$ and $\hat{r} \cdot k = n$. Hence, we can say that

$$\hat{r} = li + mj + nk \tag{2.19}$$

Thus, *the orthogonal scalar components of a unit vector are the direction cosines of the direction of the unit vector.* Now, computing the square of the magnitude of $\hat{r}$, we have

$$|\hat{r}|^2 = 1 = l^2 + m^2 + n^2 \tag{2.20}$$

We thus arrive at the familiar geometrical relation that the sum of the squares of the direction cosines of a vector is unity.

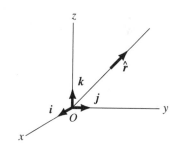

**Figure 2.23.** Unit vector $\hat{r}$ directed from $O$.

## Example 2.5

Cables $GA$ and $GB$ (see Fig. 2.24) are part of a guy-wire system support-ing two radio transmission towers. What are the lengths of $GA$ and $GB$ and the angle $\alpha$ between them?

We may directly set up the vectors $\vec{GA}$ and $\vec{GB}$ by inspecting the dia-gram. Thus on moving along the coordinate directions, it is easy to see that

$$\vec{GA} = 300j - 400i + 500k \text{ m}$$
$$\vec{GB} = 300j + 100i + 500k \text{ m}$$

Using the Pythagorean theorem, we can say for the lengths of $\vec{GA}$ and $\vec{GB}$ :

$$GA = (300^2 + 400^2 + 500^2)^{1/2} = 707 \text{ m}$$
$$GB = (300^2 + 100^2 + 500^2)^{1/2} = 592 \text{ m}$$

Now we use the dot product definition to find the angle.

$$\vec{GA} \cdot \vec{GB} = (GA)(GB) \cos \alpha$$

Therefore,

$$\cos \alpha = \frac{\vec{GA} \cdot \vec{GB}}{(GA)(GB)} = \frac{90{,}000 - 40{,}000 + 250{,}000}{(707)(592)}$$
$$= .717$$

Hence,

$$\alpha = 44.18°$$

**Figure 2.24.** Radio transmission towers.

**2.46.** Given the vectors

$$A = 10i + 20j + 3k$$
$$B = -10j + 12k$$

what is $A \cdot B$? What is $\cos(A, B)$? What is the projection of $A$ along $B$?

**2.47.** Given the vectors

$$A = 16i + 3j, \quad B = 10k - 6i, \quad C = 4j$$

compute

(a) $C(A \cdot C) + B$
(b) $-C + [B \cdot (-A)]C$

**2.48.** Given the vectors

$$A = 6i + 3j + 10k$$
$$B = 2i - 5j + 5k$$
$$C = 5i - 2j + 7k$$

what vector $D$ gives the following results?

$$D \cdot A = 20$$
$$D \cdot B = 5$$
$$D \cdot i = 10$$

**2.49.** A sailboat is tacking into a 20-knot wind. The boat has a velocity component along its axis of 6 kn but because of side slip and water currents, it has a speed a speed of .2 kn at right angles to its axis. What are the $x$ and $y$ components of the wind velocity and the boat velocity? What is the angle between the wind velocity and the sailboat velocity?

**Figure P.2.49**

**2.50.** Show that

$$\cos(A, B) = ll' + mm' + nn'$$

where $l$, $m$, $n$ and $l'$, $m'$, $n'$ are direction cosines of $A$ and $B$, respectively, with respect to the given $xyz$ reference.

**2.51.** Explain why the following operations are meaningless:

(a) $(A \cdot B) \cdot C$
(b) $(A \cdot B) + C$

**2.52.** A block $A$ is constrained to move along a 20° incline in the $yz$ plane. How far does the block have to move if the force $F$ is to do 10 ft-lb of work?

**Figure P.2.52**

**2.53.** An electrostatic field $E$ exerts a force on a charged particle of $qE$, where $q$ is the charge of the particle. If we have for $E$:

$$E = 6i + 3j + 2k \text{ dynes/coulomb}$$

what work is done by the field on a particle with a unit charge moving along a straight line from the origin to position $x = 20$ mm, $y = 40$ mm, $z = -40$ mm?

**2.54.** A force vector of magnitude 100 N has a line of action with direction cosines $l = .7$, $m = .2$, $n = .59$ relative to a reference $xyz$. The vector points away from the origin. What is the component of the force vector along a direction $a$ having direction cosines $l = -.3$, $m = .1$, and $n = .95$ for the $xyz$ reference? (*Hint:* Whenever simply a component is asked for, it is virtually always the *rectangular* component that is desired.)

45

**2.55.** What is the angle between the 1,000-N force and the axis $AB$? The force is in the diagonal plane $GCDE$.

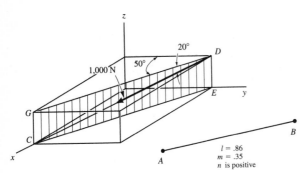

**Figure P.2.55**

**2.56.** Given a force $F = 10i + 5j + Ak$ N. If this force is to have a rectangular component of 8 N along a line having a unit vector $\hat{r} = .6i + .8k$, what should $A$ be? What is the angle between $F$ and $\hat{r}$?

**2.57.** Given a force $Ai + Bj + 20k$ N, what must $A$ and $B$ be to give a rectangular component of 10 N in the direction

$$\hat{r}_1 = .3i + .6j + .742k$$

as well as a component of 18 N in the direction

$$\hat{r}_2 = .4i + .9j + .1732k?$$

**2.58.** Find the dot product of the vectors represented by the diagonals from $A$ to $F$ and from $D$ to $G$. What is the angle between them?

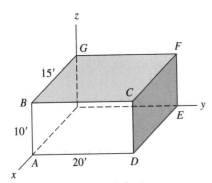

**Figure P.2.58**

**2.59.** A force $F$ is given as

$$F = 800i + 600j - 1,000k \text{ N}$$

What is the *rectangular component* along an axis $A$-$A$ equally inclined to the positive $x$, $y$, and $z$ axes?

**2.60.** What is the rectangular component of the 500-N force along the diagonal from $B$ to $A$?

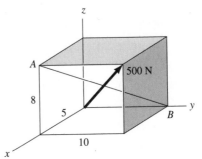

**Figure P.2.60**

**2.61.** A radio tower is held by guy wires. If $AB$ were to be moved to intersect $CD$ while remaining parallel to its original position, what is the angle between $AB$ and $CD$?

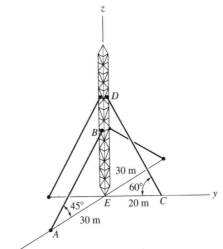

**Figure P.2.61**

**2.62.** What is the angle between the 1,000-N force and the position vector $r$?

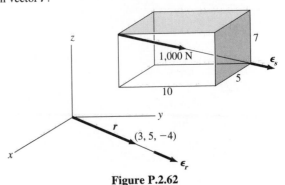

**Figure P.2.62**

46

# 2.8  Cross Product of Two Vectors

There are interactions between vector quantities that result in vector quantities. One such interaction is the moment of a force, which involves a special product of the force and a position vector (to be studied in Chapter 3). To set up a convenient operation for these situations, the *vector cross product* has been established. For the two vectors (having possibly different dimensions) shown in Fig. 2.25 as $A$ and $B$, the operation[11] is defined as

$$A \times B = C \qquad (2.21)$$

where $C$ has a magnitude that is given as

$$|C| = |A| \, |B| \, \sin \alpha \qquad (2.22)$$

**Figure 2.25.** $A \times B = C$.

The angle $\alpha$ is the smaller of the two angles between the vectors, thus making $\sin \alpha$ always positive. The vector $C$ has an orientation normal to the plane of the vectors $A$ and $B$. The sense, furthermore, corresponds to the advance of a right-hand screw rotated about $C$ as an axis while turning from $A$ to $B$ through $\alpha$—that is, from the first stated vector to the second stated vector through the smaller angle between them. In Fig. 2.25, the screw would advance upward in rotating from $A$ to $B$, whether the procedure is viewed from above or below the plane formed by $A$ and $B$. The reader can easily verify this. The description of vector $C$ is now complete, since the magnitude and direction are fully established. The line of action of $C$ is not determined by the cross product; it depends on the use of the vector $C$.

Again we remind you that the cross product, like the other vector algebraic operations, does not involve lines of action, so in taking a cross product we can move the vectors so as to come together at their tails as in Fig. 2.25.

As in the previous case, the coefficients of the vectors will multiply as ordinary scalars. This may be deduced from the nature of the definition. However, the *commutative* law breaks down for this product. We can verify, by carefully considering the definition of the cross product, that

$$(A \times B) = -(B \times A) \qquad (2.23)$$

[11]Again, we urge you to read $A \times B = C$ as "$A$ crossed into $B$ yields $C$."

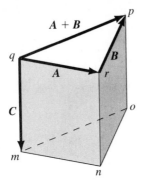

**Figure 2.26.** Prism using $A$, $B$, and $C$.

We can readily show that the cross product, like the dot product, is a distributive operation. To do this, consider in Fig. 2.26 a prism *mnopqr* with edges coinciding with the vectors $A$, $B$, $C$, and $(A + B)$. We can represent the area of each face of the prism as a vector whose magnitude equals the area of the face and whose direction is normal to the face with a sense pointing out (by convention) from the body. It will be left to the student to justify the given formulation for each of the vectors in Fig. 2.27. Since the prism is a closed surface, the net projected area in any direction must be zero, and this, in turn, means that the total area vector must be zero. We then get

$$(A + B) \times C + \tfrac{1}{2} A \times B + \tfrac{1}{2} B \times A + C \times A + C \times B = 0$$

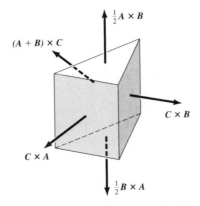

**Figure 2.27.** Area vectors for prism faces.

Noting that the second and third expressions cancel each other, we get, on rearranging the terms,

$$C \times (A + B) = C \times A + C \times B \qquad (2.24)$$

We have thus demonstrated the *distributive* property of the cross product.

Next, consider the cross product of rectangular unit vectors. Here, the product of equal vectors is zero because $\alpha$ and, consequently, $\sin \alpha$ are zero. The product $i \times j$ is unity in magnitude, and because of the right-hand-screw rule must be parallel to the $z$ axis. If the $z$ axis has been erected in a sense con-

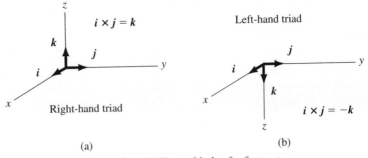

(a)  (b)

**Figure 2.28.** Different kinds of references.

sistent with the right-hand-screw rule when rotating from the $x$ to the $y$ direction, the reference is called a *right-hand triad* [see Fig. 2.28(a)] and we can write

$$i \times j = k$$

If a left-hand triad is used, the result is a $-k$ for the cross product above [see Fig. 2.28(b)]. In this text, we will use a right-hand triad as a reference. For ease in evaluation of unit cross products for such references, a simple permutation scheme is helpful. In Fig. 2.29, the unit vectors $i, j$, and $k$ are indicated on a circle in a clockwise sequence. Any cross product of a pair of unit vectors results in a positive third unit vector if going from the first vector to the second vector involves a clockwise motion on this circle. Otherwise, the vector is negative. Thus,

$$k \times j = -i, \quad k \times i = j, \quad \text{etc.}$$

Next, the cross product of two vectors in terms of their rectangular components is

$$\begin{aligned} A \times B &= (A_x i + A_y j + A_z k) \times (B_x i + B_y j + B_z k) \\ &= (A_y B_z - A_z B_y)i + (A_z B_x - A_x B_z)j + (A_x B_y - A_y B_x)k \end{aligned} \tag{2.25}$$

Another method of carrying out this long computation is to evaluate the following determinant:

$$\begin{vmatrix} A_x & A_y & A_z \\ B_x & B_y & B_z \\ i & j & k \end{vmatrix} \tag{2.26}$$

The determinant may easily be evaluated in the following manner. Repeat the first two rows below the determinant, and then form products along diagonals.

$$\begin{matrix} A_x & A_y & A_z \\ B_x & B_y & B_z \\ i & j & k \\ A_x & A_y & A_z \\ B_x & B_y & B_z \end{matrix} \tag{2.27}$$

For the products along the dashed diagonals, we must remember in this method to multiply by $-1$. We then add all six products as follows:

$$A_x B_y k + B_x A_z j + A_y B_z i - A_z B_y i - B_z A_x j - A_y B_x k$$
$$= (A_y B_z - A_z B_y)i + (A_z B_x - A_x B_z)j + (A_x B_y - A_y B_x)k$$

Clearly, this is the same result as in Eq. 2.25. It must be cautioned that this method of evaluating a determinant is correct only for $3 \times 3$ determinants. If the cross product of two vectors involves less than six nonzero components, such as in the cross product

$$(6i + 10j) \times (5j - 3k)$$

then it is advisable to multiply the components directly and collect terms, as in Eq. 2.25.

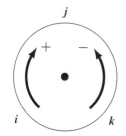

**Figure 2.29.** Permutation scheme.

# Example 2.6

A pyramid is shown in Fig. 2.30. If the height of the pyramid is 300 ft, find the angle between the outward normals to planes $ADB$ and $BDC$.[12]

We shall first find the unit normals to the aforestated planes. Then, using the dot product between these normals, we can easily find the desired angle.

To get the unit normal $n_1$ to plane $ABD$, we first compute the area vector $A_1$ for this plane. Thus, from simple trigonometry and the definition of the cross product,

$$A_1 = \tfrac{1}{2} \vec{AB} \times \vec{AD}$$

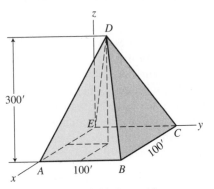

**Figure 2.30.** Pyramid.

Note next that

$$\vec{AB} = 100j \text{ ft}$$

Furthermore, we can express $\vec{AD}$ in rectangular components by moving from $A$ to $D$ along coordinate directions as follows:

$$\vec{AD} = 50j - 50i + 300k \text{ ft}$$

Hence,

$$A_1 = \tfrac{1}{2}(100j) \times (-50i + 50j + 300k)$$
$$= 15{,}000i + 2{,}500k \text{ ft}^2$$

Accordingly,

$$n_1 = \frac{A_1}{|A_1|} = \frac{15{,}000i + 2{,}500k}{\sqrt{15{,}000^2 + 2{,}500^2}} \tag{a}$$
$$= .9864i + .1644k$$

As for unit normal $n_2$ corresponding to plane $BDC$, whose area vector we denote as $A_2$, we have

$$A_2 = \tfrac{1}{2} \vec{BC} \times \vec{BD}$$

Note that

$$\vec{BC} = -100i \text{ ft}$$

And once again, moving along coordinate directions, we have for $\vec{BD}$

$$\vec{BD} = -50j - 50i + 300k \text{ ft}$$

[12] $\pi$ minus this angle is the angle between these planes.

**Example 2.6 (Continued)**

Hence,

$$A_2 = \tfrac{1}{2}(-100i) \times (-50i - 50j + 300k)$$
$$= 15{,}000j + 2{,}500k \text{ ft}^2$$

Accordingly,

$$n_2 = \frac{A_2}{|A_2|} = \frac{15{,}000j + 2{,}500k}{\sqrt{15{,}000^2 + 2{,}500^2}} \tag{b}$$
$$= .9864j + .1644k$$

Now, we use the dot product of $n_1$ and $n_2$. Thus,

$$n_1 \cdot n_2 = \cos \beta \tag{c}$$

where $\beta$ is the angle between the normals to the planes. Substituting from Eqs. (a) and (b) into (c), we get

$$\cos \beta = .0270$$

Therefore,

$$\beta = 88.5°$$

We see from this example that a plane surface can be represented as a vector, and if that plane surface is part of a closed surface, by convention the area vector is in the direction of the outward normal.

## 2.9 Scalar Triple Product

A very useful quantity is the *scalar triple product,* which for a set of vectors $A$, $B$, and $C$ is defined as

$$(A \times B) \cdot C \tag{2.28}$$

This clearly is a scalar quantity.

A simple geometric meaning can be associated with this operation. In Fig. 2.31, we have shown $A$, $B$, and $C$ as an arbitrary set of concurrent vectors. We have set up an *xyz* reference such that the $A$ and $B$ vectors are in the *xy* plane. Further, a parallelogram *abcd* in the *xy* plane is shown in the diagram. We can say that

$$|A \times B| = |A| \, |B| \sin \alpha = \text{area of } abcd$$

Furthermore, the direction of $A \times B$ is in the $z$ direction. Clearly, when we carry out Eq. 2.28, we are thus multiplying the scalar component of $C$ in the $z$ direction by the area of the aforementioned parallelogram. Thus, we have, for Eq. 2.28:

$$(A \times B) \cdot C = (\text{area of } abcd)(C_z)$$

But $C_z$ is the altitude of the parallelepiped formed by vectors $A$, $B$, and $C$ (see Fig. 2.31). We then conclude from solid geometry that *the scalar triple product is the volume of the parallelepiped formed by the concurrent vectors of the scalar triple product.*

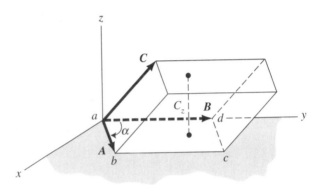

**Figure 2.31.** *A* and *B* in *xy* plane.

Using this geometrical interpretation of the scalar triple product, the reader can easily conclude that

$$(A \times B) \cdot C = - (A \times C) \cdot B = - (C \times B) \cdot A \qquad (2.29)$$

The computation of the scalar triple product is a very straightforward process. It will be left as an exercise (Problem 2.72) for you to demonstrate that

$$(A \times B) \cdot C = \begin{vmatrix} A_x & A_y & A_z \\ B_x & B_y & B_z \\ C_x & C_y & C_z \end{vmatrix} \qquad (2.30)$$

In later chapters, we shall employ the scalar triple product, although we shall not always want to associate the preceding geometric interpretation of this product.

Another operation involving three vectors is the *vector triple product* defined for vectors $A$, $B$, and $C$ as $A \times (B \times C)$. The vector triple product is a vector quantity and will appear quite often in studies of dynamics. It will be left for you to demonstrate that

$$A \times (B \times C) = B(A \cdot C) - C(A \cdot B) \qquad (2.31)$$

Notice here that the vector triple product can be carried out by using only dot products.

# Example 2.7

In Example 2.6, what is the area projected by plane $ADE$ onto an infinite plane that is inclined equally to the $x$, $y$, and $z$ axes?

The normal $n$ to the infinite plane must have three equal direction cosines. Hence, noting Eq. 2.20 for the sum of the squares of a set of direction cosines, we can say that

$$l^2 = m^2 = n^2 = \tfrac{1}{3}$$

Therefore,

$$l = m = n = \tfrac{1}{\sqrt{3}}$$

Hence,

$$n = \tfrac{1}{\sqrt{3}}i + \tfrac{1}{\sqrt{3}}j + \tfrac{1}{\sqrt{3}}k$$

The projected area then is given as

$$A_n = \left(\tfrac{1}{2}\vec{AD} \times \vec{AE}\right) \cdot n$$
$$= \left[\tfrac{1}{2}(-50i + 50j + 300k) \times (-100i)\right] \cdot \tfrac{1}{\sqrt{3}}(i + j + k)$$

The preceding result is a scalar triple product that can readily be solved as follows (disregarding the final sign):

$$A_n = \frac{1}{2\sqrt{3}}\begin{vmatrix} -50 & 50 & 300 \\ -100 & 0 & 0 \\ 1 & 1 & 1 \\ -50 & 50 & 300 \\ -100 & 0 & 0 \end{vmatrix} = \boxed{7{,}217 \text{ ft}^2}$$

## 2.10   A Note on Vector Notation

When expressing *equations,* we must at all times clearly denote scalar and vector quantities and handle them accordingly. When we are simply identifying quantities in a *discussion* or in a *diagram,* however, instead of using the vector representation, $\boldsymbol{F}$, we can just use $F$. On the other hand, $F$ will be understood to represent in an equation the magnitude of the vector $\boldsymbol{F}$. Thus, using $\boldsymbol{f}$ as the unit vector in the direction of $\boldsymbol{F}$, we can then say:

$$\boldsymbol{F} = F\boldsymbol{f}$$

$$= F[\cos{(\boldsymbol{F}, x)}\boldsymbol{i} + \cos{(\boldsymbol{F}, y)}\boldsymbol{j} + \cos{(\boldsymbol{F}, z)}\boldsymbol{k}]$$

As another example, we might want to employ the force $F$, which is shown in the coplanar diagram of Fig. 2.32(a) at a known inclination and acting at a point $a$. A correct representation of this force in a vector equation would be $F(-\cos{\alpha}\boldsymbol{i} + \sin{\alpha}\boldsymbol{j})$.

As for scalar components of any vector $\boldsymbol{F}$, we shall adopt the following understanding. The notation $F_x$, $F_y$, or $F_z$ labeling some vector component in a *diagram* will be understood to represent the *magnitude* of that particular component. Thus, in Fig. 2.32(b) the two components shown are equal in magnitude but opposite in sense. Nevertheless, they are both labeled $F_x$. However, in an equation involving these quantities, the sense must properly be accounted for by the appropriate use of signs.

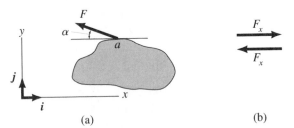

(a)                               (b)

**Figure 2.32.** Notation in diagrams.

**2.63.** If $A = 10i + 6j - 3k$ and $B = 6i$, find $A \times B$ and $B \times A$. What is the magnitude of the resulting vector? What are its direction cosines relative to the $xyz$ reference in which $A$ and $B$ are expressed?

**2.64.** What are the cross and dot products for the vectors $A$ and $B$ given as:

$$A = 6i + 3j + 4k$$
$$B = 8i - 3j + 2k?$$

**2.65.** If vectors $A$ and $B$ in the $xy$ plane have a dot product of 50 units, and if the magnitudes of these vectors are 10 units and 8 units, respectively, what is $A \times B$?

**2.66.** (a) If $A \cdot B = A \cdot B'$, does $B$ necessarily equal $B'$? Explain.
(b) If $A \times B = A \times B'$, does $B$ necessarily equal $B'$? Explain.

**2.67.** What is the cross product of the displacement vector from $A$ to $B$ times the displacement vector from $C$ to $D$?

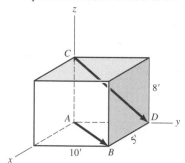

**Figure P.2.67**

**2.68.** Making use of the cross product, give the unit vector $n$ normal to the inclined surface $ABC$.

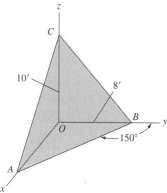

**Figure P.2.68**

**2.69.** If the coordinates of vertex $E$ of the inclined pyramid are $(5, 50, 80)$ m, what is the angle between outward normals to faces $ADE$ and $BCE$?

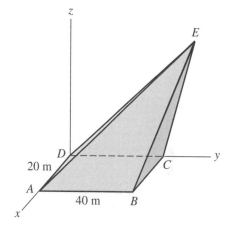

**Figure P.2.69**

**2.70.** In Problem 2.69, what is the area of face $ADE$ of the pyramid? What is the projection of the area of face $ADE$ onto a plane whose normal is along the direction $\epsilon$ where:

$$\epsilon = 0.6i - 0.8j$$

**2.71.** (a) Compute the product

$$(A \times B) \cdot C$$

in terms of orthogonal components.
(b) Compute $(C \times A) \cdot B$ and compare with the result in part (a).

**2.72.** Compute the determinant

$$\begin{vmatrix} A_x & A_y & A_z \\ B_x & B_y & B_z \\ C_x & C_y & C_z \end{vmatrix}$$

where each row represents, respectively, the scalar components of $A$, $B$, and $C$. Compare the result with the computation of $(A \times B) \cdot C$ by using the dot-product and cross-product operations.

**2.73.** In Example 2.5, what is the area vector for $GAB$ assuming a straight line connects points $A$ and $B$? Give the results in kilometers squared.

**2.74.** What is the component of the cross product $A \times B$ along the direction $n$, where

$$A = 10i + 16j + 3k$$
$$B = 5i - 2j + 2k$$
$$n = 0.8i + 0.6k$$

**2.75.** The surface *abcd* of the parallelepiped is in the *xz* plane. Compute the volume using vector analysis.

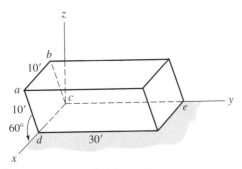

**Figure P.2.75**

**2.76.** Given the vectors

$$A = 10i + 6j$$
$$B = 3i + 5j + 10k$$
$$C = i + j - 3k$$

find

(a) $(A + B) \times C$
(b) $(A \times B) \cdot C$
(c) $A \cdot (B \times C)$

**\*2.77.** A mirror system is used to relay a laser-beam signal from mountain *M* to hill *H*. The mountain is 5,000 m high and 20,000 m NW from the mirror site *S*, while the hill is 200 m high and 15,000 m ENE from the mirror site *S*. Set up, but do not necessarily solve, the equations to find the direction of the mirror to properly relay the signal. Recall that the angle of reflection for a mirror equals the angle of incidence and that the incident ray, reflecting ray, and normal to the mirror are coplanar.

**Figure P.2.77**

## 2.11 Closure

In this chapter, we have presented symbols and notations that are associated with vectors. Also, various vector operations have been set forth that enable us to represent certain actions in nature mathematically. With this background, we shall now be able to study certain vector quantities that are of essential importance in mechanics. Some of these vectors will be formulated in terms of the operations contained in this chapter.

> *Check-Out for Sections with* †
> **2.1** What is meant by the *magnitude* of a vector? What sign must it have?
> **2.2** Can you multiply a vector $C$ by a scalar $s$? If so, describe the result.
> **2.3** What are the *law of cosines* and the *law of sines*?
> **2.4** What is meant by the *associative* law of addition?
> **2.5** Describe two ways to add any three vectors graphically.
> **2.6** How do you subtract vector $D$ from vector $F$?
> **2.7** Given a vector $D$, how would you form a *unit* vector collinear with $D$?
> **2.8** What are the scalar equations of the following vector equation?
>
> $$Di + Ej - 16k = 20i + (15 + G)k$$

**2.78.** Flight 304 from Dallas is flying NE to Chicago 900 miles away. To avoid a massive storm front, the pilot decides instead to fly due north to Topeka, Kansas, and then ENE (see Fig. P.2.4 for compass settings) to Chicago. What are the distances that he must travel from Dallas to Topeka and from Topeka to Chicago?

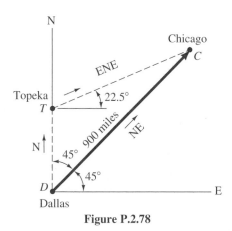

**Figure P.2.78**

**2.79.** What is the cross product between the 1,000-N force and the displacement vector $\boldsymbol{\rho}_{AB}$?

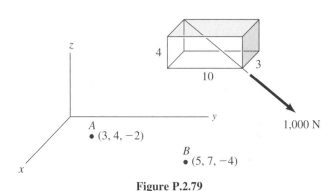

**Figure P.2.79**

**2.80.** Four members of a space frame are loaded as shown. What are the orthogonal scalar components of the forces on the ball joint at $O$? The 1,000-N force goes through points $D$ and $E$ of the rectangular parallelepiped.

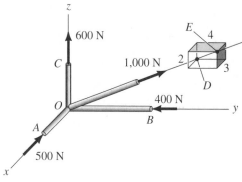

**Figure P.2.80**

**2.81.** Contractors encountered an impassable swamp while building a road from town $T$ to city $C$ 50 km SE. To avoid the swamp, they built the road SSW from $T$ and then ENE to $C$. How long is the road? (*Hint:* See the compass-settings diagram, Fig. P.2.4.)

**2.82.** Sum all forces acting on the block. Plane $A$ is parallel to the $xy$ plane. We will later study the special properties of two parallel forces (called a *couple*) that are opposite in direction and equal in magnitude.

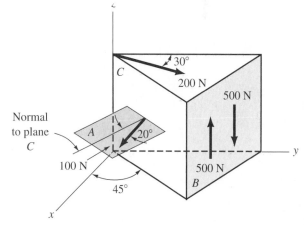

**Figure P.2.82**

**2.83.** The $x$ and $z$ components of the force $F$ are known to be 100 lb and $-30$ lb, respectively. What is the force $F$ and what are its direction cosines?

57

Figure P.2.83

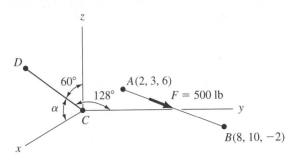

Figure P.2.86

**2.84.** A constant force given as $2i + 3k$ N moves a particle along a straight line from position $x = 10$, $y = 20$, $z = 0$ to position $x = 3$, $y = 0$, $z = -10$. If the coordinates of the $xyz$ reference are given in meter units, how much work does the force do in ft-lb?

**2.85.** Rain is falling on a ferris wheel. The velocity of the rain is constant having reached a terminal velocity of 5 ft/sec. The angular speed of the ferris wheel is constant at 0.5 RPM. At what position $\theta$ will the angle between the velocity of the occupant at $A$ and that of the rain drops be 158°? You will recall from freshman physics that the speed of a particle in circular motion is $r\omega$ with $\omega$ in radians per unit time. First use the dot-product approach and then check your result using a common sense approach.

**2.87.** The velocity of a particle of flow is given as

$$V = 10i + 16j + 2k \text{ m/sec}$$

What is the cross product $r \times V$, where $r$ is given as

$$r = 3i + 2j + 10k \text{ m}$$

Give the proper units.

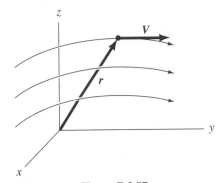

Figure P.2.87

**2.88.** A bridge truss has bar forces as shown in the cutaway sketch. What is the total force on the supporting pin at point $A$ from the members?

Figure P.2.85

Figure P.2.88

**2.86.** A force $F = 500$ lb has a line of action that goes through points $A$ and $B$. What is the angle $\delta$ between $F$ and the displacement vector $\rho_{CD}$. What is the scalar *rectangular component* of the force along axis $CD$?

**2.89.** Forces are transmitted by two members to pin $A$. If the sum of these forces is 700 lb directed vertically, what are the angles $\alpha$ and $\beta$?

400 lb    500 lb

$\beta$    $A$    $\alpha$

**Figure P.2.89**

**2.90.** A skeet shooter is aiming his gun at point $A$. What is the height $z$ of point $A$?

$z$

$A(2, 15, z)$ m

$70°$    $60°$

$O$

$y$

$x$

**Figure P.2.90**

**2.91.** If $F_1$ and the 500-N force sum vectorially to $F_T$, determine $F_1$ and $F_T$.

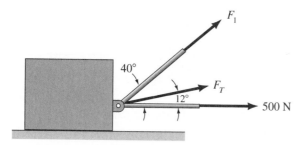

$F_1$

$40°$

$F_T$

$12°$    500 N

**Figure P.2.91**

**2.92.** The force on a charge moving through a magnetic field $B$ is given as

$$F = qV \times B$$

where $q$ = magnitude of the charge, coulombs
  $F$ = force on the body, newtons
  $V$ = velocity vector of the particle, meters per second
  $B$ = magnetic flux density, webers per meter$^2$

Suppose that an electron moves through a uniform magnetic field of $10^6$ Wb/m$^2$ in a direction inclined 30° to the field, as shown,

with a speed of 100 m/sec. What are the force components on the electron? The charge of the electron is $1.6018 \times 10^{-19}$ coulombs.

$y$

$B$

100 m/sec

$x$    $30°$

**Figure P.2.92**

**2.93.** For the line segment $AB$, determine $z_B$ and direction cosines $m$ and $n$.

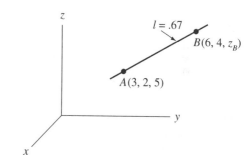

$z$

$l = .67$

$B(6, 4, z_B)$

$A(3, 2, 5)$

$y$

$x$

**Figure P.2.93**

**2.94.** Using the scalar triple product, find the area projected onto the plane $N$ from the surface $ABC$. Plane $N$ is infinite and is normal to the vector

$$r = 50i + 40j + 30k \text{ ft}$$

$z$

$C$

$10'$    $r$

Plane N

$25'$    $B$    $y$

$20'$

$A$

$x$

**Figure P.2.94**

**2.95.** A 500-lb crate is held up by three forces. Clearly the three forces should add up to a force of 500 lb going upward. What should forces $F_1$ and $F_3$ be for this condition? All forces are coplanar (in the same plane).

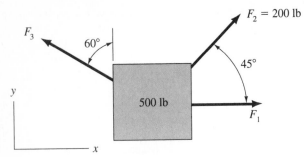

**Figure P.2.95**

**2.96.** A balloon on a gusty day goes from $A$ to $B$, then is seen at $D$ and finally is seen at $C$.

    (a) If the balloon moves along straight lines, how far has it moved?

    (b) If a bee-bee gun is to shoot down the balloon, assuming it is momentarily stationary at $C$, what are the direction cosines of the proper line of sight?

**2.97.** In Problem 2.12 what should $T_{AC}$, $T_{AB}$, and the acute angle $\theta$ between $T_{AB}$ and the vertical to minimize the tension $T_{AB}$? Take $\alpha = 36.5°$ and $W = 1,000$ N. *Hint:* Consider various possible force triangles and decide on inspection how to minimize $T_{AB}$.

**2.98.** In Problem 2.12, what should tensions $T_{AC}$, $T_{AB}$, and angle $\alpha$ be to minimize $T_{AC}$? Assume the angle between $T_{AB}$ and the vertical is equal to $60°$. See previous problem.

**2.99.** To illustrate the random motion of molecules in a liquid, a very small particle placed on the surface of the liquid is observed. The particle, if it is small enough, will jump around in a random manner giving rise to the *random walk* phenomenon described in your physics course. Suppose such a particle jumps from position $(0,0)$ mm in the $xy$ plane to $(.01, -.03)$ mm to $(-.01, -.02)$ mm to $(.01, -.01)$ mm and finally to $(.02, .01)$ mm. What is the shortest distance between end points of this movement? Solve this graphically.

**2.100.** What is the angle $\delta$ between $F$ and the displacement vector $\rho_{AB}$?

**Figure P.2.96**

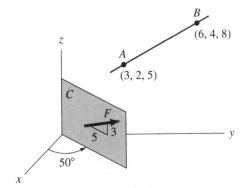

**Figure P.2.100**

# CHAPTER 3

# Important Vector Quantities

## 3.1 Position Vector

In this chapter, we shall discuss a number of useful vector quantities. Consider first the path of motion of a particle shown dashed in Fig. 3.1. As indicated in Chapter 1, the *displacement vector* $\rho$ is a directed line segment connecting any two points on the path of motion, such as points 1 and 2 in Fig. 3.1. The displacement vector thus represents the shortest movement of the particle to get from one position on the path of motion to another. The purpose of the rectangular parallelepiped shown in the diagram is to convey the magnitude and direction of $\rho$ as explained earlier. We can readily express $\rho$ between points 1 and 2 in terms of rectangular components by noting the distance in the coordinate directions needed to go from 1 to 2. Thus, in Fig. 3.1, $\rho_{12} = -2i + 6j + 3k$ m.

The directed line segment $r$ from the origin of a coordinate system to a point $P$ in space (Fig. 3.2) is called the *position vector*. The notations $R$ and $\rho$ are also used for position vectors. You can conclude from Chapter 2 that the magnitude of the position vector is the distance between the origin $O$ and point $P$. The scalar components of a position vector are simply the coordinates of the point $P$. To express $r$ in Cartesian components, we then have

$$r = xi + yj + zk \tag{3.1}$$

We can obviously express a displacement vector $\rho$ between points 1 and 2 (see Fig. 3.3) in terms of position vectors for points 1 and 2 (i.e., $r_1$ and $r_2$) as follows:

$$\rho = r_2 - r_1 = (x_2 - x_1)i + (y_2 - y_1)j + (z_2 - z_1)k \tag{3.2}$$

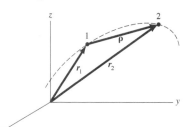

**Figure 3.1.** Displacement vector $\rho$ between points 1 and 2.

**Figure 3.2.** Position vector.

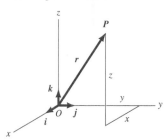

**Figure 3.3.** Relation between a displacement vector and position vectors.

61

## ■ Example 3.1

Two sets of references, *xyz* and *XYZ*, are shown in Fig. 3.4. The position vector of the origin *O* of *xyz* relative to *XYZ* is given as

$$R = 10i + 6j + 5k \text{ m} \qquad (a)$$

The position vector, $r'$, of a point *P* relative to *XYZ* is

$$r' = 3i + 2j - 6k \text{ m} \qquad (b)$$

What is the position vector $r$ of point *P* relative to *xyz*? What are the coordinates *x*, *y*, and *z* of *P*?

From Fig. 3.4, it is clear that

$$r' = R + r \qquad (c)$$

Therefore,

$$r = r' - R = (3i + 2j - 6k) - (10i + 6j + 5k)$$
$$r = -7i - 4j - 11k \text{ m} \qquad (d)$$

We can then conclude that

$$\begin{aligned} x &= -7 \text{ m} \\ y &= -4 \text{ m} \\ z &= -11 \text{ m} \end{aligned} \qquad (e)$$

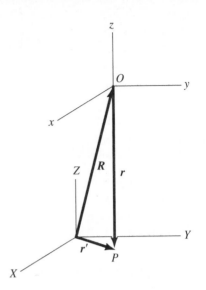

**Figure 3.4.** References *xyz* and *XYZ* separated by position vector **R**.

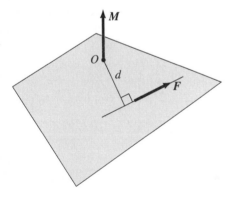

**Figure 3.5.** Moment of force **F** about *O* is *Fd*.

## 3.2   Moment of a Force About a Point

**Case A. For Simple Cases.** The moment of a force about a point *O* (see Fig. 3.5), you will recall from physics, is a vector **M** whose magnitude equals the product of the force magnitude times the perpendicular distance *d* from *O* to the line of action of the force. And the direction of this vector is perpendicular to the plane of the point and the force, with a sense determined from the familiar right-hand-screw rule.[1] The line of action of **M** is determined by the problem at hand. In Fig. 3.5, the line of action of **M** is taken for simplicity through point *O*.

**Case B. For Complex Cases.** Another approach is to employ a position vector **r** from point *O* to *any point P* along the line of action of force **F** as shown in Fig. 3.6. The moment **M** of **F** about point *O* will be shown to be given as[2]

$$M = r \times F \qquad (3.3)$$

---

[1]The sense of **M** would be that of the direction of advance of an ordinary right-hand screw at *O*, oriented normal to the plane of *O* and **F**, when this screw is turned with a sense of rotation corresponding to that of **F** around *O*.

[2]It is worth pointing out again that in determining the moment of force **F** about any point *O*, we must always take the position vector **r** going *from* point *O* to *any point along the line of action of the force F*. It is easy to make errors here.

For the purpose of forming the cross product, the vectors in Fig. 3.6 can be moved to the configuration shown in Fig. 3.7. Then the cross product between $r$ and $F$ obviously has the magnitude

$$|r \times F| = |r||F| \sin \alpha = |F||r| \sin \beta = |F| \, r \sin \beta = Fd$$

where $r \sin \beta = d$, the perpendicular distance from $O$ to the line of action of $F$, as can readily be seen in Fig. 3.7. Thus, we get the same *magnitude* of $M$ as with the elementary definition. Also, note that the *direction* of $M$ here is identical to that of the elementary definition. Thus we have the same result as for the elementary definition in all pertinent respects. We shall use either of these formulations depending on the situation at hand.

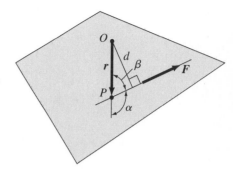

**Figure 3.6.** Put $r$ from $O$ to any point along the line of action of $F$.

The first of these formulations will be used generally for cases where the force and point are in a convenient plane, and where the perpendicular distance between the point and the line of action of the force is easily measured. As an example, we have shown in Fig. 3.8 a system of coplanar forces acting on a beam. The moment of the forces about point $A$ is then[3]

$$M_A = -(5)(1{,}000)k - (4)(600)k + (11)R_B k \text{ ft-lb}$$
$$= (11R_B - 7{,}400)k \text{ ft-lb}$$

**Figure 3.8.** Coplanar forces on a beam.

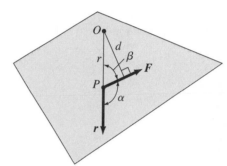

**Figure 3.7.** Move vector $r$ end $F$.

For a coplanar force system such as this, we may simply give the scalar form of the equation above, as follows:

$$M_A = 11R_B - 7{,}400 \text{ ft-lb}$$

The second formulation of the moment about a point, namely $r \times F$, is used for complicated coplanar cases and for three-dimensional cases. We shall illustrate such a case in Example 3.2 after we discuss the rectangular components of $M$.

Consider next a system of $n$ concurrent forces in Fig. 3.9 whose total moment about point $O$ (where we have established reference $xyz$) is desired. We can say that

$$M = M_1 + M_2 + M_3 + \ldots + M_n$$
$$= r \times F_1 + r \times F_2 + r \times F_3 \qquad (3.5)$$
$$+ \ldots + r \times F_n$$

[3]Please note that we still use the right-hand-screw rule in determining the signs of the respective moments.

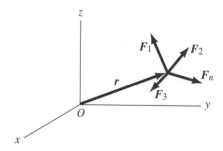

**Figure 3.9.** Concurrent forces.

Now, because of the distributive property of the cross product, Eq. 3.5 can be written

$$M = r \times (F_1 + F_2 + F_3 + \ldots + F_n) \tag{3.6}$$

We can conclude from the preceding equations that the sum of the moments about a point of a system of concurrent forces is the same as the moment about the point of the sum of the forces. This result is known as *Varignon's theorem,* which you may well recall from physics.

As a special case of Varignon's theorem, we may find it convenient to decompose a force $F$ into its rectangular components (Fig. 3.10), and then to use these components for taking moments about a point. We can then say that

$$M = r \times F = r \times (F_x i + F_y j + F_z k) \tag{3.7}$$

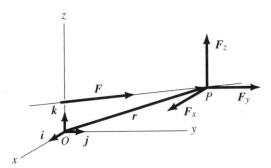

**Figure 3.10.** Decompose $F$ into components.

Now, replacing $r$ by its components, we get

$$\begin{aligned} M &= (xi + yj + zk) \times (F_x i + F_y j + F_z k) \\ &= (yF_z - zF_y)i + (zF_x - xF_z)j + (xF_y - yF_x)k \end{aligned} \tag{3.8}$$

The scalar rectangular components of $M$ are then

$$M_x = yF_z - zF_y \tag{3.9a}$$
$$M_y = zF_x - xF_z \tag{3.9b}$$
$$M_z = xF_y - yF_x \tag{3.9c}$$

As a final note, it should be apparent that, because we can choose $r$ so as to terminate anywhere along the line of action of $F$ in computing $M$, we are, in effect, stipulating that $F$ is a *transmissible* vector (defined in Chapter 1) in the computation of $M$. However, it must be emphasized that we cannot change the line of action of $F$ here.

We are now ready for the example involving the direct vectorial computation of $M$ promised just before the above discussion.

# Example 3.2

Determine the moment of the 100-lb force $F$, shown in Fig. 3.11, about points $A$ and $B$, respectively.

As a first step, let us express force $F$ vectorially. Note that the force is collinear with the vector $\boldsymbol{\rho}_{DE}$ from $D_{DE}$ to $E$, where

$$\boldsymbol{\rho}_{DE} = 8i + 4j - 4k \qquad \text{(a)}$$

To get a unit vector $\hat{\boldsymbol{\rho}}$ in the direction of $\boldsymbol{\rho}_{DE}$, we proceed as follows:

$$\hat{\boldsymbol{\rho}}_{DE} = \frac{\boldsymbol{\rho}_{DE}}{|\boldsymbol{\rho}_{DE}|} = \frac{8i + 4j - 4k}{\sqrt{8^2 + 4^2 + 4^2}} \qquad \text{(b)}$$

$$= .816i + .408j - .408k$$

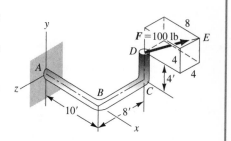

**Figure 3.11.** Find moments at $A$ and $B$.

We can then express the force $F$ in the following manner:

$$F = F\hat{\boldsymbol{\rho}}_{DE} = (100)(.816i + .408j - .408k)$$

$$= 81.6i + 40.8j - 40.8k \qquad \text{(c)}$$

To get the moment $M_A$ about point $A$, we choose a position vector from point $A$ to point $D$ which is on the line of action of force $F$. Thus, we have, for $r_{AD}$,

$$r_{AD} = 10i + 4j - 8k \text{ ft} \qquad \text{(d)}$$

and for $M_A$, we than get

$$M_A = r_{AD} \times F = (10i + 4j - 8k) \times (81.6i + 40.8j - 40.8k)$$

$$= \begin{vmatrix} 10 & 4 & -8 \\ 81.6 & 40.8 & -40.8 \\ i & j & k \\ 10 & 4 & -8 \\ 81.6 & 40.8 & -40.8 \end{vmatrix}$$

$$= (10)(40.8)k + (81.6)(-8)j + (4)(-40.8)i$$

$$\quad -(-8)(40.8)i - (-40.8)(10)j - (4)(81.6)k$$

Therefore,

$$\boxed{M_A = 163.2i - 245j + 81.6k \text{ ft-lb}} \qquad \text{(e)}$$

As for the moment about reference point $B$, we employ the position vector $r_{BD}$ from $B$ to position $D$, again on the line of action of force $F$. Thus, we have

$$r_{BD} = 4j - 8k \text{ ft}$$

Accordingly,

$$M_B = r_{BD} \times F = (4j - 8k) \times (81.6i + 40.8j - 40.8k)$$

$$= (4)(81.6)(-k) + (4)(-40.8)(i) + (-8)(81.6)(j) + (-8)(40.8)(-i) \qquad \text{(f)}$$

$$\boxed{M_B = 163.2i - 653j - 326k \text{ ft-lb}}$$

# PROBLEMS

**3.1.** What is the position vector **r** from the origin (0, 0, 0) to the point (3, 4, 5) ft? What are its magnitude and direction cosines?

**3.2.** What is the displacement vector from position (6, 13, 7) ft to position (10, −3, 4) ft?

**3.3.** A surveyor determines that the top of a radio transmission tower is at position $r_1 = (1{,}000i + 1{,}000j + 1{,}000k)$ m relative to her position. Similarly, the top of a second tower is located by $r_2 = (2{,}000i + 500j + 700k)$ m. What is the distance between the two tower tops?

**3.4.** Reference *xyz* is rotated 30° counterclockwise about its *x* axis to form reference *XYZ*. What is the position vector **r** for reference *xyz* of a point having a position vector **r′** for reference *XYZ* given as

$$r' = 6i' + 10j' + 3k' \text{ m}?$$

Use **i**, **j**, and **k** (no primes) for unit vectors associated with reference *xyz*.

**3.5.** Find the moment of the 50-lb force about the support at *A* and about support *B* of the simply supported beam.

**Figure P.3.5.**

**3.6.** Find the moment of the two forces first about point *A* and then about point *B*.
    (a) Do not use **r** × **F** format—only scalar products.
    (b) Use vector approach.

**Figure P.3.6.**

**3.7.** A particle moves along a circular path in the *xy* plane. What is the position vector **r** of this particle as a function of the coordinate *x*?

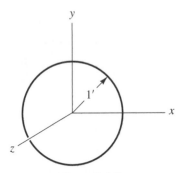

**Figure P.3.7.**

**3.8.** A particle moves along a parabolic path in the *yz* plane. If the particle has at one point a position vector $r = 4j + 2k$, give the position vector at any point on the path as a function of the *z* coordinate.

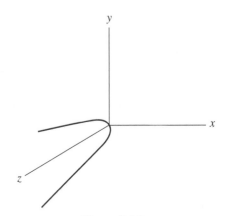

**Figure P.3.8.**

**3.9.** An artillery spotter on Hill 350 (350 m high) estimates the position of an enemy tank as 3,000 m NE of him at an elevation 200 m below his position. A 105-mm howitzer unit with a range of 11,000 m is 10,000 m due south of the spotter, and a 155-mm howitzer unit with a range of 15,000 m is 13,000 m SSE of the spotter (see Fig. P.2.4). Both gun units are located at an elevation of 150 m. Can either or both gun units hit the tank, or must an air strike be called in?

**3.10.** Find the moment of the forces about points $A$ and $B$.
(a) Use scalar approach.
(b) Use vector approach.

**Figure P.3.10.**

**3.11.** The crew of a submarine patrol plane, with three-dimensional radar, sights a surfaced submarine 10,000 yards north and 5,000 yards east while flying at an elevation of 3,000 ft above sea level. Where should the pilot instruct a second patrol plane flying at an elevation of 4,000 ft at a position 40,000 yards east of the first plane to look for confirmation of the sighting?

**3.12.** A power company lineman can comfortably trim branches 1 m from his waist at an angle of 45° above the horizontal. His waist coincides with the pivot of the work capsule. How high a branch can he trim if the maximum elevation angle of the arm is 75° and the maximum extended length is 12 m?

**Figure P.3.12.**

**3.13.** The total equivalent forces from water and gravity are shown on the dam. (We will soon be able to compute such equivalents.) Compute the moment of these forces about the toe of the dam in the right-hand corner.

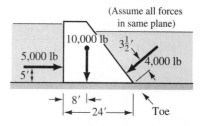

**Figure P.3.13.**

**3.14.** In an underwater "village" for research, an American flag is in place as shown. It is of plastic material and can rotate so as to be oriented parallel to the flow of water. A uniform friction force distribution from the flow is present on both faces of the flag having the value of 10 N per square meter. Also the flagpole has a uniform force from the flow of 20 N per meter of length of the flagpole. Finally there is an upward buoyant force on the flag of 30 N and on the flagpole of 8 N. What is the moment vector of these forces at the base of the flagpole?

**Figure P.3.14.**

**3.15.** Three transmission lines are placed unsymmetrically on a power-line pole. For each pole, the weight of a single line when covered with ice is 2,000 N. What is the moment at the base of a pole?

**Figure P.3.15.**

**3.16.** Compute the moment of the 1,000-lb force about points $A$, $B$, and $C$. Use the transmissibility property of force and rectangular components to make the computations simplest.

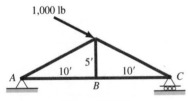

**Figure P.3.16.**

**3.17.** A truck-mounted crane has a 20-m boom inclined at 60° to the horizontal. What is the moment about the boom pivot due to a lifted weight of 30 kN? Do by vector and by scalar methods.

**Figure P.3.17.**

**3.18.** A small blimp is temporarily moored as shown in the diagram wherein $DC$ and the centerline of $AB$ are coplanar. A force $F$ from wind, weight, and buoyancy is shown acting at the centerline of the blimp. If

$$F = 5i + 10j + 18k \text{ kN}$$

what are the moment vectors from $F$ about $A$, $B$, and $C$?

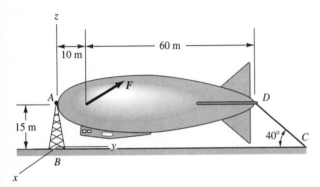

**Figure P.3.18.**

**3.19.** A force $F = 10i + 6j - 6k$ N acts at position (10, 3, 4) m relative to a coordinate system. What is the moment of the force about the origin?

**3.20.** What is the moment of the force in Problem 3.19 about the point (6, −4, −3) m?

**3.21.** Two forces $F_1$ and $F_2$ have magnitudes of 10 lb and 20 lb, respectively. $F_1$ has a set of direction cosines $l = 0.5$, $m = 0.707$, $n = -0.5$. $F_2$ has a set of direction cosines $l = 0$, $m = 0.6$, $n = 0.8$. If $F_1$ acts at point (3, 2, 2) and $F_2$ acts at (1, 0, −3), what is the sum of the moments about the origin?

**3.22.** What is the moment of a 10-lb force $F$ directed along the diagonal of a cube about the corners of the cube? The side of the cube is $a$ ft.

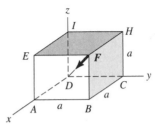

**Figure P.3.22.**

**3.23.** Three guy wires are used in the support system for a television transmission tower that is 600 m tall. Wires $A$ and $B$ are tightened to a tension of 60 kN, whereas wire $C$ has only 30 kN of tension. What is the moment of the wire forces about the base $O$ of the tower? The $y$ axis is collinear with $AO$.

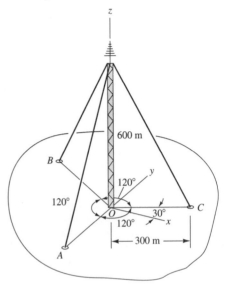

**Figure P.3.23.**

**3.24.** Cables *CD* and *AB* help support member *ED* and the 1,000-lb load at *D*. At *E* there is a ball-and-socket joint which also supports the member. Denoting the forces from the cables as $F_{CD}$ and $F_{AB}$, respectively, compute moments of the three forces about point *E*. Plane *EGD* is perpendicular to the wall. Get results in terms of $F_{CD}$ and $F_{AB}$.

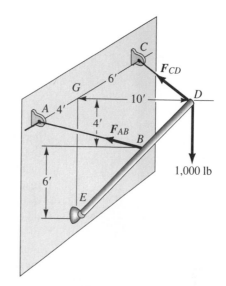

**Figure P.3.24.**

# 3.3 Moment of a Force About an Axis

**Case A. For Simple Cases.** By means of a simple situation, we shall set forth a definition of the moment of a force about an axis. Suppose that a disc is mounted on a shaft that is free to rotate in a set of bearings, as shown in Fig. 3.12. A force **F**, inclined to the plane *A* of the disc, acts on the disc. We decompose the force into two coplanar rectangular components, one normal to plane *A* of the disc and one tangent to plane *A* of the disc, that is, into forces $F_B$ and $F_A$, respectively, so as to form a plane shown tinted, normal to plane *A*.

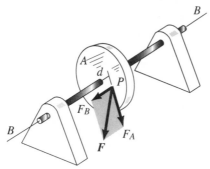

**Figure 3.12.** $F_A$ turns disc.

We know from experience that $F_B$ does not cause the disc to rotate. And we know from physics and intuition that the rotational motion of the disc is determined by the product of $F_A$ and the perpendicular distance $d$ from the centerline of the shaft to the line of action of $F_A$. You will remember from physics, this product is nothing more than the moment of force $F$ about the axis of the shaft. We shall next generalize from this simple case to the general case of taking the moment of *any* force $F$ about *any* axis.

To compute the moment (or torque) of a force $F$ about an axis $B$–$B$ (Fig. 3.13), we pass any plane $A$ perpendicular to the axis. This plane cuts $B$–$B$ at $a$ and the line of action of force $F$ at some point $P$. The force $F$ is then projected to form a rectangular component $F_B$ along a line at $P$ normal to plane $A$ and thus parallel to $B$–$B$, as shown in the diagram. The intersection of plane $A$ with the plane of forces $F_B$ and $F$ (the latter plane is shown shaded and is a plane through $F$ and perpendicular to plane $A$) gives a direction $C$–$C$ along which the other rectangular component of $F$, denoted as $F_A$, can be projected.[4] The moment of $F$ about the line $B$–$B$ is then defined as the scalar representation of the moment of $F_A$ about point $a$ with a magnitude equal to $F_A d$—a problem discussed at the beginning of the previous section (Case A). Thus in accordance with the definition, the component $F_B$, which is parallel to the axis $B$–$B$, contributes no moment about the axis, and we may say:

$$\text{Moment about axis } B\text{--}B = (F_A)(d) = |F|(\cos \alpha)(d)$$

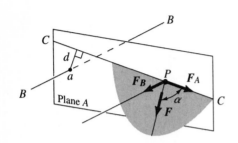

**Figure 3.13.** Formulating the moment about an axis $B$–$B$.

with an appropriate sign. The moment about an axis clearly is a scalar, even though this moment is associated with a particular axis that has a distinct direction. The situation is the same as it is with the scalar components $V_x$, $V_y$, $V_z$, etc., which are associated with certain directions but which are scalars. The reader will be quick to note that Fig. 3.13 represents a generalization of Fig. 3.12 having an axis $B$–$B$, a plane $A$ normal to this axis and finally an arbitrary force $F$. To explain further, we have redrawn Fig. 3.13 [see Fig. 3.14(a)] showing only plane $A$, axes $B$–$B$ and $C$–$C$, and the force $F_A$. In Fig. 3.14(b), we have also included the moment vector $M$. This latter diagram then takes us back to Fig. 3.5 where we first defined the moment vector of a force about a point in a most simple manner. Accordingly, we note, on the one hand, for the moment about point $a$ [see Fig. 3.14(b)], we can get the vector $M$, whereas on the other hand in Fig. 3.14(a) we can get the scalar moment $M$ about an axis $B$–$B$ at point $a$ and perpendicular to plane $A$. Thus, by taking the scalar value of $M$ in Fig. 3.14(b), we get the moment about the axis at point $a$ normal to plane $A$ as formulated in the development of Fig. 3.13.

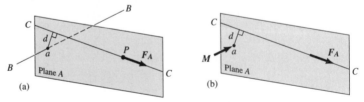

**Figure 3.14.** Comparison of Figs. 3.13 and 3.5.

---

[4]Notice that we are decomposing $F$ into only *two* rectangular components, which, to replace $F$, must be coplanar with $F$.

We can thus conclude, on considering Fig. 3.15, that the moment of $F$ about point $C$ can be considered in two ways as follows:

*Moment vector about point $C$ = $-Fd\,\mathbf{k}$*
*Torque (or moment) about $z$ axis at point $C$ = $-Fd$*

**Figure 3.15.** Consideration of the moment of $F$ about point $C$.

Before continuing, we wish to point out that $F_A$ in Fig. 3.13 can be decomposed into pairs of components in plane $A$. From Varignon's theorem we can employ these components instead of $F_A$ in computing the moment about the $B$–$B$ axis. For each force component, we multiply the force times the perpendicular distance from $a$ to the line of action of the force component using the right-hand-screw rule to determine the sense and thus the sign.

### Case B. For Complex Cases. 
When we discussed the moment of a force about a *point*, we presented a formulation useful for simple cases (i.e., a vector of magnitude $Fd$) as well as a more powerful formulation that would be needed for more complex situations (i.e., $\mathbf{r} \times \mathbf{F}$). Thus far, for moments about an *axis*, we have presented a formulation $Fd$ that is useful for simple cases,[5] and now we shall present a formulation that is needed for more complex cases. For this purpose, we have redrawn Fig. 3.13 as Fig. 3.16(a). In Fig. 3.16(b), we have shown axis $B$–$B$ of Fig. 3.16(a) as an $x$ axis and have set up coordinate axes $y$ and $z$ at *any* point $O$ anywhere on axis $B$–$B$. The coordinate distances $x$, $y$, and $z$ for the point $P$ are shown for this reference. The position vector $\mathbf{r}$ to $P$ is also shown. The force component $F_B$ of Fig. 3.16(a) now becomes force component $F_x$. And, instead of using $F_A$, we shall decompose it into components $F_y$ and $F_z$ in plane $A$ as shown in Fig. 3.16(b). We now compute the moment about the $x$ axis for force $F$ using this new arrangement which does not require $F$ to be in a plane perpendicular to plane $A$. Clearly, $F_x$ contributes no moment, as before. The force components $F_y$ and $F_z$ are in plane $A$ that is perpendicular to the axis of interest and so, as before in the case of $F_A$, we multiply each of these forces by the perpendicular distance of point $a$ to the respective lines of action of these forces.

[5]That is, for cases where the *force* and *point* in question are in a plane easily seen to be normal to the *axis* in question, thus allowing for an easy determination of the perpendicular distance between the point and the line of action of the force.

For force $F_z$, this perpendicular distance is clearly $y$, as can readily be seen from the diagram, and, for force $F_y$, this perpendicular distance is $z$. Using the right-hand-screw rule for ascertaining the sense of each of the moments, we can say:

$$\text{moment about } x \text{ axis} = (yF_z - zF_y) \tag{3.10}$$

Were we to take moments of $F$ about the origin $O$, we would get (see Eq. 3.8)

$$M = M_x i + M_y j + M_z k = r \times F$$
$$= (yF_z - zF_y)i + (zF_x - xF_z)j + (xF_y - yF_x)k \tag{3.11}$$

(a)

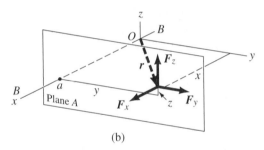

(b)

**Figure 3.16.** Moment about an axis.

Comparing Eqs. 3.10 and 3.11, we can conclude that the moment about the $x$ axis is simply $M_x$, the $x$ component of $M$ about $O$. We can thus conclude that the moment about the $x$ axis of the force $F$ is the component in the $x$ direction of the moment of $F$ about a point $O$ positioned *anywhere* along the $x$ axis. That is,

$$\text{moment about } x \text{ axis} = M_x = M_o \cdot i = (r \times F) \cdot i \tag{3.12}$$

We may generalize the preceding discussion as follows. Consider an arbitrary axis $n$–$n$ to which we have assigned a unit vector $n$ (Fig. 3.17). An arbitrary force $F$ is also shown. To get the moment $M_n$ of force $F$ about axis $n$–$n$, we choose any point $O$ along $n$–$n$. Then draw a position vector $r$ from point $O$ to any point along the line of action of $F$. This has been shown in the diagram. We can then say, from our previous discussion,

$$M_n = (r \times F) \cdot n \tag{3.13}$$

**Figure 3.17.** $M_n = (r \times F) \cdot n$.

(Notice from Eqs. 3.12 and 3.13 that the moment of a force about an axis involves a scalar triple product.) Equation 3.13 stipulates in words that:

*The moment of a force about an axis equals the scalar component in the direction of the axis of the moment vector taken about any point along the axis.*

This is the more powerful formulation that can be used for complex cases.

Note that the unit vector $n$ can have two opposite senses along the axis n, in contrast to the usual unit vectors $i$, $j$, and $k$ associated with the coordinate axes. A moment $M_n$ about the $n$ axis determined from $M \cdot n$ has a sense consistent with the sense chosen for $n$. That is, a positive moment $M_n$ has a sense corresponding to that of $n$, and a negative moment $M_n$ has a sense opposite to that of $n$. If the opposite sense had been chosen for $n$, the sign of $M \cdot n$ would be opposite to that found in the first case. However, the same physical moment is obtained in both cases.

If we specify the moments of a force about *three* orthogonal concurrent axes, we then single out *one* possible point in space for $O$ along the axes. Point $O$, of course, is the origin of the axes. These three moments about orthogonal axes then become the orthogonal scalar components of the moment of $F$ about point $O$, and we can say:

$$M = \text{(moment about the } x \text{ axis)}i +$$
$$\text{(moment about the } y \text{ axis)}j + \tag{3.14}$$
$$\text{(moment about the } z \text{ axis)}k = M_x i + M_y j + M_z k$$

From this relation, we can conclude that:

*The three orthogonal components of the moment of a force about a point are the moments of this force about the three orthogonal axes that have the point as an origin.*

You may now ask what the physical differences are in applications of moments about an axis and moments about a point. The simplest example is in the dynamics of rigid bodies. If a body is constrained so it can only spin about its axis, as in Fig. 3.12, the rotary motion will depend on the moment of the forces about the axis of rotation, as related by a scalar equation. The less familiar concept of moment about a point is illustrated in the motion of bodies that have no constraints, such as missiles and rockets. In these cases, the rotational motion of the body is related by a vector equation to the moment of forces acting on the body about a point called the *center of mass*. (The center of mass will be defined completely later.)

## Example 3.3

Compute the moment of a force $F = 10i + 6j$ N, which goes through position $r_a = 2i + 6j$ m (see Fig. 3.18), about a line going through points 1 and 2 having the respective position vectors

$$r_1 = 6i + 10j - 3k \text{ m}$$
$$r_2 = -3i - 12j + 6k \text{ m}$$

To compute this moment, we can take the moment of $F$ about either point 1 or point 2, and then find the component of this vector along the direction of the displacement vector between 1 and 2 or between 2 and 1. Mathematically, we have, using a displacement vector from point 1 to point $a$, namely $(r_a - r_1)$,

$$M_\rho = [(r_a - r_1) \times F] \cdot \hat{\rho} \qquad (a)$$

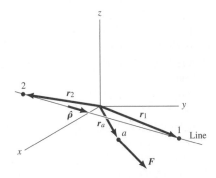

**Figure 3.18.** Find moment of $F$ about line.

where $\hat{\rho}$ is the unit vector along the line chosen to have a sense going from point 2 to point 1. The formulation above is the scalar triple product examined in Chapter 2 and we can use the determinant approach for the calculation once the components of the vectors $(r_a - r_1)$, $F$, and $\hat{\rho}$ have been determined. Thus, we have

$$r_a - r_1 = (2i + 6j) - (6i + 10j - 3k)$$
$$= -4i - 4j + 3k \text{ m}$$
$$F = 10i + 6j \text{ N}$$
$$\hat{\rho} = \frac{r_1 - r_2}{|r_1 - r_2|} = \frac{9i + 22j - 9k}{\sqrt{81 + 484 + 81}}$$
$$= .354i + .866j - .354k$$

We then have, for $M_\rho$:

$$M_\rho = \begin{vmatrix} -4 & -4 & 3 \\ 10 & 6 & 0 \\ .354 & .866 & -.354 \end{vmatrix} = \boxed{13.94 \text{ N-m}} \qquad (b)$$

Because $M_\rho$ is positive, we have a clockwise moment about the line as we look from point 2 to point 1. If we had chosen $\hat{\rho}$ to have an opposite sense, then $M_\rho$ would have been computed as $-13.94$ N-m. Then, we would conclude that $M_\rho$ is a counterclockwise moment about the line as one looks from point 1 to point 2. Note that the same physical moment is determined in both cases.

# Example 3.4

A deep submergence vessel is connected to its mother ship by a cable (Fig. 3.19). The vessel becomes snagged on some rocks and the mother ship steams ahead in a forward direction in an attempt to free the submerged vessel. The connecting cable is suspended from a crane directed up over the water 20 m above the center of mass of the mother ship and 15 m out from the longitudinal axis of the mother ship. The cable transmits a force of 200 kN. It is inclined 50° from the vertical in a vertical plane which, in turn, is oriented 20° from the longitudinal axis of the ship. What is the moment tending to cause the mother ship to roll about its longitudinal axis (i.e., the *x*-axis)?

**Figure 3.19.** Torque on a ship about a horizontal axis through the center of mass.

The position vector from the center of mass *C* to point *A* is

$$r = -15j + 20k \text{ m}$$

Next consider the 200-kN force on the cable. Notice that the cable is in a plane which was rotated 20° counterclockwise from an orientation parallel to the *xz* plane and in this rotated plane, the cable is at an angle of 50° from the vertical. We first decompose the 200-kN force into two coplanar rectangular components, one along the vertical and one along *AE*. We have then

$$F_{CABLE} = -200 \cos 50° \, k + F_{AE} = -128.6k + F_{AE} \text{ kN}$$

Next, decompose $F_{AE}$ into two coplanar rectangular components, one parallel to the *y* axis, and the other parallel to the *x* axis. We then get, noting that $|F_{AE}| = 200 \sin 50°$ kN,

$$F_{AE} = (200 \sin 50°)[-\cos 20°i - \sin 20°j] = -144i - 52.4j \text{ kN}$$

Hence,

$$F_{CABLE} = -144i - 52.4j - 128.6k \text{ kN}$$

To get the desired moment, we have the following formulation:

$$M_x = \left[ r \times F_{CABLE} \right] \cdot i$$
$$M_x = \left[ (-15j + 20k) \times (-144i - 52.4j - 128.6k) \right] \cdot i$$

Using the determinant format for the triple scalar product we get

$$M_x = \begin{vmatrix} 0 & -15 & 20 \\ -144 & -52.4 & -128.6 \\ 1 & 0 & 0 \end{vmatrix} \text{ kN-m}$$

We get on carrying out the determinant

$$M_x = 2{,}977 \text{ kN–m}$$

**3.25.** Disc $A$ has a radius of 600 mm. What is the moment of the forces about the center of the disc? What is the torque of these forces about the axis of the shaft?

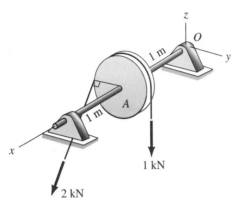

**Figure P.3.25.**

**3.26.** A force $F$ acts at position $(3, 2, 0)$ ft. It is in the $xy$ plane and is inclined at $30°$ from the $x$ axis with a sense directed away from the origin. What is the moment of this force about an axis going through the points $(6, 2, 5)$ ft and $(0, -2, -3)$ ft?

**3.27.** A force $F = 10i + 6j$ N goes through the origin of the coordinate system. What is the moment of this force $F$ about an axis going through points 1 and 2 with position vectors?

$$r_1 = 6i + 3k \text{ m}$$
$$r_2 = 16j - 4k \text{ m}$$

**3.28.** Given a force $F = 10i + 3j$ N acting at position $r = 5j + 10k$ m, what is the torque about the diagonal shown in the diagram? What is the moment about point $E$?

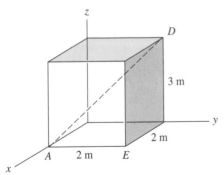

**Figure P.3.28.**

**3.29.** A blimp is moored to a tower at $A$. A force on $A$ from this blimp is

$$F = 5i + 3j + 1.8k \text{ kN}$$

What is the moment about axis $C$ on the ground? Knowledge of this moment and other moments at the base is needed to properly design the foundation of the tower.

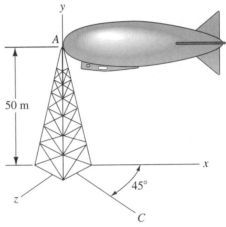

**Figure P.3.29.**

**3.30.** Compute the thrust of the applied forces shown along the axis of the shaft and the torque of the forces about the axis of the shaft.

**Figure P.3.30.**

**3.31.** What is the maximum load $W$ that the crane can lift without tipping about $A$? *Hint:* when tipping is impending, what is the supporting force at the wheel at $B$?

**Figure P.3.31.**

**3.32.** Find the moment of the 1,000-lb force about an axis going between points $D$ and $C$.

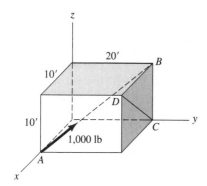

**Figure P.3.32.**

**3.33.** In Problem 3.24, what is the moment of the three indicated forces about axis $GD$?

**3.34.** The base of a fire truck extension ladder is rotated 75° counter-clockwise. The 25-m ladder is elevated 60° from the horizontal. The ladder weight is 20 kN and is regarded as concentrated at a point 10 m up from the base (the lower part of the ladder weighs much more than the upper part). A 900-N fireman and the 500-N young lady he is rescuing are at the top of the ladder. (a) What is the moment at the base of the ladder tending to tip over the fire truck? (b) What is the moment about the horizontal axis having unit vector $\hat{\rho}$ shown in the diagram?

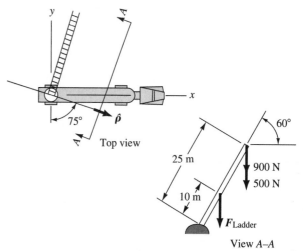

**Figure P.3.34.**

# 3.4 The Couple and Couple Moment

A special arrangement of forces that is of great importance is the *couple. The couple is formed by any two equal parallel forces that have opposite senses* (Fig. 3.20). On a rigid body, a couple has only *one* effect, a "turning" action. Individual forces or combinations of forces that do not constitute couples may "push" or "pull" as well as "turn" a body. The turning action is given quantitatively by the moment of forces about a point or an axis. We shall, accordingly, be most concerned with the moment of a couple, or what we shall call the *couple moment.*

**Figure 3.20.** A couple.

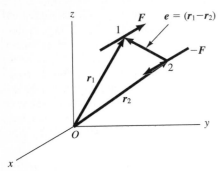

**Figure 3.21.** Compute moment of couple about $O$.

Let us now evaluate the moment of the couple about the origin. Position vectors have been drawn in Fig. 3.21 to points 1 and 2 anywhere along the respective line of action of each force. Adding the moment of each force about $O$, we have for the couple moment $M$

$$M = r_1 \times F + r_2 \times (-F)$$
$$= (r_1 - r_2) \times F \tag{3.15}$$

We can see that $(r_1 - r_2)$ is a displacement vector between points 2 and 1, and if we call this vector $e$, the formulation above becomes

$$M = e \times F \tag{3.16}$$

Since $e$ is in the plane of the couple, it is clear from the definition of a cross product that $M$ is in an orientation normal to the plane of the couple. The sense in this case may be seen in Fig. 3.22 to be directed downward, in accordance with the right-hand-screw rule. Note the use of the double arrow to represent the couple moment. Note also that the rotation of $e$ to $F$, as stipulated in the cross-product formulation, is in the same direction as the "turning" action of the two force vectors, and from now on we shall use the latter criterion for determining the sense of rotation to be used with the right-hand-screw rule.

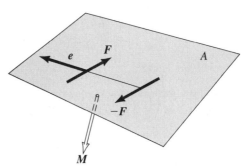

**Figure 3.22.** The couple moment $M$.

Now that the direction of couple moment $M$ has been established for the couple, we need only compute the magnitude for a complete description. Points 1 and 2 may be chosen anywhere along the lines of action of the forces without changing the resulting moment, since the forces are transmissible for taking moments. Therefore, to compute the magnitude of the couple moment vector it will be simplest to choose positions 1 and 2 so that $e$ is *perpendicular* to the lines of action of the forces ($e$ is then denoted as $e_\perp$). From the definition of the cross product, we can then say:

$$|M| = |e_\perp| |F| \sin 90° = |e_\perp| |F| = |F|d \tag{3.17}$$

where the more familiar notation, $d$, has been used in place of $|e_\perp|$ as the perpendicular distance between the lines of action of the forces.

*To summarize the preceding discussions, we may say that: The moment of a couple is a vector whose orientation is normal to the plane of the couple and whose sense is determined in accordance with the right-hand-screw rule, using the "turning" action of the forces to give the proper rotation. The magnitude of the couple moment equals the product of either force magnitude comprising the couple times the perpendicular distance between the forces.*

Note that in the computation of the moment of the couple about origin $O$, the final result in no way involved the position of point $O$. Thus, we can assume immediately that the couple has the *same* moment about every point in space. More about this in the next section.

## 3.5   The Couple Moment as a Free Vector

Had we chosen any other position in space as the origin, and had we computed the moment of the couple about it, we would have formed the same moment vector. To understand this, note that although the position vectors to points 1 and 2 will change for a new origin $O'$, the *difference* between these vectors (which has been termed $e$) does *not* change, as can readily be observed in Fig. 3.23. Since $M = e \times F$, we can conclude that *the couple has the same moment about every point in space.* The particular line of action of the vector representation of the couple moment that is illustrated in Fig. 3.24 is then of little significance and can be moved anywhere. In short, the *couple moment is a free vector.* That is, we may move this vector anywhere in space without changing its meaning, provided that we keep the direction and magnitude intact. Consequently, *for the purpose of taking moments,* we may move the couple itself anywhere in its own or a parallel plane, provided that the direction of turning is not altered—i.e., we cannot "flip" the couple over. In any of these possible planes, we can also change the magnitude of the forces of the couple to other equal values, provided that the distance $d$ is simultaneously changed so that the product $|F|d$ remains the same. Since none of these steps changes the direction or magnitude of the couple moment, all of them are permissible.

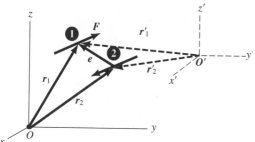

**Figure 3.23.** Vector $e$ is the same for both references.

As we pointed out earlier, the only effect of a couple on a rigid body is its turning action, which is represented quantitatively by the moment of the couple—i.e., the couple moment. Since this is its sole effect, it is only natural to represent the couple by specifications of its moment; its magnitude, then, becomes $|F|d$ and its direction that of its moment. This is the same as identifying a person by her/his job (i.e., as a teacher, plumber, etc.). Thus, in Fig. 3.24, the couple moment $C$ may be used to represent the indicated couple.

**Figure 3.24.** *C* represents couple.

## 3.6   Addition and Subtraction of Couples

Since couples themselves have zero net forces, addition per se of couples always yields zero force. For this reason, the addition and subtraction of couples is interpreted to mean addition and subtraction of the *moments* of the couples. Since couple moments are free vectors, we can always arrange to have a concurrent system of vectors. We shall now take the opportunity to illustrate many of the earlier remarks about couples by adding the two couples shown on the face of the cube in Fig. 3.25. Notice that the couple moment vectors of the couples have been drawn. Since these vectors are free, they may be moved to a convenient position and then added. The total couple moment then becomes 103.2 lb-ft at an angle of 76° with the horizontal, as shown in Fig. 3.26. The couple that creates this turning action is in a plane at right angles to this orientation with a clockwise sense as observed from below.

**Figure 3.25.** Add couples.

**Figure 3.26.** Add couple moments.

This addition may be shown to be valid by the following more elementary procedure. The couples of the cube are moved in their respective planes to the positions shown in Fig. 3.27, which does not alter the moment of the couples, as pointed out in Section 3.5. If the couple on plane $B$ is adjusted to have a force magnitude of 20 lb and if the separating distance is decreased to 5/4 ft, the couple moment is not changed (Fig. 3.28). We thus form a system of forces in which two of the forces are equal, opposite, and collinear and, since these two forces together cannot contribute moment, they may be deleted, leaving a single couple on a plane inclined to the original planes (Fig. 3.29). The distance between the remaining forces is

$$\sqrt{25 + \tfrac{25}{16}} \text{ ft} = 5.16 \text{ ft}$$

and so the magnitude of the couple moment may then be computed to be 103.2 lb-ft. The orientation of the normal to the plane of the couple is readily evaluated as 76° with the horizontal, making the total couple moment identical to our preceding result.

**Figure 3.27.** Move couples.

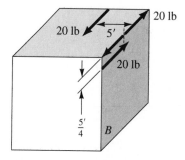

**Figure 3.28.** Change values of two forces.

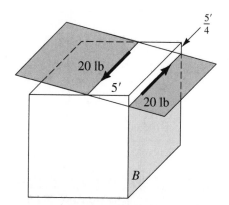

**Figure 3.29.** Eliminate collinear 20-lb forces.

A common notation for couples in a plane is shown in Fig. 3.30. The values given will be that of the couple moments.

**Figure 3.30.** Representation of couple moments in a plane.

## Example 3.5

Replace the system of forces and couple shown in Fig. 3.31 by a single couple moment. Note that the 1,000-N-m couple moment is in the diagonal plane *ABCD*. As a first step in the problem, identify a second couple moment in addition to the couple moment in the diagonal plane.

Examine the vertical forces. There is an upward sum of 1,700 N clearly not colinear with the downward 1,700-N force. These forces form the second couple. To get the couple moment for these forces, we take moments of these forces about the origin as follows:

$$C_1 = 3k \times 800j + (3k + 2i) \times (700 - 1700)j + 2i \times 200j$$
$$= 600i - 1,600k \text{ N-m}$$

As for the 1,000-N-m couple moment, we look in a direction along the $x$ axis toward the origin as shown in Fig. 3.32.

The angle $\alpha$ in Fig. 3.32 is given as follows:

$$\tan \alpha = 3/4 \qquad \therefore \ \alpha = 36.87°$$

Hence we have for the second couple moment $C_2$.

$$C_2 = -1,000 \cos 36.87°k + 1,000 \sin 36.87°j$$
$$= -800k + 600j \text{ N-m}$$

Now we can add the two couple moments to get $C_{TOTAL}$.

$$C_{TOTAL} = C_1 + C_2 = (600i - 1,600k) + (-800k + 600j)$$

$$\boxed{C_{TOTAL} = 600i + 600j - 2,400k \text{ N-m}}$$

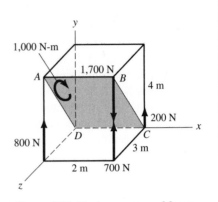

**Figure 3.31.** Replace system of forces by a single couple moment.

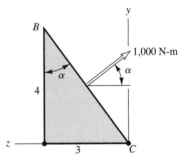

**Figure 3.32.** View along the $x$ axis.

**Figure 3.33.** To find moment of $F$ about A–A.

**Figure 3.34.** To find moment of couple about A–A.

## 3.7   Moment of a Couple About a Line

In section 3.3, we pointed out that the moment of a force $F$ about a line A–A (see Fig. 3.33) is found by first taking the moment of $F$ about *any* point $P$ on A–A and then dotting this vector into $a$, the unit vector along the line. That is,

$$M_{AA} = (r \times F) \cdot a \qquad (3.18)$$

Consider now the moment of a couple about a line. For this purpose, we show a couple moment $C$ and line A–A in Fig. 3.34. As before, we first want the moment of the couple about any point $P$ along A–A. But the moment of $C$ about *every* point in space is simply $C$ itself. Therefore, to get the moment about the line A–A all we need to do is dot $C$ into $a$. Thus,

$$M_{AA} = C \cdot a \qquad (3.19)$$

Since $C$ is a free vector, the moments of $C$ about all lines parallel to A–A must have the same value.

# Example 3.6

Consider the steering mechanism for a go-cart in Fig. 3.35. The linkages are all in a plane oriented at 45° to the horizontal. This plane is perpendicular to the steering column. In a hard turn, the driver exerts oppositely directed forces of 30 lb with each hand in order to turn the 12-in. diameter steering wheel clockwise as the driver looks at the steering wheel. What is the moment applied to each wheel about an axis normal to the ground? Assume half the transmitted torque goes to each wheel.

**Figure 3.35.** Steering mechanism of a go-cart.

The couple moment applied to the steering wheel is readily determined as

$$C = (30)(12)(.707i - .707j) = 254.5i - 254.5j \text{ in-lb}$$

The torque about the vertical axis for each wheel is now easily evaluated. Thus

$$Torque = \frac{1}{2}(254.5i - 254.5j) \cdot j = \boxed{-127.3 \text{ in-lb}}$$

## Example 3.7

In Fig. 3.36, find

(a) the sum of the forces
(b) the sum of the couples
(c) the torque of the entire system about axis $C$–$C$ having direction
     cosines $l = .46$ and $m = .63$ and going through point $A$.

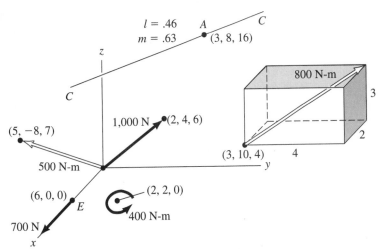

**Figure 3.36.** Force system in space.

(a)
$$\sum F = 700i + 1{,}000\left[\frac{2i + 4j + 6k}{\sqrt{2^2 + 4^2 + 6^2}}\right]$$
$$= 700i + 267.3i + 534.5j + 801.8k$$

$$\sum F = 967.3i + 534.5j + 801.8k \text{ N}$$

(b)
$$\sum C = 400k + 500\left[\frac{5i - 8j + 7k}{\sqrt{5^2 + 8^2 + 7^2}}\right]$$
$$+ 800\frac{4j - 2i + 3k}{\sqrt{4^2 + 2^2 + 3^2}} \text{ N-m}$$

## Example 3.7 (Continued)

$$\sum C = 400k + 212.8i - 340.5j + 297.9k + 594.2j$$
$$- 297.1i + 445.7k$$

$$\sum C = -84.3i + 253.7j + 1,144k \text{ N-m}$$

(c) To find $M_{CC}$ we proceed by first finding the unit vector along $C$–$C$, which we denote as $\hat{c}$. Thus from geometry

$$l^2 + m^2 + n^2 = 1$$
$$\therefore \ .46^2 + .63^2 + n^2 = 1$$
$$n = .6257$$

Hence,

$$\hat{c} = .46i + .63j + .6257k$$

We now get $M_{CC}$.

$$M_{CC} = \left[(r_E - r_A) \times (700i)\right] \bullet \hat{c} + \left[(O - r_A) \times (1,000)\frac{2i + 4j + 6k}{\sqrt{2^2 + 4^2 + 6^2}}\right] \bullet$$

$$+ \left[500\frac{5i - 8j + 7k}{\sqrt{5^2 + 8^2 + 7^2}}\right] \bullet \hat{c} + \left[800\frac{4j - 2i + 3k}{\sqrt{4^2 + 2^2 + 3^2}}\right] \bullet \hat{c} + 400k \bullet \hat{c}$$

$$M_{CC} = \{(6i - 3i - 8j - 16k) \times (700i) + (-3i - 8j - 16k)$$
$$\times (267.3i + 534.5j + 801.8k) + (212.8i - 340.5j + 297.9k)$$
$$+ (594.2j - 297.1i + 445.7k) + 400k\} \bullet (.46i + .63j + .6257k)$$

Carrying out calculations in the large bracket, we get:

$$M_{CC} = \{(5,600k - 11,200j) + (2,138i - 1,872j + 534.9k)$$
$$+ (212.8i - 340.5j + 297.9k) + (594.2j - 297.1i + 445.7k)$$
$$+ 400k\} \bullet (.46i + .63j + .6257k)$$

$$M_{CC} = 944.7 - 8,075 + 4,554 = \boxed{-2,576 \text{ N-m}}$$

**3.35.** A truck driver, while changing a tire, must tighten the nuts holding on the wheel using a torque of 80 lb-ft. If his tire tool has a length such that the forces from his hand are 22 in. apart, how much force must he exert with each hand? To remove the nuts, he exerted 70 lb with each hand. What torque did he apply?

**Figure P.3.35.**

**3.36.** Equal couples in the plane of a wheel are shown in Fig. P.3.36. Explain why they are equivalent for the purpose of turning the wheel. Are they equivalent from the viewpoint of the deformation of the wheel? Explain.

(a)    (b)

**Figure P.3.36.**

**3.37.** Oil-field workers can exert between 50 lb and 125 lb with each hand on a valve wheel (one hand on each side). If a couple moment of 100 lb-ft is required to close the valve, what diameter $d$ must the wheel have?

**Figure P.3.37.**

**3.38.** Two children push with 30 lb of force each on the rail of a 10-ft-diameter merry-go-round. What couple moment do they produce? They fasten a 20-ft-long 2-in. by 4-in. board to the merry-go-round so that the middle of the board is at the middle of the merry-go-round. What is the resulting couple moment if they push on the ends of the board? What moment about the merry-go-round axis would they generate by fastening one end of the board to the middle of the merry-go-round and both pushing in the same direction on the other end?

**3.39.** A posthole digger has a 2-ft-long handle on a 5-ft-long shaft fastened to the digging (scraping) base. From tests, we know that a couple moment of 100 lb-ft is required to dig a posthole in clay, but only 65 lb-ft is needed in sandy soil. What force $F$ must be applied in each case to dig a hole if the distance between the forces from a person's hands is 20 in.?

**Figure P.3.39.**

**3.40.** While stopping, a truck develops 350 N-m of torque at the rear axle due to the action of the brake drum on the axle. What forces are generated at the front and rear supports of the springs to which the axle is attached?

**Figure P.3.40.**

**3.41.** A couple is shown in the $yz$ plane. What is the moment of this couple about the origin? About point (6, 3, 4) m? What is the moment of the couple about a line through the origin with direction cosines $l = 0$, $m = .8$, $n = -.6$? If this line is shifted to a parallel position so that it goes through point (6, 3, 4) m, what is the moment of the couple about this line?

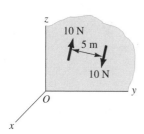

**Figure P.3.41.**

**3.42.** Given the indicated forces, what is the moment of these forces about points $A$ and $B$?

**Figure P.3.42.**

**3.43.** An eight-bladed windmill used for power generation and pumping water stops turning because a bearing on the blade shaft has "frozen up." However, the wind still blows, so each blade is subjected to a 25-lb force perpendicular to the (flat) blade surface. The force effectively acts at 2 ft from the centerline of the shaft to which the blades are attached. The blades are inclined at 60° to the axis of rotation. What is the total thrust of all the blade forces on the windmill shaft? What is the moment on the stalled shaft?

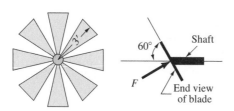

**Figure P.3.43.**

**3.44.** What is the moment of the forces shown about point $A$ and about a point $P$ having a position vector

$$r_p = 10i + 7j + 15k \text{ m?}$$

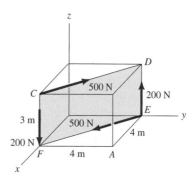

**Figure P.3.44.**

**3.45.** Find the torque about axis $A$–$A$ developed by the 100-lb force and the 3,000-ft-lb couple moment. The position vector $r_1$ is:

$$r_1 = 10i + 8j + 12k \text{ ft}$$

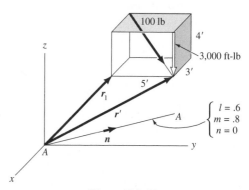

**Figure P.3.45.**

**3.46.** Find $M_{CD}$. Note that the 400-ft-lb couple moment is along the diagonal from point $A$ to point $B$.

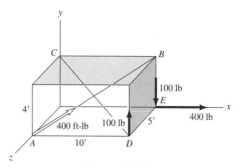

**Figure P.3.46.**

87

**3.47.** Find the torque about an axis going from $A$ to $B$.

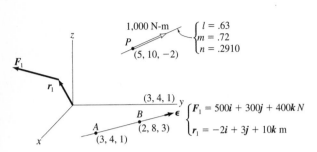

**Figure P.3.47.**

**3.48.** Given the couple moments

$$C_1 = 100i + 30j + 82k \text{ lb-ft}$$
$$C_2 = -16i + 42j \text{ lb-ft}$$
$$C_3 = 15k \text{ lb-ft}$$

what couple will restrain the twisting action of this system about an axis going from

$$r_1 = 6i + 3j + 2k \text{ ft}$$

to

$$r_2 = 10i - 2j + 3k \text{ ft}$$

while giving a moment of 100 lb-ft about the $x$ axis and 50 lb-ft about the $y$ axis?

**3.49.** Equal and opposite forces are directed along diagonals on the faces of a cube. What is the couple moment if $a = 3$ m and $F = 10$ N? What is the moment of this couple about a diagonal from $A$ to $D$?

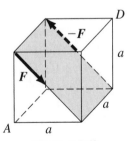

**Figure P.3.49.**

**3.50.** Find the torque about an axis going through points $A$ and $B$.

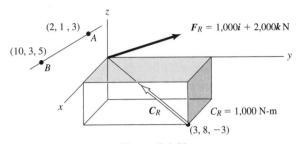

**Figure P.3.50.**

**3.51.** A force $F_1 = 10i + 6j + 3k$ N acts at position $(3, 0, 2)$ m. At point $(0, 2, -3)$ m, an equal but opposite force $-F_1$ acts. What is the couple moment? What are the direction cosines of the normal to the plane of the couple?

**3.52.** Force $F_1 = -16i + 10j - 5k$ N acts at the origin while $F_2 = -F_1$ acts at the end of a rod of length 12 m protruding from the origin with direction cosines $l = .6$, $m = .8$. What is the moment about point $P$ at

$$r_p = 3i + 10j + 15k \text{ m?}$$

What is the twist about an axis going through $P$ and having the unit vector

$$\epsilon = .2i + .8j + .566k?$$

**3.53.** Given the following data:

$$F = 100i + 300j \text{ N}$$
$$C = 200j + 300k \text{ N-m}$$
$$r_1 = 3i - 6j + 4k \text{ m}$$
$$r_2 = 8i + 3j \text{ m}$$

Find the torque about axis $A$–$A$ from $F$ and $C$.

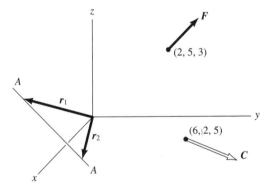

**Figure P.3.53.**

**3.54.** An oil-field pump has two valves, one on top and one on the side, that must be closed simultaneously. The valve wheels are each 27 in. in diameter and are turned with both hands by workers who can exert between 50 lb and 125 lb with each hand. If a weak worker turns the side wheel and a strong worker turns the top wheel, what is the total twisting moment (couple moment) on the pump?

**3.55.** What is the total moment about the origin of the force system shown?

**Figure P.3.55.**

**3.56.** Add the couples whose forces act along diagonals of the sides of the rectangular parallelepiped.

**Figure P.3.56.**

# 3.8 Closure

In this chapter, we have considered several important vector quantities and their properties. In particular, for rigid bodies we found we could take certain liberties with a couple without invalidating the results.

Note in particular that in the chapter on vector algebra, the line of action was of no significance. However, it should now be abundantly clear that in taking moments we *cannot* change the line of action of the force. *Only* the line of action of a *couple moment* can be changed to any parallel position for rigid bodies. Moreover, it is important to remember that we can always move a force along its line of action any time we are computing moments.

We are now ready to pursue in greater detail the important subject of equivalence of force systems for rigid body considerations. We will see that in equivalence considerations of rigid bodies, we again must be careful about what to do with lines of action. They will play a vital role in our deliberations.

**3.57.** An A-frame for hoisting and dragging equipment is held in the position shown by a cable $C$. To determine the cable force, the moment of the applied force about axis $B$–$B$ must be known. What is that moment when a 1,000-N force is applied as shown?

**Figure P.3.57.**

**3.58.** A plumber places his hands 18 in. apart on a pipe threader and can push (and pull) with 80 lb of force. What couple moment does he exert? How much could he exert if he moved his hands to the ends so that his hands are 24 in. apart? What force must he apply at the ends to achieve the same couple moment as when he held his hands 18 in. apart?

**Figure P.3.58.**

**3.59.** Find the torque about a line going from point 1 to point 2.

**Figure P.3.59.**

**3.60.** What is the moment about $A$ of the 500-N force and the 3,000-N-m couple acting on the cantilever beam?

**Figure P.3.60.**

**3.61.** A force $F = 16i + 10j - 3k$ lb goes through point $a$ having a position vector $r_a = 16i - 3j + 12k$ ft. What is the moment about an axis going through points 1 and 2 having respective position vectors given as

$$r_1 = 6i + 3j - 2k \text{ ft}$$
$$r_2 = 3i - 4j + 12k \text{ ft?}$$

**3.62.** Compute the torque about axis $C$–$C$ from the entire force system shown. The axis goes through the origin.

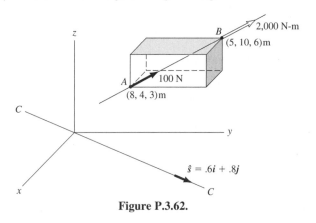

**Figure P.3.62.**

**3.63.** What is the total couple moment of the three couples shown? What is the moment of this force system about point (3, 4, 2) ft? What is the moment of this force system about the position vector $r = 3i + 4j + 2k$ ft taken as an axis? What is the total force of this system?

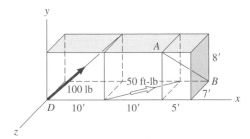

Figure P.3.63.

**3.64.** Find the torque about axis *AB*.

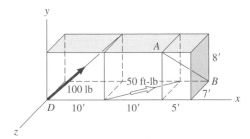

Figure P.3.64.

**3.65.** A force is developed by a liquid on a pipe any time the pipe changes the direction or the speed of flow as a result of an elbow or a nozzle. Such forces can be of considerable magnitude and must be taken into account in building design. We have shown three such forces. What moment stemming from these forces must be counteracted by the support at *O*?

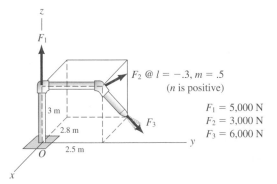

Figure P.3.65.

**3.66.** Compute the moment of the 300-lb force about points $P_1$ and $P_2$.

Figure P.3.66.

**3.67.** A tow truck is inclined at 45° to the edge *A-A* of a ravine with sides sloping at 45° to the vertical. The operator attaches a cable to a wrecked car in the ravine and starts the winch. The cable is oriented normal to *A-A* and develops a force of 15 kN. What are the moments tending to tip over the tow truck about the rear wheels (rocking backward). (*Hint:* Use the position vector from *C* to *B* in Fig. P.3.67(c).) Note that view *D–D* is normal to *A–A* and parallel to the incline.

Figure P.3.67.

91

**3.68.** A surveyor on a 100-m-high hill determines that the corner of a building at the base of the hill is 600 m east and 1,500 m north of her position. What is the position of the building corner relative to another surveyor on top of a 5,000-m-high mountain that is 10,000 m west and 3,000 m south of the hill? What is the distance from the second surveyor to the building corner?

**3.69.** Compute the moment of the 1,000-lb force about supporting points $A$ and $B$.

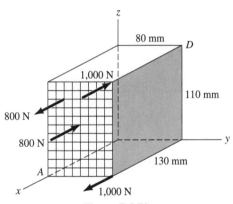

**Figure P.3.69.**

**3.70.** What is the turning action of the forces shown about the diagonal $A$–$D$?

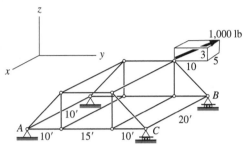

**Figure P.3.70.**

**3.71.** Find the torque of the force system about axis $A$–$B$.

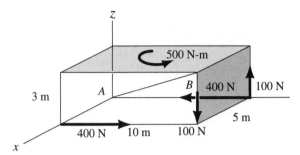

**Figure P.3.71.**

**3.72.** For the system of forces shown, what is the torque about axis $A$–$B$? *Note:* The 100-N and 50-N forces are in the $xz$ plane and are perpendicular to line $AC$.

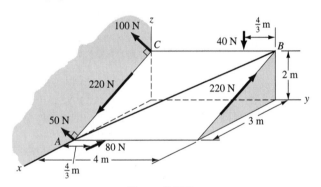

**Figure P.3.72.**

# Equivalent Force Systems

## 4.1 Introduction

In Chapter 1, we defined equivalent vectors as those that have the same capacity in some given situation. We shall now investigate an important class of situations, namely those in which a rigid-body model can be employed. Specifically, we will be concerned with equivalence requirements for force systems acting on a rigid body. Parenthetically, we will begin to see that the line of action plays a vital role in the mechanics of rigid bodies.

The effect that forces have on a rigid body is only manifested in the motion (or lack of motion) of the body induced by the forces. Two force systems, then, are equivalent if they are capable of *initiating* the same motion of the rigid body. The conditions required to give two force systems this equal capacity are:

1. Each force system must exert an equal "push" or "pull" on the body in any direction. For two systems, this requirement is satisfied if the addition of the forces in each system results in equal force vectors.
2. Each force system must exert an equal "turning" action about any point in space. This means that the moment vectors of the force systems for any chosen point must be equal.

Although these conditions will most likely be intuitively acceptable to the reader, we shall later prove them to be necessary and, for certain situations, sufficient for equivalence when we study dynamics.

As a beginning here, we shall reiterate several basic force equivalences for rigid bodies that will serve as a foundation for more complex cases. You should subject them to the tests listed above.

1. The sum of a set of concurrent forces is a single force that is equivalent to the original system. Conversely, a single force is equivalent to any complete set of its components.
2. A force may be moved along its line of action (i.e., forces are transmissible vectors).
3. The only effect that a couple develops on a rigid body is embodied in the couple moment. Since the couple moment is always a free vector, for our purposes at present the couple may be altered in any way as long as the couple moment is not changed.

Note for (1) and (2) above, we cannot change the line of action alone while maintaining equivalence.

In succeeding sections, we shall present other equivalence relations for rigid bodies and then examine perfectly general force systems with a view to replacing them with more convenient and simpler equivalent force systems. These simple replacements are often called *resultants* of the more general systems.

## 4.2  Translation of a Force to a Parallel Position

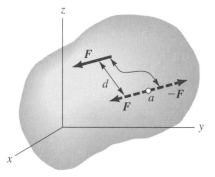

**Figure 4.1.** Insert equal and opposite forces at *a*.

In Fig. 4.1, let us consider the possibility of moving a force $F$ (solid arrow) acting on a rigid body to a parallel position at point *a* while maintaining rigid-body equivalence. If at position *a* we apply equal and opposite forces, one of which is $F$ and the other $-F$, a system of three forces is formed that is clearly equivalent to the original single force $F$. Note that the original force $F$ and the new force in the opposite sense form a couple (the pair is identified by a wavy connecting line). As usual, we represent the couple by its moment $C$, as shown in Fig. 4.2, normal to the plane $A$ of point *a* and the original force $F$. The magnitude of the couple moment $C$ is $|F|d$, where $d$ is the perpendicular distance between point *a* and the original line of action of the force. The couple moment may be moved to any parallel position, including the origin, as indicated in Fig. 4.2.

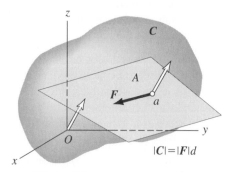

**Figure 4.2.** Equivalent system at *a*.

Thus, we see that *a force may be moved to any parallel position, provided that a couple moment of the correct orientation and magnitude is simultaneously provided.* There are, then, an infinite number of arrangements possible to get the equivalent effects of a single force on a rigid body.[1]

We now present a simple method for computing the couple moment developed on moving a force to a parallel position. Return to Fig. 4.1 and compute the moment $M$ of the original force $F$ about point $a$. We can express this as [see Fig. 4.3 (a)]

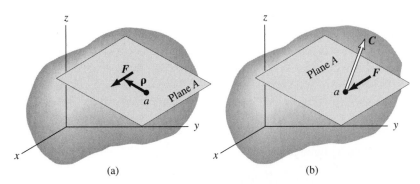

(a)               (b)

**Figure 4.3.** Couple moment  on moving $F$ is $\rho \times F$.

$$M = \rho \times F \qquad (4.1)$$

where $\rho$ is a position vector from $a$ to any point along the line of action of $F$. Now the equivalent force system, shown in Fig. 4.3 (b), must have the *same moment, $M$,* about point $a$ as the original system. Clearly, the moment about point $a$ in Fig 4.3 (b) is due only to the couple moment $C$. That is,

$$M = C \qquad (4.2)$$

Accordingly, we conclude, on comparing the previous two equations, that

$$C = \rho \times F$$

Thus, in shifting a force to pass through some new point, *we introduce a couple whose couple moment equals the moment of the force about this new point.*

We illustrate this in the following examples.

---

[1]A moment of thought will give credence to the above procedure of maintaining rigid body equivalence while moving a force to a different line of action. By moving $F$ to go through point $a$, you are eliminating the moment about $a$ that existed before the move. The couple moment inserted at the time of the move restores this lost moment.

## Example 4.1

A force $F = 6i + 3j + 6k$ lb goes through a point whose position vector is $r_1 = 2i + j + 10k$ ft (see Fig. 4.4). Replace this force by an equivalent force system, for purposes of rigid-body mechanics, going through position $P$, whose position vector is $r_2 = 6i + 10j + 12k$ ft.

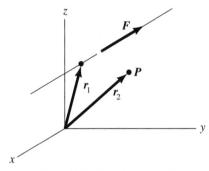

**Figure 4.4.** Move $F$ to point $P$.

The new system will consist of the force $F$ going through the position $r_2$ and, in addition, there will be a couple moment $C$ given as

$$C = \rho \times F = (r_1 - r_2) \times F$$

Inserting values, we have

$$C = [(2i + j + 10k) - (6i + 10j + 12k)]$$
$$\times (6i + 3j + 6k)$$
$$= (-4i - 9j - 2k) \times (6i + 3j + 6k)$$

$$= \begin{vmatrix} -4 & -9 & -2 \\ 6 & 3 & 6 \\ i & j & k \end{vmatrix}$$

Therefore,

$$C = -12k - 12j - 54i + 6i + 24j + 54k$$

$$\boxed{C = -48i + 12j + 42k \text{ ft-lb}}$$

Note that if the position vector $r_1$ to a point along the line of action of $F$ had not been given thus permitting us to get $\rho = (r_1 - r_2)$, we could have used *any* positon vector $\rho$ going from point $P$ to *anywhere* along the line of action of $F$.

# Example 4.2

What is the equivalent force system at position $A$ for the 100-N force shown in Fig. 4.5?

**Figure 4.5.** Find equivalent force system at $A$.

The 100-N force can be expressed vectorially as follows:

$$\boldsymbol{F} = F\frac{\vec{BE}}{\left|\vec{BE}\right|} = 100\left(\frac{-7\boldsymbol{i} - 10\boldsymbol{j} + 8\boldsymbol{k}}{\sqrt{7^2 + 10^2 + 8^2}}\right) \qquad \text{(a)}$$

$$\boldsymbol{F} = -48.0\boldsymbol{i} - 68.5\boldsymbol{j} + 54.8\boldsymbol{k} \text{ N}$$

We then have the force given above at $A$. In addition, we have a couple moment, $\boldsymbol{C}$, found using a position vector, $\boldsymbol{r}$, from $A$ to any point along the line of action of the 100-N force. Thus, choosing point $B$ for $\boldsymbol{r}$, we have

$$\boldsymbol{C} = (10\boldsymbol{i} - 8\boldsymbol{j} + 8\boldsymbol{k}) \times (-48.0\boldsymbol{i} - 68.5\boldsymbol{j} + 54.8\boldsymbol{k})$$

$$= \begin{vmatrix} 10 & -8 & 8 \\ -48.0 & -68.5 & 54.8 \\ \boldsymbol{i} & \boldsymbol{j} & \boldsymbol{k} \end{vmatrix}$$
$$\begin{matrix} 10 & -8 & 8 \\ -48.0 & -68.5 & 54.8 \end{matrix}$$

$$= (10)(-68.5)\boldsymbol{k} + (-48)(8)\boldsymbol{j} + (-8)(54.8)\boldsymbol{i}$$
$$- (8)(-68.5)\boldsymbol{i} - (54.8)(10)\boldsymbol{j} - (-8)(-48.0)\boldsymbol{k}$$

Therefore,

$$\boldsymbol{C} = 109.6\boldsymbol{i} - 932\boldsymbol{j} - 1,069\boldsymbol{k} \text{ N-m} \qquad \text{(b)}$$

The reverse of the procedure just presented may be instituted in reducing a force and a couple *in the same plane* to a *single* equivalent force. This is illustrated in Fig. 4.6 (a) where a couple composed of forces $B$ and $-B$ a distance $d_1$ apart and a force $A$ are shown in plane $N$. The moment representation of the couple is shown with force $A$ in Fig. 4.6 (b).

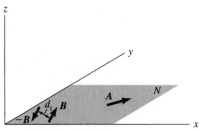

(a) Force $A$ and couple in same plane.

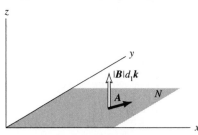

(b) Force $A$ and couple moment.

**Figure 4.6.** A coplanar force and couple.

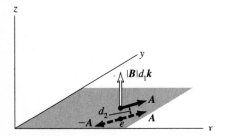

**Figure 4.7.** Equal and opposite forces placed at $e$.

Equal and opposite forces $A$ and $-A$ may next be added to the system at a specific position $e$ (see Fig. 4.7). The purpose of this step is to form another couple moment with a magnitude $|A|d_2$ equal to $|B|d_1$ and with a directon of turning opposite to the original couple moment (see Fig. 4.8) and this dictates the position of $e$. The couple moments then cancel each other out and we are left with only a single force $A$ going through point $e$. Therefore, we can always reduce a force and a couple in the same plane to a single force which clearly must have a *specific line of action* for the case at hand.

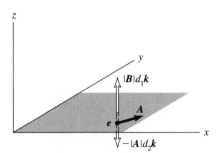

**Figure 4.8.** Adjust $d_2$ so that couple moments cancel.

We now exemplify the above procedure in the following example.

# Example 4.3

In Fig. 4.9(a), we have shown a cantilever beam supporting a single force and a couple in the $xy$ plane. We wish to reduce this system to a single force equivalent to the given system for purposes of rigid-body mechanics.

In Fig. 4.9(b), we have shown the couple moment and a point $e$ to which we shall shift the 1,000-N force. It should be clear on inspection that the couple moment accompanying this shift will have a sign opposite to the original couple moment. Our task now is to get the correct distance $d$ so as to effect a cancellation of the couple moments. Thus we require that

$$d\mathbf{i} \times (1{,}000)\,(.707\mathbf{i} - .707\mathbf{j}) + 550\mathbf{k} = \mathbf{0}$$

$$\therefore\ -(707d)\mathbf{k} + 550\mathbf{k} = \mathbf{0} \qquad \boxed{d = .778\ \text{m}}$$

Note we could have reached the above result more simply if we had resolved the force into rectangular components first. Only the vertical component has nonzero moment about $e$ and so dispensing with vectors, we can directly say

$$-707d + 550 = 0 \qquad d = .778\ \text{m}$$

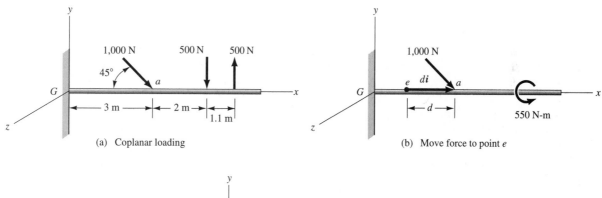

(a) Coplanar loading

(b) Move force to point $e$

(c) Single force at $e$

**Figure 4.9.** Reduction of a coplanar force and couple to a single force with a specific line of action.

# PROBLEMS

In several of the problems of this set we shall concentrate the weight of a body at its center of gravity. Most likely you are used to doing this from an earlier physics course. In Section 4.5 we shall justify this procedure.

**4.1.** Replace the 100-lb force by an equivalent system, from a rigid-body point of view, at A. Do the same for point B. Do this problem by the technique of adding equal and opposite collinear forces and also by using the cross product.

**Figure P.4.1.**

**4.2.** To back an airplane away from the boarding gate, a tractor pushes with a force of 15 kN on the nose wheels. What is the equivalent force system on the landing-gear pivot point, which is 2 m above the point where the tractor pushes?

**Figure P.4.2.**

**4.3.** Replace the 1,000-lb force by equivalent systems at points A and B. Do so by using the addition of equal and opposite collinear force components and by using the cross product.

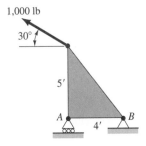

**Figure P.4.3.**

**4.4.** A parking-lot gate arms weighs 150 N. Because of the taper, the weight is concentrated at a point $1\frac{1}{4}$ m from the pivot point. What is the equivalent force system at the pivot point?

**Figure P.4.4.**

**4.5.** A plumber exerts a vertical 60-lb force on a pipe wrench inclined at 30° to the horizontal. What force and couple moment on the pipe are equivalent to the plumber's action?

**Figure P.4.5.**

**4.6.** A tractor operator is attempting to lift a 10-kN boulder. What are the equivalent force systems at A and at B from the boulder?

**Figure P.4.6.**

100

**4.7.** A small hoist has a lifting capacity of 20 kN. What are the largest and smallest equivalent force systems at $A$ for the rated maximum capacity?

**Figure P.4.7.**

**4.8.** Replace the forces by a single equivalent force.

**Figure P.4.8.**

**4.9.** Replace the forces and torques shown acting on the apparatus by a single force. Carefully give the line of action of this force.

**Figure P.4.9.**

**4.10.** A carpenter presses down on a brace-and-bit with a 150-N force while turning the brace with a 200-N force oriented for maximum twist. What is the equivalent force system on the end of the bit at $A$?

**Figure P.4.10.**

**4.11.** A force $F = 3i - 6j + 4k$ lb goes through point (6, 3, 2) ft. Replace this force by an equivalent system where the force goes through point (2, −5, 10) ft.

**4.12.** A force $F = 20i - 60j + 30k$ N goes through a point (10, − 5, 4) m. What is the equivalent system at point $A$ having position vector $r_A = 20i + 3j - 15k$ m?

**4.13.** Find the equivalent force system at the base of the cantilever pipe system stemming from force $F = 1,000$ lb.

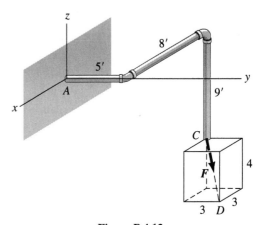

**Figure P.4.13.**

101

**4.14.** Replace the 6,000-N force and the 10,000-N-m couple moment by a single force. Where does this force cross the *x* axis?

**Figure P.4.14.**

**4.15.** In Problem 4.13, the pipe weighs 20 lb/ft. What is the equivalent force system at *A* from the weight of the pipe? [*Hint:* Concentrate the weights of the pipe sections at the respective centers of gravity (geometric centers in this case).]

**4.16.** The operator of a small boom-type crane is trying to drag a chunk of concrete. The boom is 10° above the horizontal and rotated 30° clockwise as seen from above. The cable is directed as shown in the diagram and has 60 kN of tension. What is the equivalent force system at the boom pivot point?

**Figure P.4.16.**

**4.17.** A supplementary supporting guy-wire system for a 200-m-tall tower is tightened. The cables are fastened to the ground at points 120° apart and 100 m from the tower base. What is the equivalent force system acting on the tower base when the tension is 50 kN in cable *AT*, 75 kN in *BT*, and 25 kN in *CT*?

**Figure P.4.17.**

## 4.3   Resultant of a Force System

As defined at the beginning of the chapter, a *resultant of a force system* is a simpler equivalent force system. In many computations it is desirable first to establish a resultant before entering into other computations.

For a general arrangement of forces, no matter how complex, we can always move all forces and couple moments, the latter including both those given and those formed from the movement of forces, to proceed through any single point. The result is then a system of concurrent forces at the point and a system of concurrent couple moments. These systems may then be combined into a single force and a single couple moment. Thus, in Fig. 4.10 we have shown some arbitrary system of forces and couples using full lines. The resultant force and couple moment combination at the origin of a rectangular reference is shown as dashed lines.

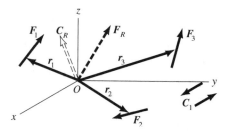

**Figure 4.10.** Resultant of general force system.

Thus, *any force system can be replaced at any point by equivalents no more complex than a single force and a single couple moment.* In special cases, which we shall examine shortly, we may have simpler equivalents such as a single force or a single couple moment. Finally, for *equilibrium of a body,* it is necessary that at any chosen point the simplest resultant system of force and couple moment acting on the body be zero vectors—a fact that will be discussed in dynamics.[2]

The methods of finding a resultant of forces involve nothing new. In moving to any new point, you will recall, there is no change in the force itself other than a shift of line of action; thus, any component of the *resultant* force, such as the *x* component, can simply be taken as the sum of the respective *x* components of all the forces in the system. We may then say for the resultant force

$$F_R = \left[ \sum_p (F_p)_x \right] i + \left[ \sum_p (F_p)_y \right] j + \left[ \sum_p (F_p)_z \right] k \qquad (4.3)$$

The couple moment accompanying $F_R$ for a chosen point $a$ may then be given as

$$C_R = [r_1 \times F_1 + r_2 \times F_2 + \ldots] + [C_1 + C_2 + \ldots] \qquad (4.4)$$

where the first bracketed quantities result from moving the noncouple forces to $a$, and the second are simply the sum of the given couple moments. The vectors $r$ are from $a$ to arbitrary points along the lines of action of the forces. In more compact form, the equation above becomes

$$C_R = \sum_p r_p \times F_p + \sum_q C_q \qquad (4.5)$$

The following example is an illustration of the procedure.

---

[2]When we refer hereafter to a resultant, we shall mean the simplest resultant.

# Example 4.4

Two forces and a couple are shown in Fig. 4.11, the couple being positioned in the *zy* plane. We shall find the resultant of the system at the origin *O*.

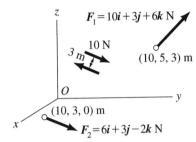

**Figure 4.11.** Find resultant at *O*.

At *O* we will have a set of two concurrent forces, which may be added to give $F_R$:

$$F_R = (10 + 6)i + (3 + 3)j + (6 - 2)k$$

$$\boxed{F_R = 16i + 6j + 4k \text{ N}}$$

The resultant couple moment at point *O* is the vector sum of the couple-moment vectors developed by moving the two forces, plus the couple moment of the couple in the *zy* plane. Thus,

$$C_R = r_1 \times F_1 + r_2 \times F_2 - 30i \text{ N-m}$$

Now

$$r_1 \times F_1 = (10i + 5j + 3k) \times (10i + 3j + 6k)$$
$$= 21i - 30j - 20k \text{ N-m}$$
$$r_2 \times F_2 = (10i + 3j) \times (6i + 3j - 2k)$$
$$= -6i + 20j + 12k \text{ N-m}$$

Hence,

$$\boxed{C_R = -15i - 10j - 8k \text{ N-m}}$$

The resultant is shown in Fig. 4.12.

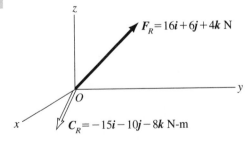

**Figure 4.12.** Resultant at *O*.

# Example 4.5

What is the resultant at $A$ of the applied loads acting in Fig. 4.13? The forces are directed to intersect the centerline of the shaft along which we

**Figure 4.13.** Find resultant at $A$; $F_1$ and $F_3$ are concurrent.

have placed the $z$ axis. We first express the loads vectorially. Thus,

$$F_1 = F_1 \hat{d}_1 = 150 \left( \frac{-10k - 3i + 4j}{\sqrt{10^2 + 3^2 + 4^2}} \right)$$

$$= -40.2i + 53.7j - 134.1k \text{ lb}$$

$$F_2 = F_2 \hat{d}_2 = 200 \left( \frac{-13k + 7i}{\sqrt{13^2 + 7^2}} \right)$$

$$= 94.8i - 176k \text{ lb}$$

$$F_3 = -100j \text{ lb}$$

$$C = -50k \text{ ft-lb}$$

We can now readily find the resultant force system at $A$. Thus,[3]

$$F_R = (-40.2 + 94.8)i + (53.7 - 100)j + (-134.1 - 176.0)k$$

$$\boxed{F_R = 54.6i - 46.3j - 310k \text{ lb}}$$

$$C_R = (-11k) \times F_1 + (-8k) \times (F_2 + F_3) + (-50k)$$
$$= -11k \times (-40.2i + 53.7j - 134.1k) + (-8k)$$
$$\times (94.8i - 176.0k - 100j) - 50k$$

$$\boxed{C_R = -209i - 316j - 50k \text{ ft-lb}}$$

---

[3]Remember that for $C$ resulting from a movement of a force, the position vector goes from the point (in this case point A) to the line of action of the force.

# 4.4  Simplest Resultants of Special Force Systems

We shall now consider special but important force systems and will establish the *simplest* resultants possible for each case. Examples will serve to illustrate the method of procedure.

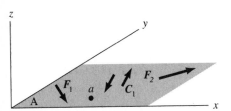

**Figure 4.14.** Coplanar force system.

**Case A. Coplanar Force Systems.** In Fig. 4.14, a system of forces and couples is shown in plane A. By moving the forces to a common point $a$ in plane A, we will form additional couples in the plane. The force portion of the equivalent system at such a point will be given as

$$F_R = \left[\sum_p (F_p)_x\right] i + \left[\sum_p (F_p)_y\right] j \tag{4.6}$$

The couple moment portion of the equivalent system can be given as (using the right hand rule for proper signs):

$$C_R = (F_1 d_1 + F_2 d_2 + \ldots)k + (C_1 + C_2 + \ldots)k \tag{4.7}$$

where $d_1$, $d_2$, etc., are perpendicular distances from point $a$ to the lines of action of the noncouple forces, and $C_1$, $C_2$, etc., are the values of the given couple moments. The resultant at $a$ is shown in Fig. 4.15.

If $F_R \neq 0$, (i.e., if $\sum_p F_x \neq 0$ and/or $\sum_p F_y \neq 0$) we can move the force

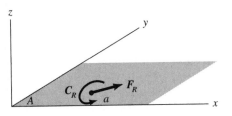

**Figure 4.15.** Resultant at point $a$.

from $a$ to yet a new parallel position so as to introduce a second couple moment to cancel $C_R$ of Fig. 4.15 in the manner described earlier in Section 4.2. Since the $x$ and $y$ directions used are arbitrary, except for the condition that they be in the plane of the forces, we can make the following conclusion. *If the force components in any direction in the plane add to other than zero, we may replace the entire coplanar system by a single force with a specific line of action.*

What happens if $\sum_p F_x = 0$ and $\sum_p F_y = 0$? Without a force at point $a$, we can no longer eliminate a couple in plane A. Thus, our second conclusion is that *if $\sum_p F_x$ and $\sum_p F_y$ are zero, the resultant must be a couple moment or be zero.*

In the coplanar case, therefore, the simplest equivalent force system must be a single force along a specific line of action, a single couple moment, or a null vector. The following example is used to illustrate the method of determining such a resultant directly without the intermediate steps followed in this discussion.

## Example 4.6

Consider a coplanar force system shown in Fig. 4.16. The *simplest* resultant is to be found. Since $\sum_p F_x$ and $\sum_p F_y$ are not zero, we know that we can replace the system by a single force, which is

$$F_R = 6i + 13j \ \text{N} \tag{a}$$

We now need to find the line of action in the plane that will make this single force equivalent to the given system. To be equivalent for rigid-body mechanics, this force without a couple moment must have the same turning action about any point or axis in space as that of the given system. Now the simplest resultant force must intercept the $x$ axis at some point $\bar{x}$.[4] We can determine $\bar{x}$ by equating the moment of the resultant force without a couple moment about the origin with that of the original system of forces and couples. Using the vector $\bar{x}i$ as a position vector from the origin to the line of action of $F_R$ (see Fig. 4.17), we accordingly have

$$\begin{aligned} \bar{x}i \times (6i + 13j) &= (8i + 2j) \\ \times (6i + 3j) &+ (5i + 3j) \times (10j) - 30k \end{aligned} \tag{b}$$

Carrying out the cross products,

$$24k - 12k + 50k - 30k = 13\bar{x}k \tag{c}$$

Hence,

$$\bar{x} = 2.46 \ \text{m}$$

By specifying the $x$ intercept, $\bar{x}$, we fully determine the line of action of the simplest resultant force. We could have also used the intercept with the $y$ axis, $\bar{y}$, for this purpose. In that case, the position vector from the origin out to the line of action is $\bar{y}j$, and we have, on equating moments about $O$ of the resultant without a couple moment with that of the original system:

$$\bar{y}j \times (6i + 13j) = (8i + 2j) \times (6i + 3j) + (5i + 3j) \times (10j) - 30k$$

$$\bar{y} = -5.35 \ \text{m}$$

[4]If the resultant force is parallel to the $x$ axis, the intercept will be at infinity.

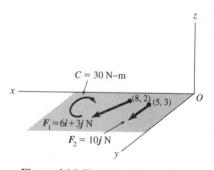

$C = 30$ N–m

$(8, 2)$ $(5, 3)$

$F_1 = 6i + 3j$ N

$F_2 = 10j$ N

**Figure 4.16.** Find simplest resultant.

$\bar{x}$

$F_R$

$\bar{x}i$

**Figure 4.17.** Simplest resultant.

## Example 4.7

Compute the *simplest* resultant for the loads shown acting on the beam in Fig. 4.18(a). Give the intercept with the x axis.

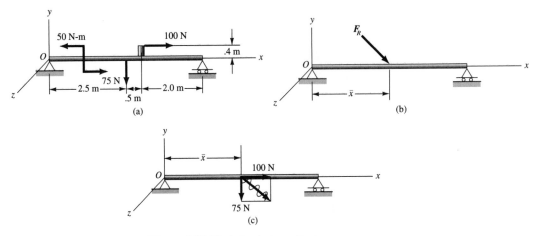

**Figure 4.18.** Find simplest resultant.

It is immediately apparent on inspection of the diagram that

$$F_R = 100i - 75j \text{ N} \tag{a}$$

Let $\bar{x}$ be the intercept with the x axis of the line of action of $F_R$ when this line of action corresponds to zero couple moment $C_R$ [see Fig. 4.18(b)]. In Fig. 4.18(c), we have decomposed $F_R$ along this line of action into rectangular components so as to permit simple calculations of moments about the origin $O$ (here we mean moments about the z axis). Accordingly, equating moments about the z axis of $F_R$ without a couple moment, with that of the original system of loads, we get,

$$-(75)(\bar{x}) = 50 - (2.5)(75) - (.4)(100)$$

$$\bar{x} = 2.37 \text{ m}$$

Thus, the simplest resultant is a force $100i - 75j$ N intercepting the beam axis at a position $\bar{x} = 2.37$ m.

As pointed out earlier, in the instance wherein $F_R = 0$, we then possibly have as the simplest resultant a couple moment normal to the plane of the coplanar force system. There is also the possibility that there is zero couple moment, in which case the forces of the coplanar force system *completely cancel* each other's effects on a rigid body. To find the couple moment for the case where $F_R = 0$, we simply take moments of the coplanar force system about *any point* in space. This moment, if it is not equal to zero, is clearly the couple-moment vector sought.

# Example 4.8

What is the simplest resultant for the forces shown acting on beam $AB$ in Fig. 4.19?

**Figure 4.19.** Coplanar loading on a simply supported beam.

Our first step will be to compute the resultant force by adding up the force vectors. Thus

$$F_R = 1,500j - 666.2i - (1,585.8)(.5)j$$
$$+ (1,585.8)(.866)i - 707.1i - 707.1j$$

Collecting terms, we have

$$F_R = (-666.2 + 1,373.3 - 707.1)i + (1,500 - 792.9 - 701.1)j = 0$$

The simplest resultant clearly must be either a couple moment or be a null vector. For this information, we shall take moments about point $A$.

$$C_R = \{[.3 - (.2)(.707)]i - (.2)(.707)j\} \times (-666.2i)$$

$$+ (.3)(1,500)k - (.6)(1,585.8)(.5)k$$

$$+ (1i + .2j) \times (-707.1i - 707.1j)$$

$$C_R = -94.20k + 450k - 475.7k$$

$$- 707.1k + 141.4k = \boxed{-685.6k \text{ N-m}}$$

We have a couple moment in the minus $z$ direction having any arbitrary line of action as the simplest resultant.

It is important that the nature of the equivalence just instituted be clearly understood. Thus, for finding the supporting force system, we can use the undeformed geometry and hence the single force replacement. However, for finding the deflection of the beam, it should be obvious that the replacement is invalid. Note, finally, that there is only one point on the beam that will allow for a single force to be equivalent to the original system for purposes of rigid-body considerations.

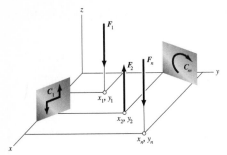

**Figure 4.20.** Parallel system of forces.

**Case B. Parallel Force Systems in Space.** Now, consider the system of $n$ parallel forces in Fig. 4.20, where the $z$ direction has been selected parallel to the forces. We also include $m$ couples whose planes are parallel to the $z$ direction because such couples can be considered to be composed of equal and opposite forces parallel to the $z$ direction. We can move the forces so that they all pass through the origin of the $xyz$ axes; the force portion of the equivalent system is then

$$F_R = \left(\sum_{p=1}^{n}(F_p)\right)k \tag{4.8}$$

The couple moment portion of the equivalent system is found by applying Eq. 4.5 to this case:

$$C_R = \sum_{p=1}^{n}\left[\left(x_p i + y_p j\right) \times F_p k\right] + \sum_{p=1}^{m}\left[\left(C_p\right)_x i + \left(C_p\right)_y j\right] \tag{4.9}$$

where $F_p$ represents the noncouple forces. Carrying out the cross product, we get

$$C_R = \sum_{p=1}^{n}\left[\left(F_p y_p\right)i - \left(F_p x_p\right)\right]j + \sum_{p=1}^{m}\left[\left(C_p\right)_x i + \left(C_p\right)_y j\right] \tag{4.10}$$

From this, we see that the couple moment must always be parallel to the $xy$ plane (i.e., perpendicular to the direction of the forces). We then have at the origin a single force and a single couple moment at right angles to each other [see Fig. 4.21(a)]. If $F_R \neq 0$, we can move $F_R$ again to another line of action in a plane $A$ perpendicular to $C_R$ [see Fig. 4.21(b)] and, choosing the proper value of $d$, ensure that $F_R d = |C_R|$ with a sense opposite to $C_R$[5] such that we eliminate the couple moment. We thus end up with a *single* force having a particular line of action specified by the intercept $\bar{x}, \bar{y}$ of the line of action of the force with the $xy$ plane. If the summation of forces should happen to be zero, the equivalent system must then be a couple moment or a null vector.

Thus, *the simplest resultant system of a parallel force system is either a force with a specific line of action, a single couple moment, or a null vector.* The following example will illustrate how we can directly determine the simplest resultant.

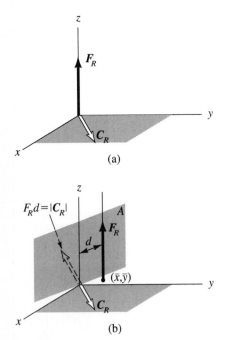

**Figure 4.21.** Simplest resultant for parallel force system.

[5]Here we urge some caution! The proper displacement vector to be used in computing the moment induced by moving $F$ is found by going from the *new* position to the *original* position. The displacement vector here accordingly must go from right to left a distance $d$. Recall in this regard that when moving a force to a line of action through a point $a$, the position vector used in computing the induced couple moment always goes *from the point to the force* [see Fig. 4.3(a)].

# ■ Example 4.9

Find the simplest resultant of the parallel force system in Fig. 4.22(a).

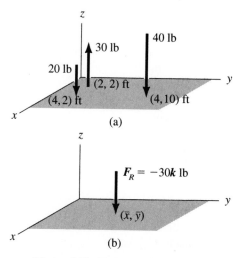

(a)

(b)

**Figure 4.22.** Find simplest resultant.

The sum of the forces is 30 lb in the negative $z$ direction. Hence, a position can be found in which a single force is equivalent to the original system. Assume that this resultant force without a couple moment proceeds through the point $\bar{x}$, $\bar{y}$ [Fig. 4.22(b)]. We can equate the moment of this resultant force about the $x$ and $y$ axes with the corresponding moments of the original system and thus form the scalar equations that yield the proper value of $\bar{x}$ and $\bar{y}$. Equating moments about the $x$ axis, we get

$$(30)(2) - (20)(2) - (40)(10) = -30\bar{y}$$

Therefore,

$$\bar{y} = 12.7 \text{ ft}$$

Equating moments about the $y$ axis, we have

$$-(30)(2) + (20)(4) + (40)(4) = 30\bar{x}$$

Therefore,

$$\bar{x} = 6 \text{ ft}$$

You can also show, as an exercise, that the same result can be reached for $\bar{x}$, $\bar{y}$ by equating moments of the resultant force without a couple moment about the origin with that of the original system about the origin.

## Example 4.10

Consider the parallel force system in Fig. 4.23(a). What is the simplest resultant?

Here we have a case where the sum of the forces is zero and so $F_R = 0$. Therefore, the simplest resultant must be a couple moment or be a null vector. To get this couple moment, $C_R$, we can take moments of the forces about *any point* in space. This moment vector then equals the desired couple moment $C_R$. One procedure is to use the origin of the reference as the point about which to take moments. Then we can say that

$$C_R = (4i + 2j) \times (-30k) + (3i + 2j) \times (40k) + (2i + 4j) \times (-10k)$$

$$= -20i + 20j \text{ N-m} \tag{a}$$

The rectangular components of $C_R$ along the x and y axes are the moments of the force system about these axes. Thus,

$$(C_R)_x = -20 \text{ N-m}$$
$$(C_R)_y = 20 \text{ N-m} \tag{b}$$

We can get the moments of the forces about the x and y axes directly and thus generate the components of the desired couple moment $C_R$. Accordingly, using the elementary definition of the moment of a force about a line as presented earlier, we have

$$(C_R)_x = -(10)(4) + (40)(2) - (30)(2) = -20 \text{ N-m}$$
$$(C_R)_y = (10)(2) - (40)(3) + (30)(4) = 20 \text{ N-m}$$

Thus, the moment of the force system about the origin, and hence about any point, is then the desired couple moment (Fig. 4.23(b))

$$C_R = -20i + 20j \text{ N-m}$$

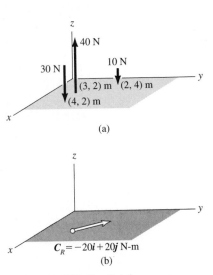

(a)

(b)

**Figure 4.23.** Parallel force system.

**Figure 4.24.** Examples of the so-called wrench. This is the simplest representation of a general force system.

Now that we have considered the concept of the simplest resultant for coplanar and parallel force systems, we wish to go back to the *general force* systems for a moment. We learned earlier that we can always replace such a system in rigid-body mechanics by a single force $F_R$ and a single couple moment $C_R$ at any chosen point. Is this always the very simplest system for rigid-body mechanics? No, it is not. To show this, decompose the couple moment $C_R$ into two rectangular components $C_\perp$ and $C_\parallel$, perpendicular to the force and collinear with the force, respectively. We can now conceivably move the force to a specific parallel position and can eliminate $C_\perp$, the component of couple moment normal to the force. However, there is nothing that we can do about the $C_\parallel$ component of couple moment collinear (or parallel) to the force. The reason for this is that any movement of the force to a parallel position *always* introduces a couple moment *perpendicular* to the force. Thus the component $C_\parallel$ cannot be affected. By eliminating $C_\perp$ we end up with the force $F_R$ and $C_\parallel$ collinear with $F_R$. This system is the simplest in the general case and it is called a *wrench* (see Fig. 4.24). However, we shall not generally use the wrench concept in this text and will work instead with the resultant force $F_R$ and the couple moment $C_R$ at any chosen point.

# PROBLEMS

*In several of the problems of this set we shall concentrate the weight of a body at its center of gravity. Most likely you are used to doing this from your previous physics course. In Section 4.5 we shall justify this procedure.*

**4.18.** Compute the resultant force system of the applied loads at positions A and B.

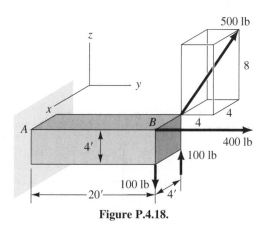

**Figure P.4.18.**

**4.19.** Compute the resultant force system at A stemming from the indicated 50-lb force. What is the twist developed about the axis of the shaft at A? The 50-lb force is normal to the wrench.

**Figure P.4.19.**

**4.20.** Find the resultant of the force system at point A. The 300-N, 200-N, and 900-N loads are at the centers of the pipe sections.

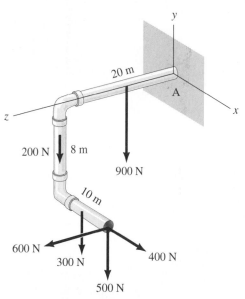

**Figure P.4.20.**

**4.21.** A 20-kN car and an 80-kN truck are stopped on a bridge. What is the resultant force system of these vehicles at the center of the bridge? At the center of the left end of the bridge? The distances given to truck and car are to respective centers of gravity where we can concentrate the weights.

**Figure P.4.21.**

113

**4.22.** Two heavy machinery crates (A weighs 20 kN and B weighs 30 kN) are placed on a truck. What is the resultant force system at the center of the rear axle? The centers of gravity of the crates, where we can concentrate the weights, are at the geometric centers.

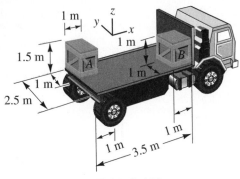

**Figure P.4.22.**

**4.23.** Replace the system of forces by a resultant at A.

**Figure P.4.23.**

**4.24.** Evaluate Forces $F_1$, $F_2$, and $F_3$ so that the resultant of the forces and torque acting on the plate is zero in both force and couple moment. (*Hint:* If the resultant is zero for one point, will it not be zero for any point? Explain why.)

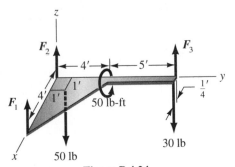

**Figure P.4.24.**

**4.25.** Find the *simplest* resultant of the forces shown acting on the beam. Give the intercept with the axis of the beam.

**Figure P.4.25.**

**4.26.** Find the *simplest* resultant of the forces shown acting on the pulley. Give the intercept with the $x$ axis.

**Figure P.4.26.**

**4.27.** A man raises a 50-lb bucket of water to the top of a bricklayer's scaffold. Also, a Jeep winch is used to raise a 200-lb load of bricks. What is the *simplest* resultant force system on the scaffold? Give the $x$ intercept. Consider the pulleys to be frictionless so that the 50-lb force and the 200-lb force are transmitted respectively to the man and to the Jeep.

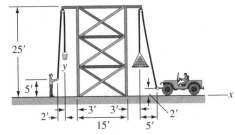

**Figure P.4.27.**

**4.28.** Find the resultant at A.

**Figure P.4.28.**

**4.29.** Compute the *simplest* resultant for the loads acting on the beam. Give the intercept with the axis of the beam.

**Figure P.4.29.**

**4.30.** Find the *simplest* resultant for the forces. Give the location of this resultant clearly.

**Figure P.4.30.**

**4.31.** Replace the system of forces acting on the rivets of the plate by the *simplest* resultant. Give the intercept of this resultant with the x axis.

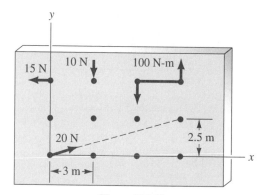

**Figure P.4.31.**

**4.32.** A parallel system of forces is such that: a 20-N force acts at position $x = 10$ m, $y = 3$ m; a 30-N force acts at position $x = 5$ m, $y = -3$m; a 50-N force acts at position $x = -2$ m, $y = 5$ m.

   (a)   If all forces point in the negative $z$ direction, give the *simplest* resultant force and its line of action.

   (b)   If the 50-N force points in the plus $z$ direction and the others in the negative $z$ direction, what is the *simplest* resultant?

**4.33.** What is the *simplest* resultant of the three forces and couple shown acting on the shaft and disc? The disc radius is 5 ft.

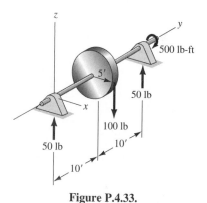

**Figure P.4.33.**

**4.34.** What is the *simplest* resultant for the system of forces? Each square is 10 mm on edge.

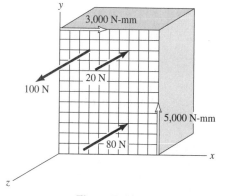

**Figure P.4.34.**

115

**4.35.** What is the simplest resultant? Where does its line of action cross the *x* axis?

**Figure P.4.35.**

**4.36.** What is the simplest resultant of the loadings shown? Be sure to give its line of action.

**Figure P.4.36.**

**4.37.** Two hoists are operated on the same overhead track. Hoist *A* has a 3,000-kN load, and hoist *B* has a 4,000-kN load. What is the resultant force system at the left end *O* of the track? Where does the *simplest* resultant force act?

**Figure P.4.37.**

**4.38.** A lo-boy trailer weighs 16,000 lb and is loaded with a 15,000-lb bulldozer *A* and a 12,000-lb front-end loader *B*. What is the simplest resultant force and where does it act? The weights of the machines and trailer act at their respective centers of gravity (C.G.).

**Figure P.4.38.**

**4.39.** Where should a 100-N force in a downward direction be placed for the *simplest* resultant of all shown forces to be at position (5, 5) m?

**Figure P.4.39.**

**4.40.** A barge must be evenly loaded so it does not list in any direction. Where can the three large machinery crates be placed (without either hanging over the edge, stacking, or standing on end)? Each crate is as tall as it is wide. Is there only one solution to this problem? Centers of gravity of crates correspond to geometric centers.

**Figure P.4.40.**

# 4.5   Distributed Force Systems

Our discussions up to now have been restricted to discrete vectors—in particular, to point forces. Vectors as well as scalars, may also be continuously distributed throughout a finite volume. Such distributions are called *vector* and *scalar fields*, respectively. A simple example of a scalar field is the temperature distribution, expressed as $T(x, y, z, t)$, where the variable $t$ indicates that the field may be changing with time. Thus, if a position $x_0$, $y_0$, $z_0$ and a time $t_0$ are specified, we can determine the temperature at this position and time provided that we know the temperature distribution function (i.e., how $T$ depends on the independent variables $x$, $y$, $z$, and $t$). A vector field is sometimes expressed in the form $F(x, y, z, t)$. A common example of a vector field is the gravitational force field of the earth—a field that is known to vary with elevation above sea level, among other factors. Note, however, that the gravitational field is virtually constant with time.

In place of the vector field, it is more convenient at times to employ three scalar fields that represent the orthogonal scalar components of a vector field at all points. Thus, for a force field we can say:

$$\text{force component in } x \text{ direction} = g(x, y, z, t)$$
$$\text{force component in } y \text{ direction} = h(x, y, z, t)$$
$$\text{force component in } z \text{ direction} = l(x, y, z, t)$$

where $g$, $h$, and $l$ represent functions of the coordinates and time. If we substitute coordinates of a special position and the time into these functions, we get the force components $F_x$, $F_y$, and $F_z$ for that position and time. The force field and its component scalar fields are then related in this way:

$$F(x, y, z, t) = g(x, y, z, t)i + h(x, y, z, t)j + l(x, y, z, t)k$$

More often, the notation for the equation above is written

$$F(x, y, z, t) = F_x(x, y, z, t)i + F_y(x, y, z, t)j + F_z(x, y, z, t)k \quad (4.11)$$

Vector fields are not restricted to forces but include other quantities such as velocity fields and heat-flow fields.

Force distributions, such as gravitational force, that exert influence directly on the elements of mass distributed throughout the body are termed *body force distributions* and are usually given per unit of mass that they directly influence. Thus, if $B(x, y, z, t)$ is such a body force distribution, the force on an element $dm$ would be $B(x, y, z, t)dm$.

Force distributions over a *surface* are called *surface force distributions*[6] $T(x, y, z, t)$ and are given per unit area of the surface directly influenced. A simple example is the force distribution on the surface of a body submerged in a fluid. In the case of a static fluid or of a frictionless fluid, the force from the fluid on an area element is always normal to the area element and directed

---

[6]Surface forces are often called *surface tractions* in solid mechanics.

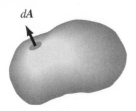

**Figure 4.25.** Area vector.

in toward the body. The force per unit area stemming from such fluid action is called *pressure* and is denoted as $p$. Pressure is a scalar quantity. The direction of the force resulting from a pressure on a surface is given by the orientation of the surface. [You will recall from Chapter 2 that an area element can be considered as a vector which is normal to the area element and directed outward from the enclosed body (Fig. 4.25).] The infinitesimal force on the area element is then given as

$$d\mathbf{f} = -p\ d\mathbf{A}$$

A more specialized, but nevertheless common, force distribution is that of a continuous load on a beam. This is often a parallel loading distribution that is symmetrical about the center plane $xy$ of a beam, as illustrated in Fig. 4.26. Various heights of bricks stacked on a beam would be an example of this kind of loading. We can replace such a loading by an equivalent coplanar distribution that acts at the center plane. The loading is given per unit length and is denoted as $w$, the *intensity of loading*. The force on an element $dx$ of the beam, then, is $w\ dx$.

**Figure 4.26.** Loading on a beam.

We have thus presented force systems distributed throughout volumes (body forces), over surfaces (surface forces or tractions), and over lines. The conclusions about resultants that were reached earlier for general, parallel, and coplanar point force systems are also valid for these distributed force systems. These conclusions are true because each distributed force system can be considered as an infinite number of infinitesimal point forces of the type used heretofore. We shall illustrate the handling of force distributions in the following examples.

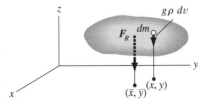

**Figure 4.27.** Gravity body force distribution.

## Case A. Parallel Body Force System—Center of Gravity.
Consider a rigid body (Fig. 4.27) whose density (mass/unit volume) is given as $\rho(x, y, z)$. It is acted on by gravity, which, for a small body, may be considered to result in a distributed parallel force field.

Since we have here a parallel system of forces in space with the same sense, we know that a single force without a couple moment along a certain line of action will be equivalent to the distribution. The gravity body force

$B(x, y, z)$ given per unit mass is $-g\mathbf{k}$. The infinitesimal force on a differential mass element dm, then , is $-g(\rho\, dv)\mathbf{k}$, where $dv$ is the volume of the element.[7] We find the resultant force on the system by replacing the summation in Eq. 4.8 with an integration, Thus,

$$F_R = -\int_V g(\rho\, dv)\mathbf{k} = -g\mathbf{k}\int_V \rho\, dv = -gM\mathbf{k}$$

where, with $g$ as a constant, the second integral becomes simply the entire mass of the body $M$.

Next, we must find the line of action of this single equivalent force without a couple moment. Let us denote the intercept of this line of action with the $xy$ plane as $\bar{x}$, $\bar{y}$ (see Fig. 4.27). The resultant at this position must have the same moments as the distribution about the $x$ and $y$ axes:

$$-F_R\bar{y} = -g\int_V y\rho\, dv \qquad F_R\bar{x} = -g\int_V x\rho\, dv,$$

Hence, we have

$$\bar{x} = \frac{\int x\rho\, dv}{M} \qquad \bar{y} = \frac{\int y\rho\, dv}{M}$$

Thus, we have fully established the simplest resultant. Now, the body is reoriented in space, keeping with it the line of action of the resultant as shown in Fig. 4.28. A new computation of the line of action of the simplest resultant for the second orientation yields a line that intersects the original line at a point $C$. It can be shown that lines of action for simplest resultants for all other orientations of the body must intersect at the same point $C$. We call this point the *center of gravity*. Effectively, we can say for rigid-body considerations that all the weight of the body can be assumed to be concentrated at the center of gravity.

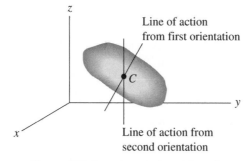

**Figure 4.28.** Location of center of gravity

---

[7]Note that $g\rho$ is the weight per unit volume, which is often given as $\gamma$, the so-called specific weight.

# Example 4.11

Find the center of gravity of the triangular block having a uniform density $\rho$ shown in Fig. 4.29.

The total weight of the body is easily evaluated as

$$F_R = g\rho\left(abc/2\right) \tag{a}$$

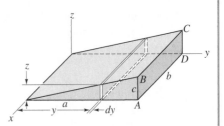

To find $\bar{y}$, we will equate the moment of $F_R$ about the $x$ axis with that of the weight distribution of the block. To facilitate the latter, we shall choose within the block *infinitesimal* elements whose weights are easily computed. Also, the moment of the weight of each element about the $x$ axis is to be likewise easily computed. Infinitesimal slices of thickness $dy$ parallel to the $xz$ plane fulfill our requirements nicely. The weight of such a slice is simply $(zb\,dy)\rho g$, where $z$ is the height of the slice (see Fig. 4.29). Because all points of the slice are at the same coordinate distance $y$ from the $x$ axis, clearly the moment of the weight of the slice is easily computed as $-y(zb\,dy)\rho g$. By letting $y$ run from 0 to $a$ during an integration, we can account for all the slices in the body. Thus, we have

**Figure 4.29.** Find center of gravity.

$$-F_R\bar{y} = -\int_0^a y(zb\,dy)\rho g \tag{b}$$

The term $z$ can be expressed with the aid of similar triangles in terms of the integration variable $y$ as follows:

$$\frac{z}{c} = \frac{y}{a}$$

$$z = \left(\frac{y}{a}\right)c \tag{c}$$

We then have for Eq. (b), on replacing $F_R$ using Eq. (a),

$$\bar{y} = \frac{1}{g\rho(abc/2)}g\int_0^a \rho y^2\,\frac{bc}{a}\,dy = \frac{2}{3}a \tag{d}$$

To find the coordinate in the $z$ direction to the center of gravity, we could reorient the body as shown in Fig 4.30. A computation similar to the preceding one would give the result that $\bar{z} = \frac{2}{3}c$. You are urged to verify this yourself.

Finally, it should be clear by inspection of Fig. 4.28 that $\bar{x} = \frac{1}{2}b$.

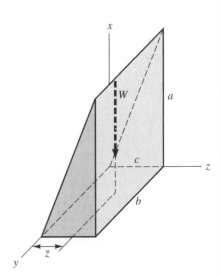

Thus we have $\boxed{\bar{x} = \dfrac{b}{2},\ \bar{y} = \dfrac{2}{3}a,\ \text{and}\ \bar{z} = \dfrac{2}{3}c}$

**Figure 4.30.** Reorientation of block.

# ■ Example 4.12

Find the center of gravity for the body of revolution shown in Fig. 4.31. The radial distance of the surface from the $y$ axis is given as

$$r = \tfrac{1}{20} y^2 \text{ ft} \qquad\qquad (a)$$

The body has constant density $\rho$, is 10 ft long, and has a cylindrical hole at the right end of length 2 ft and diameter 1 ft.

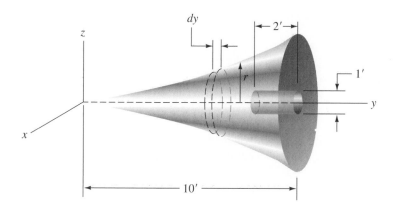

**Figure 4.31.** Body of revolution. Find center of gravity.

We need only compute $\bar{y}$, since it is clear that $\bar{z} = \bar{x} = 0$, owing to symmetry. We first compute the weight of the body. Using slices of thickness $dy$, as shown in the diagram, we sum the weight of all slices in the body assuming it is whole by letting $y$ run from 0 to 10 in an integration. We then subtract the weight of a 2-ft cylinder of diameter 1 ft to take into account the cylindrical cavity inside the body. Thus, we have, noting that the area of a circle is $\pi r^2$ or $\pi D^2/4$

$$W = \int_0^{10} \left( \pi r^2 \right) dy\, \rho g - \frac{\pi\left(1^2\right)}{4} (2)(\rho g) \qquad\qquad (b)$$

Using Eq. (a) to replace $r^2$ in terms of $y$, we get

$$W = \rho g \left( \pi \int_0^{10} \frac{y^4}{400}\, dy - \frac{\pi}{2} \right) = g\rho\pi\left( 50 - \frac{1}{2} \right) \qquad\qquad (c)$$

$$= 49.5\pi\rho g \text{ lb}$$

## Example 4.12 (Continued)

To get $\bar{y}$, we equate the moment of $W$ about the $x$ axis with that of the weight distribution. For the latter, we sum the moments about the $x$ axis of the weight of all slices, assuming first no inside cavity. Then, we subtract from this the moment about the $x$ axis of the weight of a cylinder corresponding to the cavity in the body. Because $\rho$ is constant, the center of gravity of this latter cylinder is at its geometric center so that the moment arm from the $x$ axis for the weight of the cylinder is clearly 9 ft. Thus, we have

$$-(49.5\pi\rho g)\bar{y} = -\rho g \left\{ \int_0^{10} y(\pi r^2\, dy) - \left[ \frac{\pi(1^2)}{4}(2) \right] 9 \right\}$$

$$\bar{y} = \frac{1}{49.5} \left( \int_0^{10} \frac{y^5}{400}\, dy - 4.5 \right) = \boxed{8.33 \text{ ft}}$$

Suppose as will be the case in Problem 4.43 that $\gamma$ (= $\rho g$), which is the *specific weight* (weight per unit volume), varies with $y$. That is, $\gamma = \gamma(y)$. Then for this problem, note that

$$W = \int_0^{10} \pi r^2 \gamma\, dy - \int_8^{10} \pi\left(\frac{1}{2}\right)^2 \gamma\, dy$$

also

$$-W\bar{y} = -\left[ \int_0^{10} y\pi r^2 \gamma\, dy - \int_8^{10} y\pi\left(\frac{1}{2}\right)^2 \gamma\, dy \right]$$

Note here we cannot take the short cuts used in the original problem where $\gamma$ was a constant.

# Example 4.13

A plate is shown in Fig. 4.32 lying flat on the ground. The plate is 60 mm thick and has a uniform density. The curved edge is that of a parabola with zero slope at the origin. Find the coordinates of the center of gravity.

**Figure 4.32.** Find center of gravity of plate.

The equation of a parabola oriented like that of the curved edge of the plate is

$$y = Cx^2 \qquad (a)$$

We can determine $C$ by noting that $y = 2\ m$ when $x = 3\ m$. Hence,

$$2 = C \cdot 9 \qquad (b)$$

Therefore,

$$C = \frac{2}{9}$$

The desired curve then is

$$y = \frac{2}{9}x^2$$

Therefore,

$$x = \frac{3}{\sqrt{2}}y^{1/2} \qquad (c)$$

We shall consider horizontal strips of the plate of width $dy$ (see Fig. 4.33). Using the specific weight, $\gamma$, which is weight per volume and is equal to $\rho g$, we have for the total weight $W$ of the plate:

$$W = \int_0^2 (t\, x\, dy)\gamma$$

where $t$ is the thickness. We replace $x$ using Eq. (c) to get

$$W = t\gamma \int_0^2 \left(\frac{3}{\sqrt{2}}y^{1/2}\right)dy$$

**Figure 4.33.** Use of horizontal strips.

Integrating, we get

$$W = t\gamma\frac{3}{\sqrt{2}}\left(y^{3/2}\right)\left(\frac{2}{3}\right)\Bigg|_0^2 = t\gamma\sqrt{2}(2)^{3/2} = 4t\gamma \ \text{N} \qquad (d)$$

We next take moments about the $x$ axis in order to get $\bar{y}$. Thus,

$$
\begin{aligned}
-W\bar{y} &= -\int_0^2 y(t\, x\, dy)\gamma \\
&= -\gamma t\int_0^2 (y)\left(\frac{3}{\sqrt{2}}y^{1/2}\right)dy \\
&= -\gamma t\frac{3}{\sqrt{2}}\left(y^{5/2}\right)\left(\frac{2}{5}\right)\Bigg|_0^2 \qquad (e) \\
&= -\gamma t\left(\frac{3}{\sqrt{2}}\right)\left(\frac{2}{5}\right)\left[(2^2)(2^{1/2})\right] \\
&= -\frac{24}{5}\gamma t
\end{aligned}
$$

# Example 4.13 (Continued)

Using $4t\gamma$ for $W$ from Eq. (d), we get, for $\bar{y}$:

$$\bar{y} = \tfrac{6}{5} \text{ m} \tag{f}$$

To get $\bar{x}$, we take moments about the $y$ axis, still utilizing the horizontal strips of Fig. 4.33. The center of gravity of a strip is at its center since $\gamma$ is constant and so the moment arm for a strip about the $y$ axis is $x/2$.

$$W\bar{x} = \int_0^2 \frac{x}{2}(t\,x\,dy)\gamma \tag{g}$$

Continuing with the calculations, we have

$$W\bar{x} = \frac{t\gamma}{2}\int_0^2 x^2\,dy = \frac{t\gamma}{2}\int_0^2\left(\frac{9}{2}y\right)dy$$

$$= \frac{t\gamma}{2}\frac{9}{2}\frac{y^2}{2}\bigg|_0^2 = \frac{9t\gamma}{2}$$

On replacing $W$ according to Eq. (d), we get for $\bar{x}$:

$$\bar{x} = \tfrac{9}{8} \text{ m} \tag{h}$$

Next, we proceed to get $\bar{x}$ using vertical strips as shown in Fig. 4.34. Equating moments of the weights of these vertical strips about the $y$ axis with that of the total weight $W$ at its center of gravity $\bar{x}$, we get on noting that $(2 - y)$ is the length of the strip

$$W\bar{x} = \int_0^3 x\gamma(2 - y)t\,dx = \int_0^3 x\gamma\left(2 - \frac{2}{9}x^2\right)t\,dx \tag{i}$$

$$= 2\gamma t\int_0^3\left(x - \frac{1}{9}x^3\right)dx = 2\gamma t\left(\frac{3^2}{2} - \frac{1}{9}\frac{3^4}{4}\right) = \gamma t\frac{9}{2}$$

**Figure 4.34.** Use of vertical strips.

Using $W = 4\gamma t$, we can now solve for $\bar{x}$. Thus, we again get

$$\bar{x} = \tfrac{9}{8} \text{ m}$$

Finally, it is clear that the $\bar{z}$ coordinate is zero for reference $xy$ at the center plane of the plate.

In the previous problems, we used slices of the body having a thicknesses $dy$ or $dx$. If the specific weight were a function of position, $\gamma(x, y, z)$, we could not readily use such slices, since we cannot easily express the weight of such slices in a simple manner. The reason for this is that in the $x$ and $z$ directions the dimensions of the element are finite, and so $\gamma$ would vary in these directions throughout the element. If, however, we choose an element that is infinitesimal in *all directions*, such as an infinitesimal rectangular parallelepiped having volume $dx\,dy\,dz$, then $\gamma$ can be assumed to be constant throughout the element. The weight of the element is then easily seen to be $\gamma(dx\,dy\,dz)$, where the coordinates of $\gamma$ correspond to the position of the element. We now illustrate a simple case.

# Example 4.14

Consider a block (see Fig. 4.35) wherein the specific weight $\gamma$ at corner $A$ is 200 lbf/ft$^3$. The specific weight in the block does not change in the $x$ direction. However, it decreases linearly by 50 lbf/ft$^3$ in 10 ft in the $y$ direction, and increases linearly by 50 lbf/ft$^3$ in 8 ft in the $z$ direction, as has been shown in the diagram. What are the coordinates $\bar{x}$, $\bar{y}$ of the center of gravity for this block?

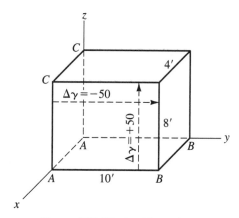

**Figure 4.35.** Block with varying $\gamma$.

We must first express $\gamma$ at any position $P(x, y, z)$. Using simple proportions, we can say

$$\gamma = 200 - \frac{y}{10}(50) + \frac{z}{8}(50) \qquad \text{(a)}$$
$$= 200 - 5y + 6.25z \text{ lbf/ft}^3$$

We shall first compute the weight of the block (i.e., the resultant force of gravity). We do not use an infinitesimal slice or rectangular rod of the block, as we have done heretofore. With the specific weight varying with both $y$ and $z$, it would not be an easy matter to compute the weight and moment of a slice or a rod. Instead, we shall use an infinitesimal rectangular parallelepiped having volume $dx\, dy\, dz$, located at a position having coordinates $x$, $y$, and $z$ as has been shown in Fig. 4.36(a). Because of the vanishingly small size of this element, the specific weight $\gamma$ can be considered constant inside the element, and so the weight $dW$ of the element can be given as[8]

$$dW = \gamma(dx\, dy\, dz) = (200 - 5y + 6.25z)\, dx\, dy\, dz$$

---

[8]We are deleting higher order quantities.

## Example 4.14 (Continued)

To include the weight of *all* such elements in the block, we first let $x$ "run" from 0 to 4 ft while holding $y$ and $z$ fixed. The rectangular parallelepiped of Fig. 4.36(a) then becomes a rectangular rod as shown in Fig. 4.36(b). Having run its course, $x$ is no longer a variable in this summation process. Next, let $y$ "run" from 0 to 10 while holding $z$ constant. The rectangular rod of Fig. 4.36(b) then becomes an infinitesimal slice, as shown in Fig. 4.36(c). The variable $y$ has thus run its course and is no longer a variable. This leaves only the variable $z$, and now we let $z$ "run" from 0 to 8. Clearly, we cover the entire block by this process.

We can do this mathematically by a process called *multiple integration*. We perform three integrations, paralleling the three steps outlined in the previous paragraph. Thus, we can formulate $W$ as follows:

$$-W\boldsymbol{k} = \int_0^8 \int_0^{10} \int_0^4 (200 - 5y + 6.25z)\, dx\, dy\, dz\, (-\boldsymbol{k})$$

We first consider integration with respect to $x$ holding $y$ and $z$ constant. That is

$$\int_0^4 (200 - 5y + 6.25z)\, dx$$

As in the first step set forth in the previous paragraph, to go from a rectangular parallelepiped to a rectangular rod, we integrate with respect to $x$ from $x = 0$ to $x = 4$ while holding $y$ and $z$ constant. Thus,

$$\int_0^4 (200 - 5y + 6.25z)\, dx = (200x - 5yx + 6.25z\,x)\Big|_0^4$$
$$= 800 - 20y + 25z$$

With $x$ no longer a variable (since it has run its course), the equation for $W$ becomes

$$W = \int_0^8 \int_0^{10} (800 - 20y + 25z)\, dy\, dz$$

Now, we hold $z$ constant and integrate with respect to $y$ from 0 to 10. (This takes us from a rectangular rod to a slice.) Thus,

$$\int_0^{10} (800 - 20y + 25z)\, dy = \left(800y - 20\frac{y^2}{2} + 25zy\right)\Big|_0^{10}$$
$$= 8{,}000 - 1{,}000 + 250z$$

Now $y$ has run its course, and we have

$$W = \int_0^8 (7{,}000 + 250z)\, dz$$

(a)

(b)

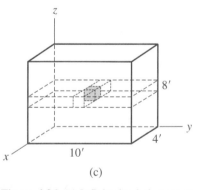

(c)

**Figure 4.36.** (a) Infinitesimal element at $P(x, y, z)$; (b) $x$ runs from 0 to 4, while $z$ and $y$ are fixed, to form rectangular rod; (c) $y$ runs from 0 to 10, while holding $z$ fixed, to form slice.

## Example 4.14 (Continued)

By integrating with respect to $z$, we sum up all the slices, and we have covered the entire block. Thus,

$$W = \left(7{,}000z + 250\,\frac{z^2}{2}\right)\Bigg|_0^8 = 64{,}000 \text{ lb}$$

To get $\bar{y}$, we equate the moment about the $x$ axis of the resultant force without a couple moment with the moment of the distribution. Thus, using multiple integration as described above:

$$-(64{,}000)\bar{y} = -\int_0^8\int_0^{10}\int_0^4 y(200 - 5y + 6.25z)\,dx\,dy\,dz$$

Therefore, integrating with respect to $x$, then $y$, and then $z$ as before, we have

$$
\begin{aligned}
64{,}000\bar{y} &= \int_0^8\int_0^{10}(200yx - 5y^2x + 6.25yzx)\Big|_0^4\,dy\,dz \\
&= \int_0^8\int_0^{10}(800y - 20y^2 + 25yz)\,dy\,dz \\
&= \int_0^8\left(800\,\frac{y^2}{2} - \frac{20y^3}{3}x + \frac{25y^2}{2}z\right)\Bigg|_0^{10}\,dz \\
&= \int_0^8(40{,}000 - 6{,}667 + 1{,}250z)\,dz \\
&= \left(33{,}333z + 1{,}250\,\frac{z^2}{2}\right)\Bigg|_0^8 = 307{,}000
\end{aligned}
$$

and

$$\boxed{\bar{y} = 4.79 \text{ ft}}$$

Because $\gamma$ does not depend on $x$, we can directly conclude by inspection that $\bar{x} = 2$ ft.

Next, in the case of a body made up of simple shapes (subbodies) such as cones, spheres, cylinders, and cubes, we can find the center of gravity of the body by using the centers of gravity of the known subbody shapes. Thus, we can say on taking moments about the $y$ axis that

$$W_{\text{total}}(\bar{x}) = \sum_i W_i(\bar{x})_i \tag{4.12}$$

where $W_i$ is the weight of the $i$th subbody and where $(\bar{x})_i$ is the $x$ coordinate to the center of gravity of the $i$th subbody. Bodies made up of simple subbodies are called *composite bodies*.

## Example 4.15

Find the weight and center of gravity of a large steam turbine for power generation (Fig. 4.37) needed for earthquake safety calculations. The specific weights of each turbine component are shown. Cylinder 2 has a radius of 5 m and is 14 m long. Half of this cylinder is embedded in the block 1.

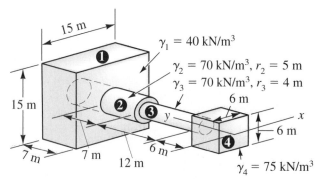

$\gamma_1 = 40$ kN/m$^3$

$\gamma_2 = 70$ kN/m$^3$, $r_2 = 5$ m
$\gamma_3 = 70$ kN/m$^3$, $r_3 = 4$ m

$\gamma_4 = 75$ kN/m$^3$

**Figure 4.37.** A composite body.

The weights of each component of this model are first calculated.

$$W_1 = \gamma_1 V_1 = [40][(7)(15)(15) - (7)(\pi)(5^2)] = 41 \text{ MN}$$

$$W_2 = \gamma_2 V_2 = [70][(14)(\pi)(5^2)] = 77 \text{ MN}$$

$$W_3 = \gamma_3 V_3 = [70][(12)(\pi)(4^2)] = 42.2 \text{ MN}$$

$$W_4 = \gamma_4 V_4 = [75][(6)(6)(6)] = 16.2 \text{ MN}$$

$$W_{TOTAL} = \sum W_i = 41 + 77 + 42.2 + 16.2 = \boxed{176.4 \text{ MN}}$$

The center of gravity lies along the axis of the shaft. Taking moments about the $x$ axis we have

$$176.4\bar{y} = (41)(28.5) + (77)(25) + (42.2)(12) + (16.2)(3) = 3,648.5 \text{ MN}$$

$$\boxed{\bar{y} = 20.68 \text{ m}}$$

## Case B. Parallel Force Distribution over a Plane Surface—

**Center of Pressure.**  Let us now consider a normal pressure distribution over a *plane* surface $A$ in the $xy$ plane in Fig. 4.38. The vertical ordinate is taken as a pressure ordinate, so that over the area $A$ we have a pressure distribution $p(x, y)$ represented by the pressure surface. Since in this case there is a parallel force system with one sense of direction, we know that the simplest resultant is a single force, which is given as

$$F_R = -\int p\, dA = -\left(\int p\, dA\right)k \qquad (4.13)$$

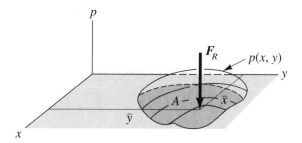

**Figure 4.38.** Pressure distribution.

The position $\bar{x}, \bar{y}$ can be computed by equating the moments about the $y$ and $x$ axes of the resultant force without a couple moment with the corresponding moments of the distribution. Solving for $\bar{x}$ and $\bar{y}$,

$$\bar{x} = \frac{\int px\, dA}{\int p\, dA}$$

$$\bar{y} = \frac{\int py\, dA}{\int p\, dA}$$

Since we know that $p$ is a function of $x$ and $y$ over the surface, we can carry out the preceding integrations either analytically or numerically. The point thus determined is called the *center of pressure*.

(In later chapters, we shall consider distributed frictional forces over plane and curved surfaces. In these cases, the simplest resultant is not necessarily a single force as it was in the special case above.)

## Example 4.16

A plate $ABCD$ on which both distributed and point force systems act is shown in Fig. 4.39. The pressure distribution is given as

$$p = -4y^2 + 100 \text{ psf} \qquad (a)$$

Find the simplest resultant for the system.

To get the resultant force, we consider a strip $dy$ along the plate as shown in Fig. 4.39. The reason for using such a strip is that the pressure $p$ is uniform along this strip, as can be seen from the diagram. Hence, the force from the pressure on the strip is simply $p\,dA = p(dy)(5)$. Thus, we can say that

$$F_R = -\int_0^5 p(5)(dy) - 500$$

$$= -\int_0^5 (-4y^2 + 100)(5)\,dy - 500 \qquad (b)$$

$$F_R = (20\frac{y^3}{3} - 500y)\Big|_0^5 - 500 = \boxed{-2{,}167 \text{ lb}}$$

**Figure 4.39.** Find simplest resultant.

To get the position $\bar{x}$, $\bar{y}$ of the resultant force $F_R$ without a couple moment, we equate moments of $F_R$ about the $x$ and $y$ axes with that of the original system. Thus, starting with the $x$ axis, we have using strip $dy$ as before:

$$-2{,}167\bar{y} = -\int_0^5 yp(5dy) - (500)(2)$$

$$= -\int_0^5 5y(-4y^2 + 100)\,dy - 1{,}000$$

$$= (20\frac{y^4}{4} - 500\frac{y^2}{2})\Big|_0^5 - 1{,}000 = -4{,}125$$

Therefore,

$$\boxed{\bar{y} = 1.904 \text{ ft}}$$

Now, considering the $y$ axis, we still use the strips $dy$ because $p$ is uniform along such strips. However, the force $df = p\,dA = p(5)(dy)$ may be considered acting at the center of the strip, and accordingly has a moment arm about the $y$ axis equal to $5/2$ for each strip. Hence, we can say that

$$2{,}167\bar{x} = \int_0^5 \frac{5}{2} p(5dy) + (500)(2) - \frac{500}{12}$$

$$= \frac{25}{2}\int_0^5 (-4y^2 + 100)\,dy + 1{,}000 - 41.7 = 5{,}125$$

Therefore,

$$\boxed{\bar{x} = 2.36 \text{ ft}}$$

In a stationary liquid, the pressure at the surface of the liquid is transmitted uniformly throughout the liquid. In addition, there is superposed a linearly increasing pressure with depth resulting from gravity acting on the liquid. Thus, if we have $p_{atm}$ at the surface (often called the free surface), the pressure in the liquid is

$$p = p_{atm} + \gamma y$$

where $\gamma$ is the specific weight of the liquid and $y$ is the depth below the surface. Thus, for a given liquid the pressure is uniform at any constant distance below the free surface. With this in mind, examine the following example.

# Example 4.17

In Fig. 4.40 find the force on the door $AB$ from water whose specific weight is 9,806 N/m$^3$ and on whose free surface there is atmospheric pressure equal to 101,325 N/m$^2$ ($\equiv$ 101,325 Pa).[9] Also find the center of pressure.

The pressure on the door $AB$ is then

$$p = p_{atm} + (\gamma)(y) = p_{atm} + (\gamma)(s)(\sin 45°)$$

where $s$ is the distance from $O$ along the inclined wall $OR$. The resultant force is then

$$F = \int_5^9 \left[ p_{atm} + (\gamma)(s)(\sin 45°) \right](2)(ds)$$

$$= \int_5^9 \left[ 101,325 + (9,806)(.707)s \right](2)ds$$

$$= (2)\left[ 101,325s + (9,806)(.707)\frac{s^2}{2} \right]_5^9$$

$$\therefore \quad \boxed{F = 1.199 \times 10^6 \text{ N}}$$

To get the center of pressure equate moments about $O$ with that of the resultant. Thus using the notation $\bar{s}$ to locate the center of pressure we have

$$1.199 \times 10^6 \, \bar{s} = \int_5^9 s\left[ p_{atm} + (\gamma)(s)(\sin 45°) \right](2) \, ds$$

$$= \left[ 101,325 \frac{s^2}{2} + (9,806)(.707)\frac{s^3}{3} \right]_5^9 (2)$$

$$\therefore \quad \boxed{\bar{s} = 7.061 \text{ m}}$$

Clearly the *total* force on the door from inside and outside would include the contribution of the atmospheric pressure on the outside. We can easily determine this force by deleting $p_{atm}$ in the preceding calculations.[10]

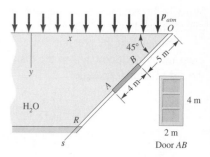

**Figure 4.40.** Door $AB$ is exposed on one side to water and on the other side to air.

---

[9]The unit of pressure in the SI system is the *pascal*, where 1 pascal = 1 Pa = 1 N/m$^2$.

[10]We have touched here on the subject of *hydrostatics*. For a treatment of this subject that may be similar to what you will study in your upcoming course in fluid mechanics, see Shames, I.H., *Mechanics of Fluids*, 3rd Edition, 1992, McGraw-Hill, Inc., Chapter 3

We finish this series of examples with a multiple integration problem.

## ■*Example 4.18

What is the simplest resultant and the center of pressure for the pressure distribution shown in Fig. 4.41?

Notice that the pressure varies linearly in the $x$ and $y$ directions. The pressure at any point $x$, $y$ in the distribution can be given as follows with the aid of similar triangles:

$$p = \left(\frac{y}{10}\right)(20) + \left(\frac{x}{5}\right)(30) \qquad \text{(a)}$$

$$= 2y + 6x \text{ Pa}$$

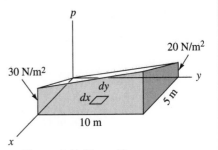

**Figure 4.41.** Nonuniform pressure distribution.

We cannot employ a convenient strip here along which the pressure is uniform, as in Example 4.16. For this reason we consider rectangular area element $dx\, dy$ to work with (see Fig. 4.41). For such a small area, we can assume the pressure as constant so that $p\, dx\, dy$ is the force on the element. To find the resultant force, we must integrate over the $10 \times 5$ rectangle. This integration involves two variables and is again a case of *multiple integration*. Thus, we can say that

$$F_R = -\int_0^{10}\int_0^5 p\, dx\, dy$$

wherein we first integrate with respect to $x$ while holding $y$ constant and then integrate with respect to $y$ (in this way we cover the entire $10 \times 5$ rectangular area). Thus, we have

$$F_R = -\int_0^{10}\int_0^5 (2y + 6x)\, dx\, dy$$

$$= -\int_0^{10}\left(2yx + \frac{6x^2}{2}\right)\Bigg|_0^5 dy$$

$$= -\int_0^{10} (10y + 75)\, dy$$

$$= -\left[\frac{10y^2}{2} + 75y\right]\Bigg|_0^{10} = -1{,}250 \text{ N}$$

$$\boxed{F_R = -1{,}250 \text{ N}}$$

To find $\bar{y}$ for $F_R$ without a couple moment, we equate moments of $F_R$ about the $x$ axis with that of the distribution. Thus,

$$-\bar{y}(1{,}250) = -\int_0^{10}\int_0^5 py\, dx\, dy$$

## Example 4.18 (Continued)

Therefore,

$$\bar{y} = \frac{1}{1,250} \int_0^{10} \int_0^5 (2y + 6x)y \, dx \, dy$$

$$= \frac{1}{1,250} \int_0^{10} \left( 2y^2 x + \frac{6x^2}{2} y \right) \Bigg|_0^5 dy$$

$$= \frac{1}{1,250} \int_0^{10} \left( 10y^2 + 75y \right) dy$$

$$= \frac{1}{1,250} \left( \frac{10y^3}{3} + 75 \frac{y^2}{2} \right) \Bigg|_0^{10}$$

$$\boxed{\bar{y} = 5.67 \text{ m}}$$

As for $\bar{x}$, we proceed as follows:

$$\bar{x}(1,250) = \int_5^{10} \int_0^5 px \, dx \, dy$$

Therefore,

$$\bar{x} = \frac{1}{1,250} \int_0^{10} \int_0^5 (2y + 6x)x \, dx \, dy$$

$$= \frac{1}{1,250} \int_0^{10} \left( 2y \frac{x^2}{2} + \frac{6x^3}{3} \right) \Bigg|_0^5 dy$$

$$= \frac{1}{1,250} \int_0^{10} \left( 25y + 250 \right) dy$$

$$= \frac{1}{1,250} \left( 25 \frac{y^2}{2} + 250y \right) \Bigg|_0^{10}$$

$$\boxed{\bar{x} = 3.00 \text{ m}}$$

The center of pressure is thus at (3.00, 5.67) m.

## Case C. Coplanar Parallel Force Distribution. As we pointed out earlier, this type of loading may be considered for beams loaded symmetrically over the longitudinal midplane of the beam. The loading is represented by an intensity function $w(x)$ as shown in Fig. 4.26. This coplanar parallel force distribution can be replaced by a single force given as

$$F_R = -\int w(x) \, dx \boldsymbol{j}$$

We find the position of $F_R$ without a couple moment by equating moments of $F_R$ and the distribution $w$ about a convenient point of the beam, usually one of the ends. Solving for $\bar{x}$, we get

$$\bar{x} = \frac{\int x w(x) \, dx}{\int w(x) \, dx}$$

## ■ Example 4.19

A simply supported beam is shown in Fig. 4.42 supporting a 1,000-lb point force, a 500 lb-ft couple, and a coplanar, parabolic, distributed load $w$ lb/ft. Find the simplest resultant of this force system.

To express the intensity of loading for the coordinate system shown in the diagram, we begin with the general formulation

$$w^2 = ax + b \tag{a}$$

Note from the diagram that when $x = 25$ we have $w = 0$, and when $x = 65$, we have $w = 50$. Subjecting Eq. (a) to these conditions, we can determine $a$ and $b$. Thus,

$$0 = a(25) + b \tag{b}$$

$$2,500 = a(65) + b \tag{c}$$

Subtracting, we can get $a$ as follows:

$$-2,500 = -40a$$

Therefore,

$$a = 62.5$$

From Eq. (b), we get

$$b = -(25)(62.5) = -1,562.5$$

Thus, we have

$$w^2 = 62.5x - 1,562.5 \tag{d}$$

Summing forces, we get for $F_R$,

$$F_R = -1,000 - \int_{25}^{65} \sqrt{62.5x - 1,562.5} \, dx \tag{e}$$

To integrate this, we may change variables as follows:

$$\mu = 62.5x - 1,562.5 \tag{f}$$

Therefore,

$$d\mu = 62.5 \, dx$$

Substituting into the integral in Eq. (e), we have[11]

$$F_R = -1,000 - \int_0^{2,500} \mu^{1/2} \frac{d\mu}{62.5}$$

$$= -1,000 - \frac{1}{62.5} \mu^{3/2} \left(\frac{2}{3}\right) \Big|_0^{2,500}$$

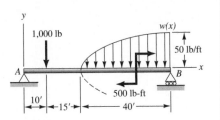

**Figure 4.42.** Find simplest resultant.

---

[11]Do not forget to change the limits for $\mu$. Thus, from Eq. (f), the upper limit is (62.5)(65) − 1,562.5 = 2,500, whereas the lower limit is (62.5)(25) − 1,562.5 = 0.

■ **Example 4.19 (Continued)**

$$= -1,000 - \frac{1}{62.5}(2,500)^{3/2}\left(\frac{2}{3}\right)$$

$$F_R = \boxed{-2,333 \text{ lb}}$$

We now compute $\bar{x}$ for the resultant without a couple as follows:

$$-2,333\bar{x} = -(10)(1,000) - \int_{25}^{65} x\sqrt{62.5x - 1,562.5} \, dx - 500 \qquad (g)$$

We can evaluate the integral most readily by consulting the mathematical formulas in Appendix I. We find the following formula (No 6):

$$\int x\sqrt{a + bx} \, dx = -\frac{2(2a - 3bx)\sqrt{(bx + a)^3}}{15b^2}$$

In our case $b = 62.5$ and $a = -1,562.5$, so the indefinite integral for our case is

$$\int x\sqrt{62.5x - 1,562.5} \, dx = -\frac{(2)(-3,125 - 187.5x)\sqrt{(62.5x - 1,562.5)^3}}{(15)(3,906)}$$

Putting in limits, we have

$$\int_{25}^{65} x\sqrt{62.5x - 1,562.5} \, dx = -\frac{(2)(-3,125 - 187.5x)\sqrt{(62.5x - 1,562.5)^3}}{(15)(3,906)}\Bigg|_{25}^{65}$$

$$= 65,333 - 0 = 65,333$$

Going back to Eq. (g), we can now solve easily for $\bar{x}$. Thus,

$$\bar{x} = -\frac{1}{2,330}\left[-(10)(1,000) - 65,300 - 500\right]$$

$$\boxed{\bar{x} = 32.5 \text{ ft}}$$

Before closing, it will be pointed out that, for a loading function $w(x)$, the resultant $\int_0^x w \, dx$, equals the *area* under the loading curve. This fact is particularly useful for the case of a triangular loading function such as is shown in Fig. 4.43. Hence, we can say on inspection that the resultant force has the value

$$F_R = \frac{1}{2}(5)(1,000) = 2,500 \text{ N}$$

Furthermore, you can readily show that the *simplest* resultant has a line of action that is (2/3) × (length of loading) from the toe of the loading.[12] Thus, $F_R$ without a couple moment is at a position (2/3)(5) to the right of $a$ (see Fig. 4.43). You are urged to use this information when needed.

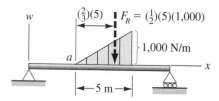

**Figure 4.43.** Triangular loading resultant.

[12] In Chapter 8, you will learn that the simplest resultant force for a distribution $w(x)$ goes through the *centroid* of the area under $w(x)$. The centroid will be carefully defined at that time.

**4.41.** A force field is given as

$$F(x, y, z, t) = (10x + 5)i + (16x^2 + 2z)j + 15k \text{ N}$$

What is the force at position (3, 6, 7) m? What is the difference between the force at this position and that at the origin?

**4.42.** A magnetic field is developed such that the body force on the rectangular parallelepiped of metal is given as

$$f = (.01x + \tfrac{1}{8})k \text{ oz/lbm}$$

If the mass density of the metal is 450 lbm/ft³, what is the *simplest* resultant body force from such a field? Note that at the earth's surface the number of pounds mass equals the number of pounds force of weight.

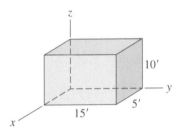

**Figure P.4.42.**

**4.43.** A body of revolution has a variable specific weight such that $\gamma = (36 + .01x^2)$ kN/m³ with $x$ in meters. A hole of diameter 3 m and length 6 m is cut from the body as shown. Where is the center of gravity?

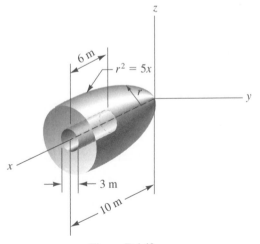

**Figure P.4.43.**

**4.44.** The specific weight $\gamma$ of the material in the solid cylinder varies linearly as one goes from face $A$ to face $B$. If

$$\gamma_A = 400 \text{ lbf/ft}^3, \qquad \gamma_B = 500 \text{ lbf/ft}^3$$

what is the position of the center of gravity of the cylinder?

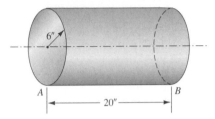

**Figure P.4.44.**

**4.45.** The specific weight of the material in a right circular cone is constant. Where is the center of gravity of the cone? *Hint:* Rotate cone 90° so that gravity is perpendicular to the $z$ axis. Use concept of similar triangles to show that $r/R = (h - z)/h$ and solve for $r$ needed for the integration.

**Figure P.4.45.**

**4.46.** Show that the center of gravity of the right triangular plate of thickness $t$ is at $x = a/3$ and $y = b/3$.

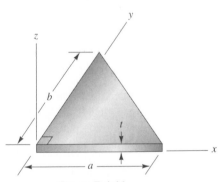

**Figure P.4.46.**

**4.47.** Show that the volume and center of gravity of the conical frustum are, respectively,

$$V = \frac{\pi h}{3}\left(r_2^2 + r_1 r_2 + r_1^2\right)$$

and

$$\bar{z} = \frac{h}{4}\frac{3r_2^2 + r_1^2 + 2r_1 r_2}{r_2^2 + r_1^2 + r_1 r_2}$$

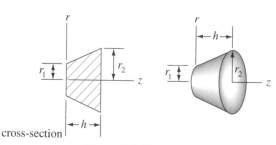

Figure P.4.47.

**4.48.** Find the center of gravity of the plate bounded by a straight line and a parabola.

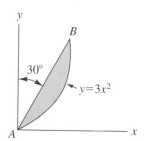

Figure P.4.48.

**4.49.** A massive radio-wave antenna for detection of signals from outer space is a body of revolution with a parabolic face (see the diagram). These antennas may be carved from rock in a valley away from other disturbing signals. What would the antenna weigh if made from concrete (23.6 kN/m³) for location in a remote desert area?

Figure P.4.49.

**4.50.** In Problem 4.49 find the distance from the ground to the center of gravity if the total weight is $2.37 \times 10^8$ kN.

**\*4.51.** A plate of thickness 30 mm has a specific weight $\gamma$ that varies linearly in the $x$ direction from 26 kN/m³ at $A$ to 36 kN/m³ at $B$, and varies in the $y$ direction as the square of $y$ from 26 kN/m³ at $A$ to 40 kN/m³ at $C$. Where is the center of gravity of the plate?

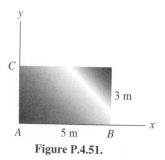

Figure P.4.51.

**4.52.** You are looking down on a plate with a hole in it as shown. The thickness has a constant value equal to $t$ and the specific weight $\gamma$ is constant. Find the coordinates $\bar{x}, \bar{y}$ of the *center of gravity*.

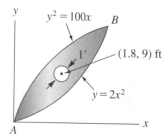

Figure P.4.52.

**4.53.** Find the center of gravity of the plate having uniform thickness and uniform specific weight. You are looking down onto the top of the plate.

Figure P.4.53.

137

**4.54.** The top view of a plate is shown. Find center of gravity coordinates $\bar{x}$, $\bar{y}$.

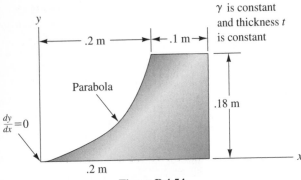

Figure P.4.54.

**4.55.** The thin circular rod has a weight of $w$ N/m. What is the $y$ coordinate of its center of gravity? The rod forms one-half of a circle.

Figure P.4.55.

**4.56.** Find $\bar{y}$ for the center of gravity of the horizontal plate with a hole in it.

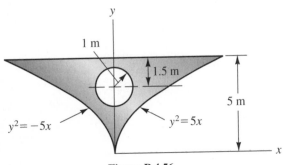

Figure P.4.56.

**\*4.57.** Suppose in Problem 4.42 that

$$f = (.01x + .2y + .3z)\mathbf{k} \text{ oz/lbm}$$

Find the *simplest* resultant for $\rho = 450$ lbm/ft³. Find the proper line of action.

**4.58.** After a fast stop and swerve to the left, the load of sand (specific weight = 15 kN/m³) in a dump truck is in the position shown. What is the simplest resultant force on the truck from the sand and where does it act? If the truck was full (with a level top) before the stop, how much sand spilled? Use the results of Problem 4.46.

Figure P.4.58.

*In Problems 4.59 through 4.62 use the known positions of centers of gravity of simple shapes.*

**4.59.** An I-beam cantilevered out from a wall weighs 30 lb/ft and supports a 300-lb hoist. Steel (487 lb/ft³) cover plates 1 in. thick are welded on the beam near the wall to increase the carrying capacity of the beam. What is the moment at the wall due to the weight of the reinforced beam and the hoisted load of 4,000 lb at the outermost position of the hoist? What is the *simplest* resultant force and its location?

Figure P.4.59.

**4.60.** The bulk materials trailer weighs 10,000 lb and is filled with cement ($\gamma = 94$ lb/ft$^3$) in the front compartment (sections 1 and 2), and half-filled with water ($\gamma = 62.5$ lb/ft$^3$) in the rear compartment (sections 3 and 4). What is the *simplest* resultant force, and where does it act? What is the resultant when the water is drained? Use the center of gravity and volume results from Problem 4.47 (conical frustum).

**Figure P.4.60.**

**4.61.** Find the coordinates $(x, y)_{CG}$ of the center of gravity of the loaded conveyor system. The centers of gravity of the crates $C$ and $D$ are at their geometric centers. $W_E$ is the weight of the frame whose C.G. is at its geometric center.

$W_A = 200$ lb
$W_B = 100$ lb
$W_C = 80$ lb
$W_D = 300$ lb
$W_E = 1,000$ lb
$W_{BELT} = 5$ lb/ft

**Figure P.4.61.**

**4.62.** Find the center of gravity of the body shown. It has a constant specific weight throughout. Cone and cylinder are on block surfaces.

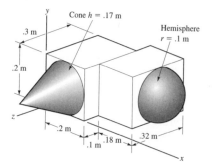

**Figure P.4.62.**

**4.63.** Find the *simplest* resultant of a normal pressure distribution over the rectangular area with sides $a$ and $b$. Give the coordinates of the center of pressure.

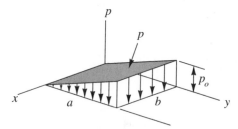

**Figure P.4.63.**

**4.64.** Find the *simplest* resultant acting on the vertical wall $ABCD$. Give the coordinates of the center of pressure. The pressure varies such that $p = E/(y + 1) + F$ psi, with $y$ in feet, from 10 psi to 50 psi, as indicated in the diagram. E and F are constants.

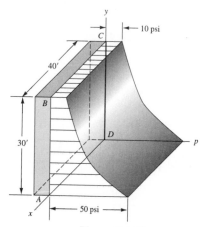

**Figure P.4.64.**

139

**4.65.** One floor of a warehouse is divided into four areas. Area 1 is stacked high with TV sets such that the distributed load is $p = 120$ lb/ft$^2$. Area 2 has refrigerators with $p = 65$ lb/ft$^2$. Area 3 has stereos stacked so that $p = 80$ lb/ft$^2$. Area 4 has washing machines with $p = 50$ lb/ft$^2$. What is the simplest resultant force and where does it act?

**Figure P.4.65.**

**4.66.** Consider a pressure distribution $p$ forming a hemispherical surface over a domain of radius 5 m. If the maximum pressure is 5 Pa, what is the *simplest* resultant from this pressure distribution?

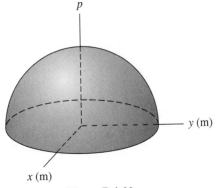

**Figure P.4.66.**

**4.67.** A rectangular tank contains water. If the tank is rotated clockwise 10° about an axis normal to the page, what torque is required to maintain the configuration? Width of tank is 1 ft.

**Figure P.4.67.**

**4.68.** (a) Find the torque about axis $AB$ from the wrench.
(b) Find the torque about axis $AB$ from the distributed loads. (*Hint:* Look down from above to help view problem.)

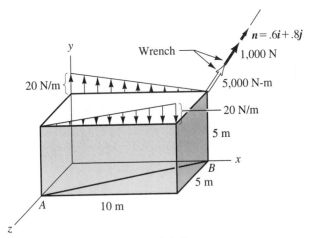

**Figure P.4.68.**

**4.69.** For the system of forces shown, determine the torque about the axis going from $A$ to $B$. *Note:* The 100-N force and the triangular load distribution are in the $yz$ plane and the 300-N-m couple is on the top face of the rectangular box.

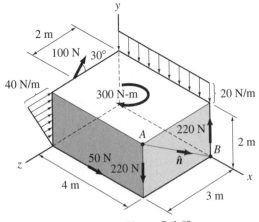

**Figure P.4.69.**

**4.70.** A *manometer* is a simple pressure measuring device. One such manometer called a *U tube* is shown in the diagram. The tank contains water including the tube to level $M$ where mercury is present. $M$ and $N$ are at the same level. What is the gage pressure (i.e., the pressure above atmosphere) at point $a$ in the tank for the following data:

$$d_1 = .2 \text{ m} \quad d_2 = .6 \text{ m} \quad \gamma_{H_2O} = 9,806 \text{ N/m}^3$$
$$\gamma_{Hg} = (13.7)(\gamma_{H_2O}) \text{ N/m}^3$$

[*Hint: M* and *N* are at the same level and are joined by the same fluid, namely mercury. Hence, the pressures there are equal.]

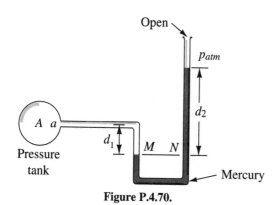

**Figure P.4.70.**

**4.71.** Imagine a liquid which when stationary stratifies in such a way that the specific weight is proportional to the square root of the pressure. At the free surface, the specific weight is known and has the value $\gamma_0$. What is the pressure as a function of depth from the free surface? What is the resultant force on one face $AB$ of a rectangular plate submerged in the liquid? The width of the plate is $b$.

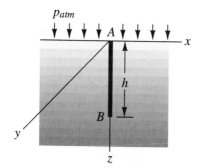

**Figure P.4.71.**

**4.72.** (a) Calculate the force on the door from all fluids inside and outside. The specific gravity of the oil is 0.8. This means that the oil has a specific weight $\gamma$ which is 0.8 times that of water ($\gamma_{water} = 62.4$ lb/ft³). Note that a uniform pressure on the surface of a liquid extends undiminished throughout the depth of the liquid.

(b) Determine the distance from the surface of the oil to the *simplest* resultant force on the door from all fluids.

**Figure P.4.72.**

**4.73.** At what height $h$ will the water cause the door to rotate clockwise? The door is 3 m wide. Neglect friction and the weight of the door.

**Figure P.4.73.**

**4.74.** Find the force on the door from the inside and outside pressures. Give the position of the resultant force above the base of the door. The specific gravity $S$ of the oil is 0.7, i.e., $\gamma_{oil} = (0.7)\gamma_{H_2O}$.

**Figure P.4.74.**

**4.75.** Find the total force on door $AB$ from fluids. The specific gravity $S$ of a fluid is $\gamma_{fluid}/\gamma_{H_2O}$. Take $S_{oil} = 0.6$. Find the position of this force from the bottom of the door.

**Figure P.4.75.**

**4.76.** What is the simplest resultant force from the water and where does it act on the 60-m-high 800-m-long straight earthfill dam? (Water weighs 9,806 N/m³.)

**Figure P.4.76.**

**4.77.** A block 1 ft thick is submerged in water. Compute the simplest resultant force and the center of pressure on the bottom surface. Take $\gamma = 62.4$ lb/ft³.

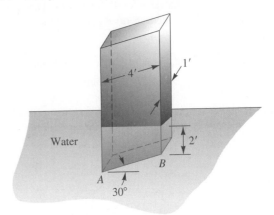

**Figure P.4.77.**

**\*4.78.** What is the resultant force from water and where does it act on the 40-m-high circular concrete dam between two walls of a rocky gorge? (Water weighs 9,806 N/m³.)

**Figure P.4.78.**

**4.79.** The weight of the wire $ABCD$ per unit length, $w$, increases linearly from 4 oz/ft at $A$ to 20 oz/ft at $D$. Where is the center of gravity of the wire?

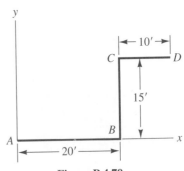

**Figure P.4.79.**

**4.80.** Find the center of gravity of the wire. The weight per unit length increases as the square of the length of wire from a value of 3 oz/ft at $A$ until it reaches the value of 8 oz/ft at $C$. It then decreases 1 oz/ft for every 10 ft of length.

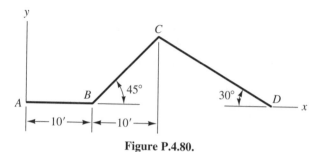

**Figure P.4.80.**

**4.81.** What is the center of gravity coordinate $y_c$ for a thin circular rod shown in the diagram? It has a weight of $w$ N/m. The rod is placed symmetrically about the $y$ axis. The angle $\phi$ is in degrees.

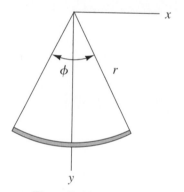

**Figure P.4.81.**

**4.82.** At what distance $\bar{x}$ from point $A$ can the system of thin rods be suspended so that it will just balance? Use the formula developed in the previous problem for the radial distance to the center of gravity of a thin circular rod, namely $\dfrac{360r}{\pi\phi}\sin\dfrac{\phi}{2}$.

**Figure P.4.82.**

## 4.6 Closure

We now have the tools that enable us to replace, for purposes of rigid body mechanics, any system of forces by a resultant consisting of a force and a couple moment. These tools will prove very helpful in our computations. More important at this time, however, is the fact that in considering conditions of equilibrium for rigid bodies we need only concern ourselves with this resultant to reach conclusions valid for any force system, no matter how complex. From this viewpoint, we shall develop the fundamental equations of statics in Chapter 5 and then employ them to solve a large variety of problems.

**4.83.** Replace the force and couples acting on the plate by a single force. Give the intercept of the line of action of this force with the vertical edge $BC$ of the plate.

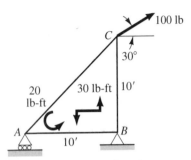

**Figure P.4.83.**

**4.84.** A 100-kN bridge pier supports a 10-m segment length of roadway weighing 150 kN and a 150-kN truck. The truck is located at the same position along the roadway as the pier. What is the equivalent force system acting on the base of the bridge pier when the truck is (a) in the center of the outside lane and (b) in the center of the inside lane?

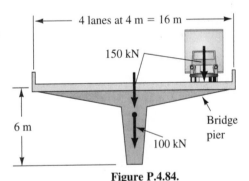

**Figure P.4.84.**

**4.85.** Find the center of gravity of a flat plate having constant specific weight $\gamma$ and thickness $t$. You are looking down on the plate from above. Proceed as follows:
(a) Give weight of plate in terms of $\gamma t$.
(b) Determine $\bar{y}$.
(c) Determine $\bar{x}$. Use vertical strips and be careful with integration limits.

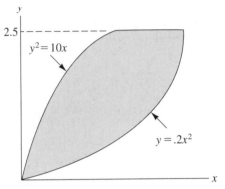

**Figure P.4.85.**

**4.86.** A heavy duty off-the-road dump truck is loaded with iron ore that weighs 51 kN/m³. What is the *simplest* resultant force on the truck and where does it act?

**Figure P.4.86.**

**4.87.** Find the *center of gravity* of the *system* of two plates shown (not separately) in terms of the specific weight $\gamma$ and thickness $t$, both of which are uniform and the same for both plates. You are looking down on the plates.

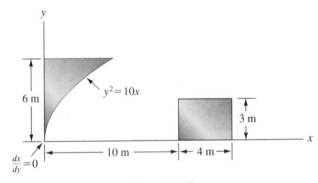

**Figure P.4.87.**

**4.88.** A Jeep weighs 11 kN and has both a front winch and a rear power take-off. The tension in the winch cable is 5 kN. The power take-off develops 300 N-m of torque $T$ about an axis parallel to the $x$ axis. If the driver weighs 800 N, what is the resultant force system at the indicated center of gravity of the Jeep where we can consider the weight of the Jeep to be concentrated?

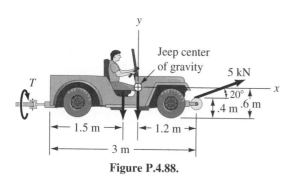

**Figure P.4.88.**

**4.89.** What is the *simplest* resultant for the forces and couple acting on the beam?

**Figure P.4.89.**

**4.90.** A parabolic body of revolution has cut out of it a second parabolic body of revolution starting at $A$ and forming a sharp edge at $B$ with zero slope at $A$.

    (a) What is $r'$ as a function of $x$ for the cut-out body of revolution?

    (b) Set up an integral for computing $W$ (weight) and then the center of gravity coordinate $\bar{x}$.

*Note:* $\gamma$ varies with $x$. Do not solve the integral.

$$\gamma = 10^4\,(x^2 + 3x)\,\tfrac{N}{m^3}$$

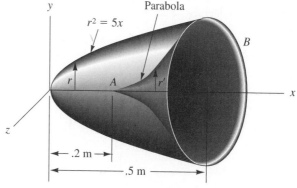

**Figure P.4.90.**

**4.91.** Find the torque about axis $OB$ from the system of forces.

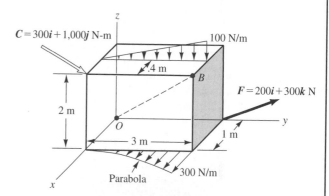

**Figure P.4.91.**

**4.92.** A rectangular plate shown as $ABC$ can rotate about hinge $B$. What length $l$ should $BC$ be so there is zero torque about $B$ from the water, air, and weight of the plate? Take this weight as 1,000 N/m of length. The width is 1 m. $\gamma_{H_2O} = 9,806$ N/m$^3$.

**Figure P.4.92.**

145

**4.93.** An open rectangular tank of water is partially filled with water. The dimensions are shown.

    (a) Determine the force on the bottom of the tank from the water.

    (b) Determine the force on the door shown at the side of the tank. Indicate the position of this force from the bottom of the tank.

Note that the atmospheric pressure develops equal and opposite forces on both sides of the door and hence yields no net force.

**Figure P.4.93.**

**4.94.** Sandbags are piled on a beam. Each bag is 1 ft wide and weighs 100 lb. What is the simplest resultant force and where does it act? What linear mathematical function of the distributed load can be used to represent the sandbags over the left 3 ft of the beam?

**Figure P.4.94.**

**4.95.** A cantilever beam is subjected to a linearly varying load over part of its length. What is the *simplest* resultant force, and where does it act? What is the moment at the supported end?

**Figure P.4.95.**

**4.96.** Find the torque about an axis along position vector $r = 3i + 4j + 2k$ m. Note we have a pressure distribution on *ABED* and a coplanar loading in the *yz* plane along the *y* axis.

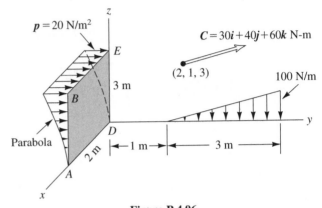

**Figure P.4.96.**

**4.97.** Compute the *simplest* resultant force for the loads acting on the cantilever beam.

**Figure P.4.97.**

**4.98.** Find the resultant force system at $A$ for the forces on the bent cantilever beam. $BC$ is parallel to $z$ axis.

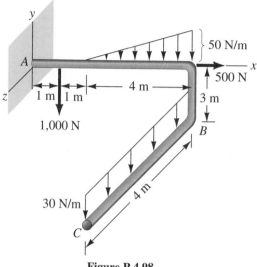

**Figure P.4.98.**

**4.100.** The L-shaped concrete post supports an elevated railroad. The concrete weighs 150 lb/ft³. What is the simplest resultant force from the weight and the load and where does it act? Load acts at center of top surface.

**Figure P.4.100.**

**4.99.** (a) Find equations describing both parabolas.
  (b) Using *vertical* strips (and a composite body approach), find weight of plate in terms of $\gamma t$.
  (c) Using vertical strips, find $\bar{x}$ of C.G.

**4.101.** Explain why the system shown can be considered a system of parallel forces. Find the *simplest* resultant for this system. The grid is composed of 1-m squares.

**Figure P.4.99.**

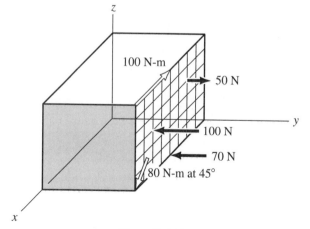

**Figure P.4.101.**

147

**4.102.** A plate of thickness $t$ has as the upper edge a parabolic curve with infinite slope at the origin. Find the $x$, $y$ coordinates of the center of gravity for this plate.

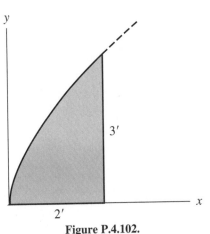

**Figure P.4.102.**

**4.103.** A rectangular tank contains a liquid. At the top of the liquid there is a pressure of $.1380$ N/mm² absolute. What is the simplest resultant force in the inside surface of the door $AB$? Where is the center of pressure relative to the bottom of the door? Take $\gamma = 8{,}190$ N/m³ for the liquid.

**Figure P.4.103.**

**4.104.** The specific weight of the material in a right circular cone varies directly as the square of the distance $y$ from the base. If $\gamma_0 = 50$ lb/ft³ is the specific weight at the base, and if $\gamma' = 70$ lb/ft³ is the specific weight at the tip, where is the center of gravity of the cone? (See hint in Problem 4.45.)

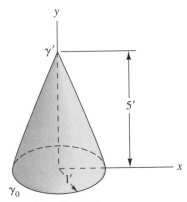

**Figure P.4.104.**

**\*4.105.** A block has a rectangular portion removed (darkened region). If the specific weight is given as

$$\gamma = (2.0x + y + 3xyz) \text{ kN/m}^3$$

find $x$ for the center of gravity.

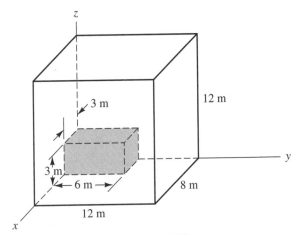

**Figure P.4.105.**

**4.106.** Compute the *simplest* resultant for the loads shown acting on the simply supported beam. Give the line of action.

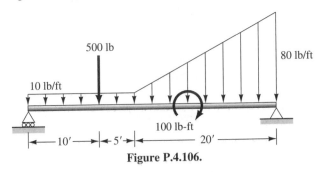

**Figure P.4.106.**

**4.107.** Find the center of gravity of the plate.

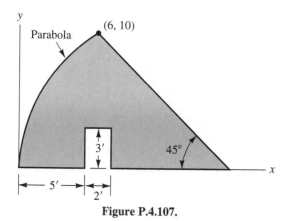

**Figure P.4.107.**

**4.108.** Find the center of gravity of the truss. All members have the same weight per unit length.

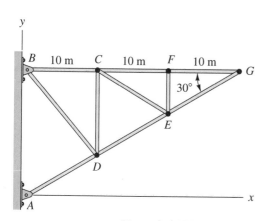

**Figure P.4.108.**

**\*4.109.** The pressure $p_0$ at the corner $O$ of the plate is 50 Pa and increases linearly in the $y$ direction by 5 Pa/m. In the $x$ direction, it increases parabolically starting with zero slope so that in 20 m the pressure has gone from 50 Pa to 500 Pa. What is the simplest resultant for this distribution? Give the coordinates of the center of pressure.

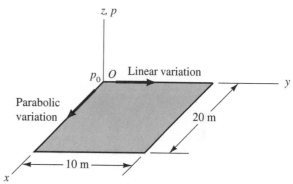

**Figure P.4.109.**

**4.110.** A sluice-gate door in a dam is 3 m wide and 3 m high. The water level in the dam is 4 m above the top of the door. The gate is opened until the water level falls 4 m. What is the simplest resultant force on the closed door at both water levels? Where do the forces act (i.e., where is the "center of pressure" in each case)? Water weighs 9,806 N/m³.

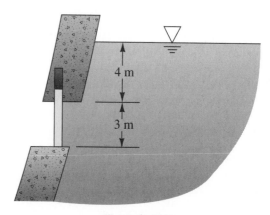

**Figure P.4.110.**

149

**4.111.** A cylindrical tank of water is rotated at constant angular speed $\omega$ until the water ceases to change shape. The result is a free surface which, from fluid mechanics considerations, is that of a paraboloid. If the pressure varies directly as the depth below the free surface, what is the resultant force on a quadrant of the base of the cylinder? Take $\gamma = 62.4$ lb/ft³. [*Hint:* Use circular strip in quadrant having area $1/4(2\pi r \, dr)$.]

**4.112.** Find the $x$ and $y$ coordinates of the center of gravity of the bodies shown. These consist of:
  (a) A plate $ABC$ whose thickness $t = 50$ mm.
  (b) A rod $D$ of diameter .3 m and length 3 m.
  (c) A block $F$ whose thickness (not shown) is .3 m.
The density of the three bodies is the same.

**Figure P.4.111.**

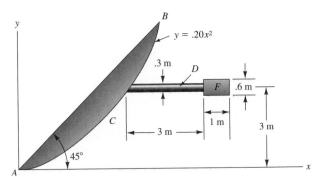

**Figure P.4.112.**

# CHAPTER 5

# Equations
# of Equilibrium

## 5.1 Introduction

You will recall from Section 1.10 that a *particle* in equilibrium is one that is stationary or that moves uniformly relative to an inertial reference. A *body* is in equilibrium if all the particles that may be considered to comprise the body are in equilibrium. It follows, then, that a rigid body in equilibrium cannot be rotating relative to an inertial reference. In this chapter, we shall consider bodies in equilibrium for which the rigid-body model is valid. For these bodies, there are certain simple equations that relate all the surface and body forces, or their equivalents, that act on the body. With these equations, we can sometimes ascertain the value of a certain number of unknown forces. For instance, in the beam shown in Fig. 5.1, we know the loads $F_1$ and $F_2$ and also the weight $W$ of the beam, and we want to determine the forces transmitted to the earth so that we can design a foundation to support the structure properly. Knowing that the beam is in equilibrium and that the small deflection of the beam will not appreciably affect the forces transmitted to the earth, we can write rigid-body equations of equilibrium involving the unknown and known forces acting on the beam and thus arrive at the desired information.

**Figure 5.1.** Loaded beam.

Note in the beam problem above that a number of steps are implied. First, there is the singling out of the beam itself for discussion. Then, we express certain equations of equilibrium for the beam, which we take as a rigid body. Finally, there is the evaluation of the unknowns and interpretation of the results. In this chapter, we will carefully examine each of these steps.

Of critical importance is the need to be able to isolate a body or part of a body for analysis. Such a body is called a *free body*. We will first carefully investigate the development of free-body diagrams. We urge you to pay special heed to this topic, since *it is the most important step in the solving of mechanics problems*. An incorrect free-body diagram means that all ensuing

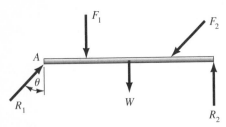

**Figure 5.2.** Free-body diagram of beam.

**Figure 5.3.** Smooth spheres in equilibrium.

**Figure 5.4.** Free-body diagrams.

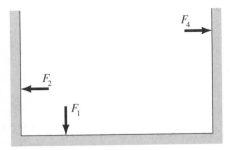

**Figure 5.5.** This is *not* a free-body diagram.

work, no matter how brilliant, will lead to wrong results. More than just a means of attacking statics problems, the free-body concept is your first exposure to the overridingly important topic of *engineering analysis* in general.[1] We now examine this critical step.

## 5.2  The Free-Body Diagram

Since the equations of equilibrium for a particular body actually stem from the dynamic considerations of the body, we must be sure to include *all* the forces (or their equivalents) acting *on* this body, because they all affect the motion of the body and must be accounted for. To help identify all the forces and so ensure the correct use of the equations of equilibrium, we isolate the body in a simple diagram and show *all* the forces from the *surroundings* that act *on* the body. Such a diagram is called a *free-body diagram*. When we isolate the beam in our problem from its surroundings, we get Fig. 5.2. On the left end, there is an unknown force from the ground that has a magnitude denoted as $R_1$ and a direction denoted as $\theta$, with a line of action going through a known point $A$. [We may also use components $(R_1)_x$ and $(R_1)_y$ as unknowns rather than $R_1$ and $\theta$.] The right side involves a force in the vertical direction with an unknown magnitude denoted as $R_2$. The direction is vertical because the beam is on rollers to allow for thermal expansion or thermal contraction and to thus allow unhindered stretching or shrinking of the beam in the axial direction. As a result, the ground exerts a negligibly small horizontal force there. Once all the forces acting on the beam have been identified, including the three unknown quantities $R_1$, $R_2$, and $\theta$, we can solve for these unknowns by using three equations of equilibrium.

Consider now the hard spheres shown in Fig. 5.3 in a condition of equilibrium with surfaces smooth enough to permit us to neglect friction completely. The contact forces thus must be in a direction normal to the surface of contact. The free bodies of the spheres are shown in Fig. 5.4. Notice that $F_3$ is the magnitude of the force from sphere $B$ on sphere $A$, while the reaction, also shown as $F_3$ according to Newton's third law, is the magnitude of the force from sphere $A$ on sphere $B$.

You might be tempted to consider a portion of the container as a free body in the manner shown in Fig. 5.5. But even if this diagram did clearly depict a body (which it does not!), it would not qualify as a free body, since all the forces acting on the body have not been shown.

In engineering problems, bodies are often in contact in a number of standard ways. In Fig. 5.6, you will find the types of forces transmitted from body $M$ to body $N$ for body connections that are often found in practice. (These are not free-body diagrams, since all the forces on any given body have not been shown.)

---

[1]The author has found through many years of experience that the absence of a free-body diagram in a student's work on a particular problem signifies that:
1. There will most likely be errors in the analysis of the problem, or
2. Even worse, the student does not have a good grasp of the problem.

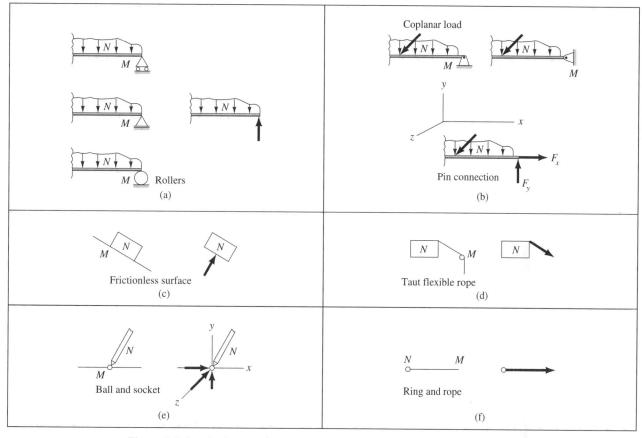

**Figure 5.6.** Standard connections.

In general, to ascertain the nature of the force system that a body $M$ is capable of transmitting to a second body $N$ through some connector or support, we may proceed in the following manner. Mentally move the bodies relative to each other in each of three orthogonal directions. In those directions where relative motion is impeded or prevented by the connector or support, there can be a force component at this connector or support in a free-body diagram of either body $M$ or $N$. Next, mentally rotate bodies $M$ and $N$ relative to each other about the orthogonal axes. In each direction about which relative rotation is impeded or prevented by the connector or support, there can be a couple-moment component at this connector or support in a free-body diagram of body $M$ or $N$. Now as a result of equilibrium considerations of body $M$ or $N$, certain force and couple-moment components that are capable of being generated at a support or connector will be zero for the particular loadings at hand. Indeed, one can often readily recognize this by inspection. For instance, consider the pin-connected beam loaded in a coplanar manner shown partially in Fig. 5.7. If we mentally move the beam relative to the ground in the $x$, $y$, and $z$ directions, we get resistance from the pin for each direction, and so the ground at $A$ can transmit force components $A_x$, $A_y$, and

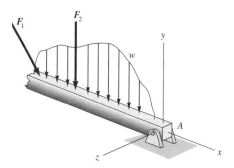

**Figure 5.7.** Pin connection.

$A_z$. However, because the loading is coplanar in the $xy$ plane, the force component $A_z$ must be zero and can be deleted. Next, mentally rotate the beam relative to the ground at $A$ about the three orthogonal axes. Because of the smooth pin connection, there is no resistance about the $z$ axis and so $M_z = 0$. But there is resistance about the $x$ and $y$ axes. However, the coplanar loading in the $xy$ plane cannot exert moments about the $x$ and $y$ axes, and so the couple moments $M_x$ and $M_y$ are zero. All told, then, we just have force components $A_x$ and $A_y$ at the pin connection, as has been shown earlier in Fig. 5.6(b), wherein we relied on physical reasoning for this result.

## 5.3   Free Bodies Involving Interior Sections

Let us consider a rigid body in equilibrium as shown in Fig. 5.8. Clearly, every portion of this body must also be in equilibrium. If we consider the body as two parts $A$ and $B$, we can present either part in a free-body diagram. To do this, we must include on the portion chosen to be the free body the forces *from the other part* that arise at the common section (Fig. 5.9). The surface between both sections may be any curved or plane surface, and over it there will be a continuous force distribution. In the general case, we know that such a distribution can be replaced by a single force and a single couple moment (at any chosen point) and this has been done in the free-body diagram of parts $A$ and $B$ in Fig. 5.9. Notice that Newton's third law has been observed.

**Figure 5.8.** Rigid body in equilibrium.

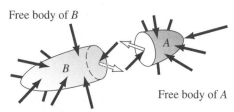

Free body of $B$

Free body of $A$

**Figure 5.9.** Free bodies of parts $A$ and $B$.

As a special case, consider a beam with one end embedded in a massive wall (cantilever beam) and loaded within the $xy$ plane (Fig. 5.10). A free body of the portion of the beam extending from the wall is shown in Fig. 5.11. Because of the geometric symmetry of the beam about the $xy$ plane and the fact that the loads are in this plane, the exposed forces in the cut section can be considered coplanar. Hence, these exposed forces can be replaced by a single force and a single couple moment in the center plane, and it is the usual practice to decompose the force into components $F_y$ and $F_x$. Although a line of action for the force can be found that would enable us to eliminate the couple moment, it is desirable in structural problems to work with an equivalent system that has the force passing through the center of the beam cross section, and thus to have a couple moment. In the next section, we will see how $F$ and $C$ can be ascertained.

**Figure 5.10.** Cantilever beam.

**Figure 5.11.** Free-body diagram of cantilever beam.

# Example 5.1

As a further illustration of a free-body diagram, we shall now consider the frame[2] shown in Fig. 5.12, which consists of members connected by frictionless pins. The force systems acting on the assembly and its parts will be taken as coplanar. We shall now sketch free-body diagrams of the assembly and its parts.

***Free-body diagram of the entire assembly.***    The magnitude and direction of the force at $A$ from the wall onto the assembly is not known. However, we know that this force is in the plane of the system. Therefore, two components are shown at this point (Fig. 5.13). Since the direction of the force $C$ is known, there are then three unknown scalar quantities, $A_y$, $A_x$, and $C$, for the free body.

***Free-body diagram of the component parts.***    When two members are pinned together, such as members $DE$ and $AB$ or $DE$ and $BC$, we usually consider the pin to be part of one of the bodies. However, when more than two members are connected at a pin, such as members $AB$, $BC$, and $BF$ at $B$, we often isolate the pin and consider that all members act on the pin rather than directly on each other, as illustrated in Fig. 5.14. Notice four sets of forces that form pairs of reactions have been enclosed with dashed lines. The fifth set is the 1,000-lb force on the pin and on the member $BF$.

**Figure 5.12.** A frame.

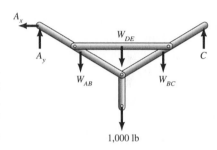

**Figure 5.13.** Free-body diagram of frame.

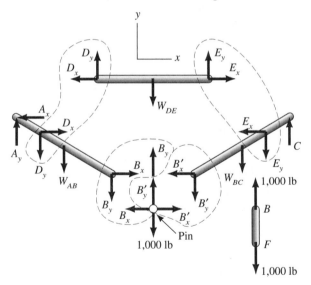

**Figure 5.14.** Free-body diagrams of parts.

[2]A *frame* is a system of connected straight or bent, long, slender members where some of the connecting pins are not at the ends of the members as is the case for structures that we will study later called *trusses*.

### ■ Example 5.1 (Continued)

Do not be concerned about the proper *sense* of an unknown force component that you draw on the free-body diagram, for you may choose either a positive or negative sense for these components. When the values of these quantities are ascertained by methods of statics, the proper sense for each component can then be established; but, having chosen a sense for a component, you must be sure that the *reaction* to this component has the *opposite* sense—else you will violate Newton's third law.

***Free-body diagram of portion of the assembly to the right of M–M.*** In making a free body of the portion to the right of section *M–M* (see Fig. 5.15a), we must remember to put in the weight of the portions of the members remaining *after* the cut has been made. At the two cuts made by *M–M* we must replace coplanar force distributions by resultants, as in the case of the previously considered cantilever beam. This is accomplished by inserting two force components usually normal and tangential to the cross section and a couple moment as was done for the cantilever beam. Note in Fig. 5.15(b) that there are seven unknown scalar quantities for this free-body diagram. They are $C_1$, $C_2$, $F_1$, $F_2$, $F_3$, and $F_4$. Apparently, the number of unknowns varies widely for the various free bodies that may be drawn for the system. For this reason, you must choose the free-body diagram that is suitable for your needs with some discretion in order to effectively solve for the desired unknowns.

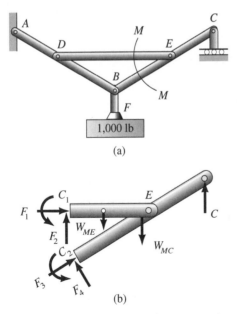

(a)

(b)

**Figure 5.15.** Cutting a frame to expose forces in certain members.

# Example 5.2

Draw a free-body diagram of the beam $AB$ and the frictionless pulley in Fig. 5.16 (a). The weight of the pulley is $W_D$, and the weight of the beam is $W_{AB}$.

**Figure 5.16.** (a) Beam AB: (b) Free-body diagram of AB; (c) Free-body diagram of Pulley D

The free-body diagram of beam $AB$ is shown in Fig. 5.16(b). The weight of the beam has been shown at its center of gravity. Components $B_x$ and $B_y$ are forces from the pulley $D$ acting on the beam through the pin at $B$. The free-body diagram of the pulley is shown in Fig. 5.16(c).

Some students may be tempted to put the weight of the pulley at $B$ in the free-body diagram of beam $AB$. The argument given is that this weight "goes through $B$." To put the pulley weight at $B$ on free body $AB$ is strictly speaking an error! The fact is that the weight of the pulley is a body force acting throughout the *pulley* and *does not* act on the *beam BD*. It so happens that the simplest resultant of this body force distribution on $D$ goes through a *position* corresponding to pin $B$. This does not alter the fact that this weight acts *on the pulley* and *not on the beam*. The beam can only feel forces $B_x$ and $B_y$ transmitted from the pulley to the beam through pin $B$. These forces are related to the pulley weight as well as the tension in the cord around the pulley through equations of equilibrium for the free body

## Example 5.3

We finish this series of free-body diagrams with a three-dimensional case. In Fig. 5.17(a) we show a structure having ball-joint connections at $A$ and $D$, a fixed-end support at $B$ and, finally, at $C$ the column $B$ resting directly on the foundation. Draw the free-body diagram for the structure.

We show the free body diagram for this case in Fig. 5.17(b).

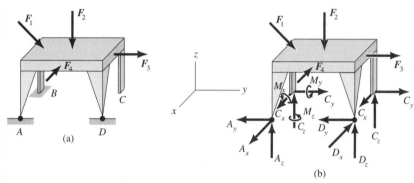

**Figure 5.17.** Three-dimensional structure with a free-body diagram.

# *5.4   Looking Ahead—Control Volumes

In *rigid-body mechanics* we use the free body wherein we isolate a body or a portion of a body and we identify all the external forces acting on the body so that we can employ Newton's law.

In *fluid mechanics*, we may either make a free body of some chosen chunk of fluid (here it is called a system), but more likely it will be more profitable to identify some volume in space involving fluid flow through the volume. Such a volume is called a *control volume*. Here, as in the case of a free body, we must specify *all* the *external forces* such as tractions on the bounding surfaces of the control volume and, in addition, body forces on the material inside the control volume. This identification and force specification is needed to ensure, in appropriate equations, that Newton's law and other laws are satisfied for the fluid and other bodies inside the control volume at any time $t$.

Thus, you must develop sensitivity at this early stage of your studies in depicting external forces for a free body. The same care will be needed in your upcoming courses in fluid mechanics.

**5.1.** Draw the free-body diagram of the gas-grill lid when it is lifted at the handle to a 45° open position.

**5.3.** Draw a free-body diagram of the A-frame.

Figure P.5.1.

**Figure P.5.3.**

**5.2.** A large antenna is supported by three guy wires and rests on a large spherical ball joint. Draw the free-body diagram of the antenna.

**5.4.** Draw complete free-body diagrams for the member $AB$ and for cylinder $D$. Neglect friction at the contact surfaces of the cylinder. The weights of the cylinder and the member are denoted as $W_D$ and $W_{AB}$, respectively.

**Figure P.5.2.**

**Figure P.5.4.**

**5.5.** Draw free-body diagrams of the plate *ABCD* and the bar *EG*. Assume that there is no friction at the pulley *H* or at the contact surface *C*.

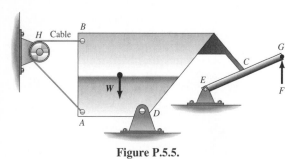

**Figure P.5.5.**

**5.6.** Draw the free-body diagram of one part of the two-piece posthole digger.

**Figure P.5.6.**

**5.7.** Draw a free-body diagram for each member of the system. Neglect the weights of the members. Replace the distributed load by a resultant.

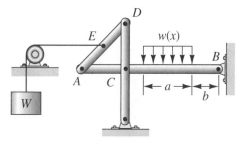

**Figure P.5.7.**

**5.8.** Draw the free-body diagrams for the oars of a rowboat when the rower pushes with one hand and pulls with the other (i.e., turns the boat).

**Figure P.5.8.**

**5.9.** Draw a free-body diagram of the beam. Replace all distributions by simplest equivalent force systems. Neglect the weight of the beam.

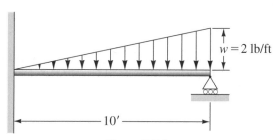

**Figure P.5.9.**

**5.10.** Two cantilever beams are pinned together at *A*. Draw free-body diagrams of each cantilever beam.

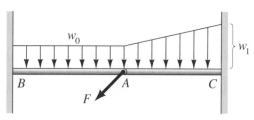

**Figure P.5.10.**

**5.11.** Draw free-body diagrams of each part of the tree-branch trimmer.

**Figure P.5.11.**

**5.12.** Draw free-body diagrams for the two booms and the body $E$ of the power shovel. Consider the weight of each part to act at a central location. (Regard the shovel and payload as concentrated forces, $W_S$ and $W_{PL}$, respectively.)

**Figure P.5.12.**

**5.13.** Draw the free-body diagram for the bulldozer, $B$, hydraulic ram, $R$, and tractor, $T$. Consider the weight of each part, $B$, $R$, and $T$.

**Figure P.5.13.**

**5.14.** Draw a free-body diagram of the whole apparatus and of each of its parts: $AB$, $AC$, $BC$, and $D$. Include the weights of all bodies. Label forces.

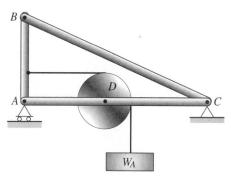

**Figure P.5.14.**

**5.15.** Draw a free-body diagram of members $CG$, $AG$, and the disc $B$. Include as the only weight that of disc $B$. Label all forces. (*Hint:* Consider the pin at $G$ as a separate free body.)

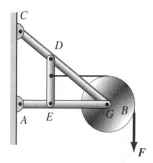

**Figure P.5.15.**

**5.16.** Draw the free-body diagram of the horizontally bent cantilevered beam. Use only $xyz$ components of all vectors drawn.

**Figure P.5.16.**

## 5.5   General Equations of Equilibrium

For every free-body diagram, we can replace the system of forces and couples acting on the body by a single force and a single couple moment at a point $a$. The force will have the same magnitude and direction, no matter where point $a$ is chosen to move the entire system by methods discussed earlier. However, the couple-moment vector will depend on the point chosen. We will prove in dynamics that:

> *The necessary conditions for a rigid body to be in equilibrium are that the resultant force $F_R$ and the resultant couple moment $C_R$ for any point a be zero vectors.*

That is,

$$F_R = 0 \tag{5.1a}$$
$$C_R = 0 \tag{5.1b}$$

We shall prove in dynamics, furthermore, that the conditions above are *sufficient* to maintain an *initially stationary* body in a state of equilibrium. These are the fundamental equations of statics. You will remember from Section 4.3 that the resultant $F_R$ is the sum of the forces moved to the common point, and that the couple moment $C_R$ is equal to the sum of the moments of all the original forces and couples taken about this point. Hence, the equations above can be written

$$\sum_i F_i = 0 \tag{5.2a}$$
$$\sum_i \boldsymbol{\rho}_i \times F_i + \sum_i C_i = 0 \tag{5.2b}$$

where the $\boldsymbol{\rho}_i$'s are displacement vectors from the common point $a$ to any point on the lines of action of the respective forces. From this form of the equations of statics, we can conclude that for equilibrium to exist, *the vector sum of the forces must be zero and the moment of the system of forces and couples about any point in space must be zero.*

Now that we have summed forces and have taken moments about a point $a$, we will demonstrate that we cannot find another *independent* equation by taking moments about a *different* point $b$. For the body in Fig. 5.18, we have initially the following equations of equilibrium using point $a$:

$$F_1 + F_2 + F_3 + F_4 = 0 \tag{5.3}$$
$$\boldsymbol{\rho}_1 \times F_1 + \boldsymbol{\rho}_2 \times F_2 + \boldsymbol{\rho}_3 \times F_3 + \boldsymbol{\rho}_4 \times F_4 = 0 \tag{5.4}$$

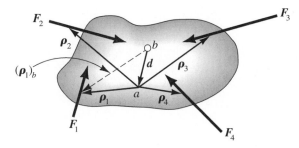

**Figure 5.18.** Consider moments about point $b$.

The new point $b$ is separated from $a$ by the position vector $d$. The position vector (shown dashed) from $b$ to the line of action of the force $F_1$ can be given in terms of $d$ and the displacement vector $\rho_1$ as follows. Similarly for $(\rho_2)_b$, which is not shown, and others.

$$(\rho_1)_b = (d + \rho_1)$$
$$(\rho_2)_b = (d + \rho_2), \quad \text{etc.}$$

The moment equation for point $b$ can then be given as

$$(\rho_1 + d) \times F_1 + (\rho_2 + d) \times F_2 + (\rho_3 + d) \times F_3 + (\rho_4 + d) \times F_4 = 0$$

Using the distributive rule for cross products, we can restate this equation as

$$(\rho_1 \times F_1 + \rho_2 \times F_2 + \rho_3 \times F_3 + \rho_4 \times F_4)$$
$$+ d \times (F_1 + F_2 + F_3 + F_4) = 0 \tag{5.5}$$

Since the expression in the second set of parentheses is zero, in accordance with Eq. 5.3, the remaining portion degenerates to Eq. 5.4, and thus we have not introduced a new equation. Therefore, *there are only two independent vector equations of equilibrium for any single free body.*

We shall now show that instead of using Eqs. 5.3 and 5.4 as the equations of equilibrium, we can instead use Eqs. 5.4 and 5.5. That is, instead of summing forces and then taking moments about a point for equilibrium, we can instead take moments about *two* points. Thus, if Eq. 5.4 is satisfied for point $a$, then for point $b$ we end up in Eq. 5.5 with

$$d \times (F_1 + F_2 + F_3 + F_4) = 0 \tag{5.6}$$

If point $b$ can be any point in space making $d$ arbitrary, then the above equation indicates that the vector sum of forces is zero. If a point $b$ happens to be chosen making Eq. (5.6) identically $0 = 0$, (and hence useless), choose another point $b$. We then have equilibrium since $F_R = 0$ and $C_R = 0$.

Using the vector Eqs. 5.2, we can now express the scalar equations of equilibrium. Since, as you will recall, the rectangular components of the moment of a force about a point are the moments of the force about the orthogonal axes at the point, we may state these equations in the following manner:

$$\sum_i (F_x)_i = 0 \quad \text{(a)} \qquad \sum_i (M_x)_i = 0 \quad \text{(d)}$$

$$\sum_i (F_y)_i = 0 \quad \text{(b)} \qquad \sum_i (M_y)_i = 0 \quad \text{(e)} \qquad (5.7)$$

$$\sum_i (F_z)_i = 0 \quad \text{(c)} \qquad \sum_i (M_z)_i = 0 \quad \text{(f)}$$

From this set of equations, it is clear that *no more than six unknown scalar quantities in the general case can be solved by methods of statics for a single free body.*[3]

We can easily express *any number* of scalar equations of equilibrium for a free body by selecting references that have different axis directions, along which we can sum forces and about which we can take moments. However, in choosing six *independent* equations, we will find that the remaining equations will be dependent on these six. That is, these equations will be sums, differences, etc., of the independent set and so will be of no use in solving for desired unknowns other than for purposes of checking calculations.

## 5.6   Problems of Equilibrium I

We shall now examine problems of equilibrium in which the rigid-body assumption is valid. To solve such problems, we must find the value of certain unknown forces and couple moments. We first draw a free-body diagram of the entire system or portions thereof to clearly *expose* pertinent unknowns for analysis. We then write the equilibrium equations in terms of the unknowns along with the known forces and geometry. As we have seen, for any free body there is a limited number of independent scalar equations of equilibrium. Thus, at times we must employ several free-body diagrams for portions of the system to produce enough independent equations to solve all the unknowns.

For any free body, we may proceed by expressing two basic vector equations of statics. After carrying out such vector operations as cross products and additions in the equations, we form scalar equations. These scalar equations are then solved simultaneously (together with scalar equations from other free-body diagrams that may be needed) to find the unknown forces and

---

[3]Keep in mind that we can also take moments about two sets of axes just as we could take moments about two points for the vector equations of equilibrium. The two sets of axes like the two points must be chosen properly so as to yield *independent* scalar equations. This can readily be done. If your second point and associated axes do not yield three additional independent equations, select another point and axes until you have your six independent equations of equilibrium.

couple moments. We can also express the scalar equations immediately by using the alternative scalar equilibrium relations that we formulated in previous sections. In the first case, we start with more compact vector equations and arrive at the expanded scalar equations by the formal procedures of vector algebra. In the latter case, we evaluate the expanded scalar equations by carrying out arithmetic operations on the free-body diagram as we write the equations. Which procedure is more desirable? It all depends on the problem and the investigator's skill in vector manipulation. It is true that many statics problems submit easily to a direct scalar approach, but the more challenging problems of statics and dynamics definitely favor an initial vector approach. In this text, we shall employ the particular procedure that the occasion warrants.

In statics problems, we must assign a sense to each component of an unknown force or couple moment in order to write the equations. If, on solving the equations, *we obtain a negative sign for a component, then we have guessed the wrong sense for that component*. Nothing need be redone should this occur. Continue with the remainder of the problem, retaining the minus sign (or signs). At the end of the problem, report the correct sense of your force components and couple-moment components.

We shall now solve and discuss a number of problems of equilibrium. These problems are divided into four classes of force systems:

1. Concurrent.
2. Coplanar.
3. Parallel.
4. General.

The type of *simplest* resultant for each special system of forces is most useful in determining the number of scalar equations available in a given problem. The procedure is to classify the force system, note what simplest resultant force system is associated with the classification, and then consider the number of scalar equations necessary and sufficient to guarantee this resultant to be zero. The following cases exemplify this procedure.

**Case A. Concurrent System of Forces.**   In this case, since the simplest resultant is a single force at the point of concurrency, the only requirement for equilibrium is that this force be zero. We can ensure this condition if the orthogonal components of this force are separately equal to zero. Thus, we have *three* equations of equilibrium of the form

$$\sum_i (F_x)_i = 0, \qquad \sum_i (F_y)_i = 0, \qquad \sum_i (F_z)_i = 0 \qquad (5.8)$$

As was pointed out in the general vector discussion, there are other ways of ensuring a zero resultant. Suppose that the moments of the concurrent force system are zero about three nonparallel axes: $\alpha$, $\beta$, and $\gamma$. That is,

$$\sum_i (M_\alpha)_i = 0, \qquad \sum_i (M_\beta)_i = 0, \qquad \sum_i (M_\gamma)_i = 0 \qquad (5.9)$$

Any one of the following three conditions must then be true:

**1.** The resultant force $F_R$ is zero.
**2.** $F_R$ cuts all three axes (see Fig. 5.19).
**3.** $F_R$ cuts two axes and is parallel to the third (see Fig. 5.20).

**Figure 5.19.** $F_R$ cuts three axes.

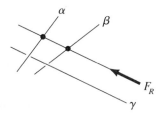

**Figure 5.20.** $F_R$ cuts two axes and is parallel to third.

We can guarantee condition 1 and thus equilibrium if we select axes $\alpha$, $\beta$, and $\gamma$ so that no straight line can intersect all three axes or can cut two axes and be parallel to the third. Then we can use Eqs. 5.9 as the equations of equilibrium under the aforestated conditions rather than using Eqs. 5.8. What happens if an axis used violates these conditions? The resulting equation will either be an *identity* $0 = 0$ or will be dependent on a previous independent equation of equilibrium for one of the axes. No harm is done. One should use other axes until three independent equations are found.

Similarly, one can sum forces in one direction and take moments about two axes. Setting these equal to zero can yield three independent equations of equilibrium. If not, use other axes.

The essential conclusion to be drawn is that *there are three independent scalar equations of equilibrium for a concurrent force system.* For such systems it is most likely that you will always sum forces rather than take moments. However, for other force systems that we shall undertake, there will be ample opportunity to profitably employ alternate forms of equations other than those that we shall at first prescribe. The important thing to remember is that, just as in the concurrent force systems, only a definite number of equations for a given free body will be independent. Simply writing more equations beyond this number will only lead to identities and equations that will be of no use for solving for the desired unknowns.

# ■ Example 5.4 ■

What are the tensions in cables $AC$ and $AB$ in Fig. 5.21? The system is in equilibrium. The following data apply:

$$W = 1,000 \text{ N} \qquad \beta = 50° \qquad \alpha = 37°$$

**Figure 5.21.** Derrick holding a beam.

Immediately it should be clear from observation of the diagram that

$$T_{AD} = 1,000 \text{ N (tension)}$$

A suitable free body that exposes the desired unknowns is the ring $A$, which may be considered as a particle for this computation because of its comparatively small size (Fig. 5.22). Physical intuition indicates that the cables are in tension and hence pulling away from $A$ as we have indicated in the diagram although, as mentioned previously, it is *not* necessary to recognize at the outset the correct sense of an unknown force. The force system acting on the particle must be a concurrent system. Here it is also coplanar as well, and therefore we may solve for only two unknowns. Hence, we can proceed to the scalar equations of equilibrium. Thus,

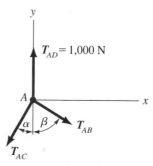

**Figure 5.22.** Free-body diagram of pin $A$.

$$\underline{\sum F_y = 0} \qquad 1,000 - T_{AC} \cos 37° - T_{AB} \cos 50° = 0$$

$$\therefore .7986 T_{AC} + .6428 T_{AB} = 1,000 \qquad \text{(a)}$$

$$\underline{\sum F_x = 0} \qquad -T_{AC} \sin \alpha - T_{AB} \sin \beta = 0$$

$$\therefore T_{AC} = 1.2729 T_{AB} \qquad \text{(b)}$$

Solving for $T_{AC}$ and $T_{AB}$ from Eqs. (a) and (b), we get

$$\boxed{T_{AB} = 602.6 \text{ N}} \qquad \boxed{T_{AC} = 767.1 \text{ N}}$$

Since the signs for $T_{AC}$ and $T_{AB}$ are positive, we have chosen the correct senses for the forces in the free-body diagram.

## ■ Example 5.4 (Continued)

Another way of arriving at the solution, is to consider the *force polygon* that was discussed in Section 2.3. Because the forces are in equilibrium, the polygon must close; that is, the head of the final force must coincide with the tail of the initial force. In this case, we have a triangle, as shown in Fig. 5.23 approximately to scale. We can now use the law of sines.

$$\frac{T_{AB}}{\sin 37°} = \frac{1,000}{\sin 93°} \qquad\qquad T_{AB} = 602.6 \text{ N}$$

$$\frac{T_{AC}}{\sin 50°} = \frac{1,000}{\sin 93°} \qquad\qquad T_{AC} = 767.1 \text{ N}$$

**Figure 5.23.** Force polygon.

The force polygon may thus be used to good advantage when three concurrent coplanar forces are in equilibrium.

As a final alternative, let us now initiate the computation for the unknown tensions from the basic *vector* equations of statics. First, we must express all forces in vector notation.

$$T_{AC} = T_{AC} (-\sin 37° \, i - \cos 37° \, j)$$

$$T_{AB} = T_{AB} (\sin 50° \, i - \cos 50° \, j)$$

We get the following equation when the vector sum of the forces is set equal to zero:

$$T_{AC} (-.6018i - .7986j) + T_{AB} (.7660i - .6428j) + 1,000j = 0$$

Choosing point *A*, the point of concurrency, we clearly see that the sum of moments of the forces about this point is zero, so the second basic equation of equilibrium is intrinsically satisfied. We now regroup the terms of the preceding equation in the following manner:

$$(-.6018T_{AC} + .7660T_{AB})i + (-.7986T_{AC} - .6428T_{AB} + 1,000)j = 0$$

To satisfy this equation, each of the quantities in parentheses must be zero. This gives the scalar equations (a) and (b) stated earlier, from which the scalar quantities $T_{AB}$ and $T_{AC}$ can be solved.

The three alternative methods of solution are apparently of equal usefulness in this simple problem. However, the force polygon is only of practical use for three concurrent coplanar forces, where the trigonometric properties of a triangle can be directly used. The other methods can be readily extended to more complex concurrent problems.

# Example 5.5

Find the forces in cables $DB$ and $CB$ in Fig. 5.24. The 500-N force is parallel to the $y$ axis. Consider $B$ to be a ball joint located in the $xz$ plane. Rod $AB$ is a compression member, with a ball joint at $A$.

In Fig. 5.25 we have indicated the forces acting on joint $B$. Clearly we have a three-dimensional concurrent force system with three unknowns. We can readily determine the unknown forces here by simply setting the sum of the forces equal to zero. However, since we only want the forces in the two cables, we shall proceed by setting the moment of the forces about point $A$ of rod $AB$ equal to zero, thereby not including the force in member $AB$.

Denoting the force in $BC$ as $T_C$ and the force in $BD$ as $T_D$, we proceed now to establish the rectangular components of these forces.

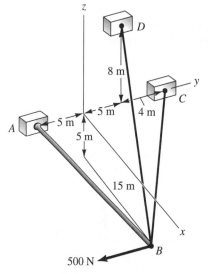

Figure 5.24. Rod $AB$ and cables $CB$ and $DB$ support a 500-lb force.

$$T_C = T_C\left(\frac{-15i + 9j + 5k}{\sqrt{15^2 + 9^2 + 5^2}}\right) = T_C(-.824i + .495j + .275k)\ N$$

$$T_D = T_D\left(\frac{-15i + 5j + 13k}{\sqrt{15^2 + 5^2 + 13^2}}\right) = T_D(-.733i + .244j + .635k)\ N$$

The position vector that we shall use for the moment about point $A$ is $r_{AB}$, which is

$$r_{AB} = 15i + 5j - 5k\ m$$

We now set the moments about point $A$ equal to zero.

$$\sum M_A = 0$$

$$(15i + 5j - 5k) \times [T_C(-.824i + .495j + .275k) +$$

$$T_D(-.733i + .244j + .635k) - 500j] = 0$$

We simplify the calculations further by noting that cable $BD$ has a direction inclined to the plane $ACB$ in which the other three forces lie. This can only mean that the force $T_D$ must have a zero value. Hence, deleting this force in the above equation and carrying out the cross products, it is easy to get the remaining nonzero force. We thus have

$$T_C = 649\ N \qquad T_D = 0\ N$$

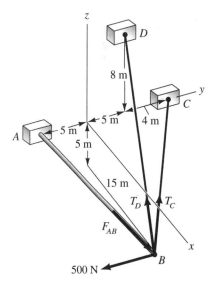

Figure 5.25. Free-body diagram of joint $B$.

## Example 5.5 (Continued)

Finally, as indicated at the outset, we could proceed by summing forces in the coordinate directions. The resulting scalar equations for all three unknown forces are

$$-0.824T_C - 0.733T_D + 0.905T_A = 0$$

$$0.495T_C + 0.244T_D + 0.302T_A = 500$$

$$0.275T_C + 0.635T_D - 0.302T_A = 0$$

We may now solve the simultaneous equations using *Cramer's rule*. Thus we calculate first the determinant of the coefficients of the unknowns. Thus

$$\begin{vmatrix} -0.824 & -0.733 & 0.905 \\ 0.495 & 0.244 & 0.302 \\ 0.275 & 0.635 & -0.302 \end{vmatrix} = 0.272$$

To calculate $T_C$ we proceed as follows:

$$T_C = \dfrac{\begin{vmatrix} 0 & -0.733 & 0.905 \\ 500 & 0.244 & 0.302 \\ 0 & 0.635 & 0.302 \end{vmatrix}}{0.272} = 649 \text{ N}$$

Note that the first column of the determinant consists of the right side terms of the set of simultaneous equations in place of the coefficients of the desired unknown. We can solve for the other unknowns similarly. We then would have the compressive force in member *AB* which is 591 N.

---

**Case B. Coplanar Forces System.**   We have shown that the simplest resultant for a coplanar force system (see Fig. 4.14) is a single force or a single couple moment normal to the plane. Thus, to ensure that the resultant force is zero, we require for a coplanar system in which all forces are in the *xy* plane:

$$\sum_i (F_x)_i = 0, \qquad \sum_i (F_y)_i = 0 \tag{5.10}$$

To ensure that the resultant couple moment is zero, we require for moments about any axis parallel to the *z* axis:

$$\sum_i (M_z)_i = 0 \tag{5.11}$$

We conclude that there are *three* scalar equations of equilibrium for a coplanar force system. Other combinations, such as two moment equations for two axes parallel to the *z* axis and a single force summation, if properly chosen, may be employed to give the three independent scalar equations of equilibrium, as was discussed in case A.

## Example 5.6

**Figure 5.26.** A car is being towed up an incline at a constant speed.

A car shown in Fig. 5.26 is being towed at a steady speed up an incline having an angle of 15°. The car weighs 3,600 lb. The center of gravity is located as is shown in the diagram. Calculate the supporting force on each wheel and force $T$.

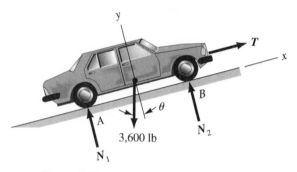

**Figure 5.27.** Free-body diagram of the car.

A free-body diagram of the car is shown in Fig. 5.27. The forces $N_1$ and $N_2$ are the total forces, respectively, for the rear wheels and the front wheels. Note, because the wheels are rotating at constant speed, there are no friction forces present. We have thus formed on this free body a *coplanar* force system involving three unknowns and hence the unknowns are solvable by rigid-body statics. Using axes tangent and normal to the incline we have

## ▪ Example 5.6 (Continued)

$$\underline{\sum F_x = 0} \quad T - W\sin\theta = 0$$

$$\therefore T = (3{,}600)(\sin 15°) = 931.7 \text{ lb}$$

$$\boxed{T = 931.7 \text{ lb}}$$

$$\underline{\sum F_y = 0} \quad N_1 + N_2 - W\cos\theta = 0$$

$$\therefore N_1 + N_2 = (3{,}600)(\cos 15°) = 3{,}477 \text{ lb} \qquad (1)$$

$$\underline{\sum M_A = 0} \quad (N_2)(10) + (W\sin\theta)(1) - (W\cos\theta)(5) - (T)(1.5) = 0$$

$$\therefore N_2 = \frac{1}{10}[-(3{,}600)(\sin 15°)(1) + (3{,}600)(\cos 15°)(5)$$

$$+(931.7)(1.5)] = 1{,}785 \text{ lb}$$

From Eq.(1), we can now get $N_1$. Thus

$$N_1 = 3{,}477 - N_2 = 3{,}477 - 1{,}785 = 1{,}692 \text{ lb}$$

Clearly each rear wheel has acting on it a normal force of $N_1/2 = 846.1$ lb and each front wheel has a normal force of $N_2/2 = 892.5$ lb.

$$\therefore \quad \boxed{\begin{array}{l} \text{Rear wheel support force} = 846.1 \text{ lb} \\ \text{Front wheel support force} = 892.5 \text{ lb} \end{array}}$$

We may now check this solution by using a redundant equation of equilibrium. Thus

$$\underline{\sum M_B \overset{?}{=} 0}$$

$$-(N_1)(10) + (W\sin\theta)(1) - (W\cos\theta)(5) - (T)(1.5) = 0$$

$$\therefore -(1{,}692)(10) + (3{,}600)(\sin 15°)(1) +$$

$$(3{,}600)(\cos 15°)(5) - (931.7)(1.5)] = 0$$

$$.8634 \approx 0$$

We have here a roundoff error, which we can accept for the accuracy of the calculations taken in this problem.

# ■ Example 5.7 ■

A frame is shown in Fig. 5.28 in which the frictionless pulley at $D$ has a mass of 200 kg. Neglecting the weights of the bars, find the force transmitted from one bar to the other at joint $C$.

**Figure 5.28.** Loaded frame.

To expose force components $C_x$ and $C_y$, we form the free body of bar $BD$. This is shown as F.B.D. I in Fig. 5.29. It is clear that for this free body we have six unknowns and only three independent equations of equilibrium.[4] The free-body diagram of the bent bar $AC$ is then drawn (F.B.D. II in Fig. 5.29). Here, we have three more equations but we bring in three

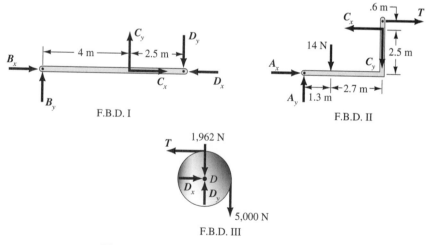

**Figure 5.29.** Free-body diagrams of frame parts.

[4] It should be noted that it is possible to have situations wherein there are more unknowns than independent equations of equilibrium for a given free body, but wherein some of the unknowns—perhaps the desired ones—can be still determined by the equations available. However, not all the unknowns of the free body can be solved. Accordingly, be alert for such situations, so as to minimize the work involved. In this case, we must consider other free-body diagrams.

## Example 5.7 (Continued)

more unknowns. Finally, the free-body diagram of the pulley (F.B.D. III in Fig. 5.29) gives three more equations with no additional unknowns. We now have nine equations available and nine unknowns and can proceed with confidence. Since only two of the unknowns are desired, we shall take select scalar equations from each of the free-body diagrams to arrive at the components $C_x$ and $C_y$ most quickly.

*From F.B.D. III:*

$$\sum M_D = 0:$$

$$(T)(.6) - (5,000)(.6) = 0$$

Therefore,

$$T = 5,000 \text{ N}$$

$$\sum F_x = 0:$$

$$-T + D_x = 0$$

Therefore,

$$D_x = 5,000 \text{ N}$$

$$\sum F_y = 0:$$

$$-1,962 - 5,000 + D_y = 0$$

Therefore,

$$D_y = 6,962 \text{ N}$$

*From F.B.D. I:*

$$\sum M_B = 0:$$

$$(4)(C_y) - (6.5)(D_y) = 0$$

Therefore,

$$C_y = 11,313 \text{ N}$$

*From F.B.D. II:*

$$\sum M_A = 0:$$

$$-(1.3)(14) - (T)(3.1) - C_y(4) + C_x(2.5) = 0$$

Therefore,

$$C_x = 24,300 \text{ N}$$

We can give the force at $C$ (transmitted from bar AC to bar BD) as

$$C = 24,300i + 11.313j \text{ N}$$

# PROBLEMS

**5.17.** In a tug of war, when team *B* pulls with a 400-lb force, how much force must team *C* exert for a draw? With what force does team *A* pull?

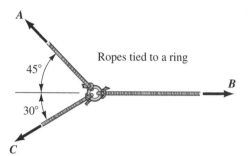

**Figure P.5.17.**

**5.18.** Find the tensile force in cables *AB* and *CB*. The remaining cables ride over frictionless pulleys *E* and *F*.

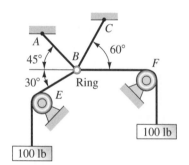

**Figure P.5.18.**

**5.19.** Find the force transmitted by wire *BC*. The pulley *E* can be assumed to be frictionless in this problem.

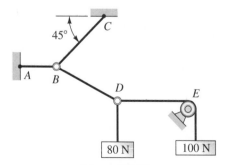

**Figure P.5.19.**

**5.20.** Find the tensions in the three cables connected to *B*. The entire system of cables is coplanar. The roller at *E* is free to turn without resistance.

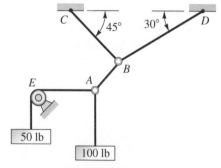

**Figure P.5.20.**

**5.21.** A 700-N circus performer causes a .15-m sag in the middle of a 12-m tightrope with a 5,000-N initial tension. What additional tension is induced in the cable? What is the cable tension when the performer is 3 m from the end and the sag is .12 m?

**Figure P.5.21.**

**5.22.** A 27-lb mirror is held up by a wire fastened to two hooks on the mirror frame. (a) What is the force on the wall hook and the tension in the wire? (b) If the wire will break at a tension of 32 lb, must the wall hook be moved (i.e., the wire lengthened or shortened and the 4-in. rise distance changed)? If so, to what point?

**Figure P.5.22.**

**5.23.** Explain why equilibrium of a concurrent force system is guaranteed by having $\sum_i (F_y)_i = 0$, $\sum_i (M_d)_i = 0$, and $\sum_i (M_e)_i = 0$. Axes $d$ and $e$ are not parallel to the $xz$ plane. Moreover, the axes are oriented so that the line of action of the resultant force cannot intersect both axes.

**5.24.** Cylinders $A$ and $B$ weigh 500 N each and cylinder $C$ weighs 1,000 N. Compute all contact forces.

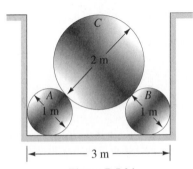

**Figure P.5.24.**

**5.25.** A block having a mass of 500 kg is held by five cables. What are the tensions in these cables? Lower cables are identical and are identically connected at ends.

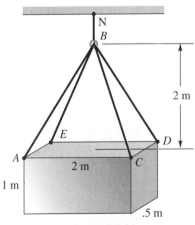

**Figure P.5.25.**

**5.26.** Gear $D$ has a weight of 300 N while members $AB$ and $BC$ are light. What torque $T$ is necessary for equilibrium?

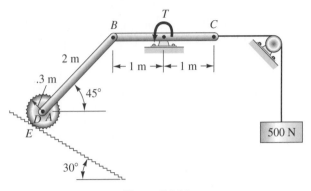

**Figure P.5.26.**

**5.27.** Find the supporting forces at $A$.

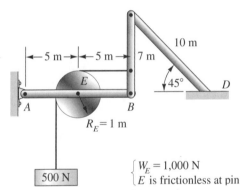

$\begin{cases} W_E = 1{,}000 \text{ N} \\ E \text{ is frictionless at pin} \end{cases}$

**Figure P.5.27.**

**5.28.** Find the supporting forces at $A$ and $B$. At $D$ there is a cylinder weighing 300 N.

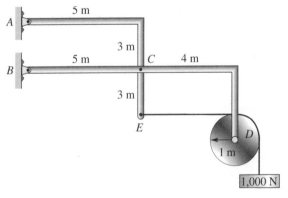

**Figure P.5.28.**

**5.29.** Find the supporting force systems for the beams shown. Note that there is a *pin connection* at C. Neglect the weights of the beams.

**Figure P.5.29.**

**5.30.** Find the supporting force systems at A and B. The length of CB is 8 m.

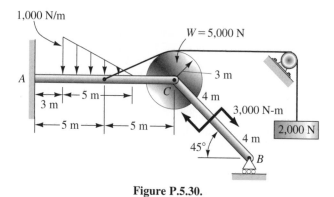

**Figure P.5.30.**

**5.31.** What are the supporting forces at A and D for the frame shown? What are the forces in members AB, BE, and BC?

**Figure P.5.31.**

**5.32.** What are the supporting forces for the frame? Neglect all weights except the 10-kN weight. Disregard friction.

**Figure P.5.32.**

**5.33.** Find the supporting forces at E and F. Pulleys A and B offer no rotational resistance from friction at the bearings.

Data

$W_A = 100$ N
$W_B = 100$ N
$W_C = 200$ N
$D_A = 2$ m
$D_B = 2$ m

**Figure P.5.33.**

177

**5.34.** What are the tensions in cables *AB*, *BC*, and *BD*? Points *A* and *B* are in the *yz* plane.

**Figure P.5.34.**

**5.35.** An elastic cord *AB* is just taut before the 1,000-N force is applied. If it takes 5.0 N/mm of elongation of the cord, what is the tension *T* in the cord after the 1,000-N force is applied? Set up the equation for *T* but do not solve.

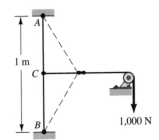

**Figure P.5.35.**

**5.36.** A thin hoop of radius 1 m and weight 500 N rests on an incline. What friction force *f* at *A* is needed for this configuration? What is the tension in wire *CB*?

**Figure P.5.36.**

**5.37.** A stepped cylinder is pulled down an incline by a force *F*, which is increased from zero to 20 N very slowly while always maintaining the 30° inclination shown. If the cylinder is in equilibrium when *F* = 0, how far does *O* move as a result of *F* after equilibrium has been established with *F* = 20 N? The stepped cylinder has a mass of 10 kg. There is no slipping at the base. The force from the spring is *K* times the extension of the spring. For this spring, *K* = 5 N/mm. *F* remains parallel to the position shown.

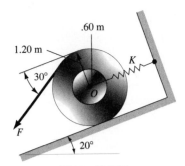

**Figure P.5.37.**

**5.38.** A 10-kg ring is supported by a smooth surface *E* and a wire *AB*. A body *D* having a mass of 3 kg is fixed to the ring at the orientation shown. What is the tension in the wire *AB*? What is its orientation $\alpha$? Point *A* is directly above point *O*.

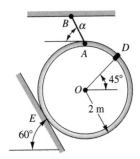

**Figure P.5.38.**

**5.39.** Find the supporting forces on the beam *EF* and the supporting forces at *A*, *B*, *C*, and *D*.

**Figure P.5.39.**

**5.40.** What is the supporting force system at $A$ for the cantilever beam? Neglect the weight of the beam.

**Figure P.5.40.**

**5.41.** In Problem 5.40, find the force system transmitted through the cross section at $B$.

**5.42.** A cantilever beam $AB$ is pinned at $B$ to a simply supported beam $BC$. For the loads given, find the supporting force system at $A$. Determine force components that are normal and tangential to the cross-section of beam $AB$. Neglect the weights of the beams.

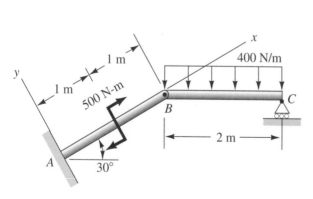

**Figure P.5.42.**

**5.43.** Find the supporting force system at $A$.

**Figure P.5.43.**

**5.44.** A light bent rod $AD$ is pinned to a straight light rod $CB$ at $C$. The bent rod supports a uniform load. A spring is stretched to connect the two rods. The spring has a spring constant of $10^4$ N/m, and its unstretched length is .8 m. Find the supporting forces at $A$ and $B$. The force in the spring is $10^4$ times the elongation in meters.

**Figure P.5.44.**

179

**5.45.** Light rods *AD* and *BC* are pinned together at *C* and support a 300-N and a 100-N load. What are the supporting forces at *A* and *B*?

**Figure P.5.45.**

**5.47.** Solve for the supporting forces at *A* and *C*. *AB* weighs 100 lb, and *BC* weighs 150 lb.

**Figure P.5.47.**

**5.46.** A light rod *CD* is held in a horizontal position by a strong elastic band *AB* (shock cord), which acts like a spring in that it takes $10^3$ N per meter of elongation of the band. The upper part of the band is connected to a small wheel free to roll on a horizontal surface. What is the angle $\alpha$ needed to support a 200-N load as shown?

**Figure P.5.46.**

**5.48.** What torque *T* is needed to maintain the configuration shown for the compressor if $p_1 = 5$ psig? The system lies horizontally.

**Figure P.5.48.**

**5.49.** Work Problem 5.48 for the system oriented vertically with *BC* weighing 3 lb and *CD* weighing 5 lb.

**5.50.** Neglecting friction, find the angle $\beta$ of line *AB* for equilibrium in terms of $\alpha_1$, $\alpha_2$ $W_1$, and $W_2$.

**5.52.** Find the supporting forces at *A* and *G*. The weight of *W* is 500 N and the weight of *C* is 200 N. Neglect all other weights. The cord connecting *C* and *D* is vertical.

Figure P.5.52.

Figure P.5.50.

**5.53.** What torque *T* is needed for equilibrium if cylinder *B* weighs 500 N and *CD* weighs 300 N?

**5.51.** If the rod *CD* weighs 20 lb, what torque *T* is needed to maintain equilibrium? The system is in a vertical plane. Cylinder *A* weighs 10 lb and cylinder *B* weighs 5 lb. Disregard friction. At *D* there is a slot.

Figure P.5.51.

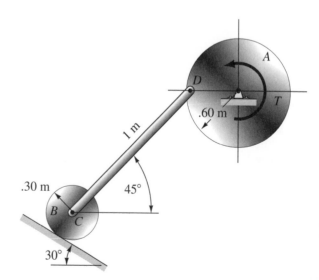

Figure P.5.53.

**5.54.** A bar $AB$ is pinned to two identical planetary gears each of diameter .30 m. Gear $E$ is pinned to bar $AB$ and meshes with the two planetary gears, which in turn mesh with stationary gear $D$. If a torque $T$ of 100 N-m is applied to bar $AB$, what external torque is needed to be applied to the upper planetary gear to maintain equilibrium? The system is horizontal.

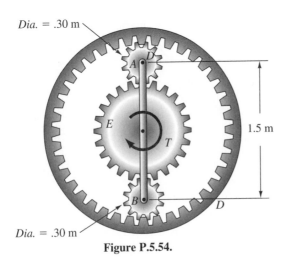

Dia. = .30 m

Dia. = .30 m

**Figure P.5.54.**

**5.55.** In Problem 5.54, equilibrium is maintained by applying a torque on gear $E$ rather than the upper planetary gear. What is this torque?

**5.56.** A Bucyrus–Erie transit crane is holding a chimney having a weight of 20 kN. The chimney is held by a cable that goes over a pulley at $A$, then goes over a second pulley at $D$, and then to a winch at $K$. The position of boom $AH$ (on top) is maintained by two separate cables, one from $A$ to $B$, and the other from $B$ to pulley $C$. Find the tensions in cables $AB$ and $BC$. Note that $BC$ is oriented 30° from the vertical for the setup shown. Consider only the weight of the load and neglect friction.

**Figure P.5.56.**

**5.57.** What are the forces at the arm connection at $B$ and the cable tensions when the power shovel is in the position shown? Arm $AC$ weighs 13,000 N, arm $DF$ weighs 11,000 N, and the shovel and payload together weigh 9,000 N and act at the center of gravity as shown. $B$ is at the same elevation as $G$.

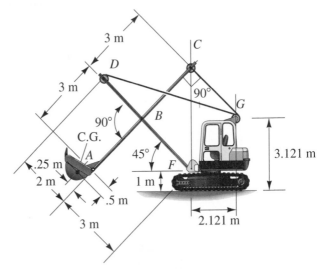

**Figure P.5.57.**

182

# 5.7 Problems of Equilibrium II

**Case C. Parallel Forces in Space.** In the case of parallel forces in space (see Fig. 4.20), we already know that the simplest resultant can be either a single force or a couple moment. If the forces are in the $z$ direction, then

$$\sum_i (F_z)_i = 0 \tag{5.12}$$

ensures that the resultant force is zero. Also,

$$\sum_i (M_x)_i = 0, \qquad \sum_i (M_y)_i = 0 \tag{5.13}$$

guarantees that the resultant couple moment is zero, where the $x$ and $y$ axes may be chosen in any plane perpendicular to the direction of the forces.[5] Thus, three independent scalar equations are available for equilibrium of parallel forces in space.

A summary of the special cases discussed thus far is given below. For even simpler systems such as the concurrent-coplanar and the parallel-coplanar systems, clearly, there is one less equation of equilibrium.

|  | *Summary for Special Cases* | |
|---|---|---|
| *System* | *Simplest Resultant* | *Number of Equations for Equilibrium* |
| Concurrent (three-dimensional) | Single force | 3 |
| Coplanar single couple moment | Single force or | 3 |
| Parallel (three-dimensional) | Single force or single couple moment | 3 |

[5]For parallel forces in the $z$ direction, a simplest resultant consisting of a couple moment only must have this couple moment parallel to the $xy$ plane (see diagram). Recall from Chapter 3 that the orthogonal $xyz$ components of $\mathbf{C}_R$ equals the torque of the system about these axes. Hence, by setting $\sum_i (M_x)_i = \sum_i (M_y)_i = 0$, we are ensuring that $\mathbf{C}_R = \mathbf{0}$.

## Example 5.8

Determine the forces required to support the uniform beam in Fig. 5.30 shown loaded with a couple, a point force, and a downward parabolic distribution of load having zero slope at the origin. The weight of the beam is 100 lb.

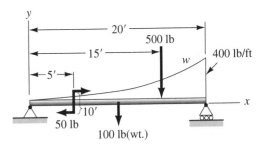

**Figure 5.30.** Find supporting forces.

Since a couple can be rotated without affecting the equilibrium of the body, we can orient the couple so that the forces are vertical. Accordingly, we have here a beam loaded by a system of parallel coplanar loads. Clearly, the supporting forces must be vertical, as shown in Fig. 5.31, where we have a free-body diagram of the beam. Since there are only two unknown quantities, we can handle the problem by statical consideration of this free body.

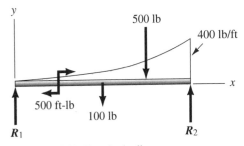

**Figure 5.31.** Free-body diagram.

The equation for the loading curve must be $w = ax^2 + b$, where $a$ and $b$ are to be determined from the loading data and the choice of reference. With an $xy$ reference at the left end, as shown, we then have the conditions:

1. When $x = 0$, $w = 0$.

2. When $x = 20$, $w = 400$.

To satisfy these conditions, $b$ must be zero and $a$ must be unity; the loading function is thus given as

$$w = x^2 \text{ lb/ft}$$

## Example 5.8 (Continued)

In this problem, we shall again work directly with the scalar equations. By summing moments about the left and right ends of the beam, we can then solve for the unknowns directly:

$$\sum M_1 = 0:$$

$$-500 - (10)(100) - (15)(500) - \int_0^{20} xy \ dx + 20R_2 = 0$$

Replacing $y$ by $x^2$, then integrating and canceling terms, we get

$$-9,000 - \left.\frac{x^4}{4}\right|_0^{20} + 20R_2 = 0$$

By inserting limits and solving, we get one of the unknowns:

$$R_2 = 2,450 \text{ lb}$$

Next,

$$\sum M_2 = 0:$$

$$-20R_1 - 500 + (10)(100) + (5)(500) + \int_0^{20}(20 - x)y \ dx = 0$$

Replacing $y$ by $x^2$, integrating and then solving for $R_1$, we have

$$R_1 = 817 \text{ lb}$$

As a check on these computations, we can sum forces in the vertical direction. The result must be zero (or as close to zero as the accuracy of our calculations permits):

$$\sum F_y = 0:$$

$$R_1 + R_2 - 100 - 500 - \int_0^{20} x^2 \ dx = 0$$

$$3,267 - 600 - \left.\frac{x^3}{3}\right|_0^{20} = 0$$

Therefore,

$$2,667 - 2,667 = 0$$

Always take the opportunity to check a solution in this manner (i.e., by using a redundant equilibrium equation). In later problems, we shall rely heavily on calculated reactions (supporting forces); thus, we must make sure they are correct.

## Example 5.9

In Fig. 5.32, find the supporting forces at A, B, and D. Note the pin connection at C. Also note that at E we have a welded connection.

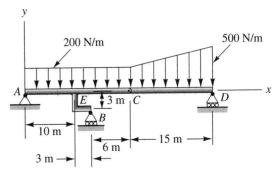

**Figure 5.32.** Members AC and CD are pinned at C.

The free-body diagram for the entire system is shown in Fig. 5.33.
We have a coplanar system of forces for this free body and so we have
only three independent equations of equilibrium. However we have here
four unknowns. One of the unknowns, namely $A_x$, can be seen by inspection to be zero leaving now a coplanar parallel system with three
unknowns but with only two equations of equilibrium.

**Figure 5.33.** Free-body diagram I. (F.B.D. I)

We shall next consider the free body of member CD. This is shown
in Fig. 5.34 where we have simplified the distributed loading in order to
better facilitate the ensuing computations. Clearly, $C_x$ must be zero. Thus,
for F.B.D. II in Fig. 5.34, we will then have only two unknowns for which
we have two equations of equilibrium. By taking moments about point C
in Fig. 5.34, we can directly determine $D_y$.

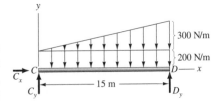

**Figure 5.34.** Free-body diagram II.
(F.B.D. II)

# Example 5.9 (Continued)

$$\sum M_C = 0:$$

$$(D_y)(15) - (200)(15)\left(\frac{15}{2}\right) - \left(\frac{1}{2}\right)(15)(300)\left[\left(\frac{2}{3}\right)(15)\right] = 0$$

$$D_y = 3,000 \text{ N}$$

We may now go back to Fig. 5.33 to solve for the remaining two unknowns. Thus, simplifying the loading between $C$ and $D$ as we have done in Fig. 5.34, we have

$$\sum F_y = 0:$$

$$A_y + B_y + 3,000 - (200)(34) - \frac{1}{2}(300)(15) = 0 \qquad \text{(a)}$$

$$\sum M_B = 0:$$

$$-A_y(13) + (3,000)(21) - (200)(34)\left(\frac{34}{2} - 13\right)$$

$$-\frac{1}{2}(300)(15)\left[6 + \left(\frac{2}{3}\right)(15)\right] = 0 \qquad \text{(b)}$$

From Eq. (b), we get

$$A_y = -15.38 \text{ N}$$

$$\text{(c)}$$

And from Eq. (a), we have

$$B_y = 6,065 \text{ N}$$

Note that $A_y$ was negative indicating that we had inserted the wrong sense for this force in F.B.D. I. We did not make any changes in the ensuing calculations while going to Eq. (a) (valid for F.B.D. I) to calculate $B_y$ (i.e., we used the result from Eq. (c) with the negative sign of $A_y$ intact).

Also, if you were tempted in F.B.D. I to include the force $C_y$ at pin $C$ in this diagram you were flirting with the prime error in the statics of rigid bodies—namely, including a force which is *internal* to the particular *free body* drawn.

## Case D. General Force Systems.

The simplest resultant in the general case is a force and a couple moment. Six equations of equilibrium can be given for each free-body diagram. We now examine two examples for this case.

## Example 5.10

A derrick is shown in Fig. 5.35 supporting a 1,000-lb load. The vertical beam has a ball-and-socket connection into the ground at $d$ and is held by guy wires $ac$ and $bc$. Neglect the weight of the members and guy wires, and find the tensions in the guy wires $ac$, $bc$, and $ce$.

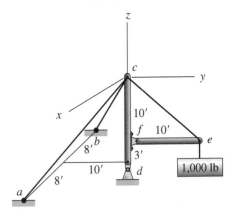

**Figure 5.35.** Loaded derrick.

If we select as a free body both members and the interconnecting guy wire $ce$, we shall expose two of the desired unknowns (Fig. 5.36). Note that this is a general three-dimensional force system with only five unknowns.[6] Although all these unknowns can be solved by statical considerations of this free body, you will notice that, if we take moments about point $d$, we will involve in a vector equation only the desired unknowns $T_{bc}$ and $T_{ac}$. Accordingly, all unknown forces need not be computed for this free-body diagram. You should always look for such short cuts in situations such as this.

To determine the unknown tension $T_{ce}$, we must employ another free-body diagram. Either the vertical or horizontal member will expose this unknown in a manner susceptible to solution. The latter has been selected and is shown in Fig. 5.37. Note that we have here a coplanar force system with three unknowns. Again, you can see that, by taking moments about point $f$, we will involve only the desired unknown.

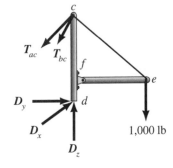

**Figure 5.36.** Free-body diagram 1.

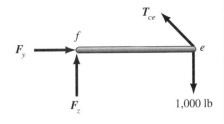

**Figure 5.37.** Free-body diagram 2.

[6]It should be clear on inspection of Fig. 5.35 that, due to symmetry, forces in the two supporting cables must be equal. Using this information is tantamount to using one of the equations of equilibrium. However, for practice we will not use this information and we will solve for both of these cables and demonstrate their equality. Also, note that there is a sixth equation of equilibrium that is identically satisfied. To see this, look at moments of the forces about axis $cd$ in Fig. 5.36. Why is the total moment identically equal to zero about this axis, thus denying us an equation to help solve for unknowns? Is the derrick completely constrained? Explain.

## Example 5.10 (Continued)

The vector $T_{ac}$ may then be given as

$$T_{ac} = T_{ac}\left[\frac{1}{\sqrt{333}}(8i - 10j - 13k)\right] \tag{a}$$

Similarly, we have for $T_{bc}$,

$$T_{bc} = T_{bc}\left[\frac{1}{\sqrt{333}}(-8i - 10j - 13k)\right] \tag{b}$$

Using the free-body diagram in Fig. 5.36, we now set the sum of moments about point $d$ equal to zero. Thus, employing the relations above, we get

$$13k \times \frac{T_{ac}}{\sqrt{333}}(8i - 10j - 13k) + 13k \times \frac{T_{bc}}{\sqrt{333}}(-8i - 10j - 13k)$$
$$+10j \times (-1,000k) = 0 \tag{c}$$

When we make the substitution of variable $t_1 = T_{ac}/\sqrt{333}$ and $t_2 = T_{bc}/\sqrt{333}$, the preceding equation becomes

$$[130(t_1 + t_2) - 10,000]i + [104(t_1 - t_2)]j = 0 \tag{d}$$

The scalar equations,

$$130(t_1 + t_2) - 10,000 = 0$$
$$104(t_1 - t_2) = 0$$

can now be readily solved to give $t_1 = t_2 = 38.5$. Hence, we get $T_{ac} = 38.5\sqrt{333} = 702$ lb and $T_{bc} = 38.5\sqrt{333} = 702$ lb.[7]

$$\therefore \quad \begin{array}{l} T_{ac} = 702 \text{ lb} \\ T_{bc} = 702 \text{ lb} \end{array}$$

Turning finally to free-body diagram 2 in Fig. 5.37, we see that, in summing moments about $f$, the horizontal component of the tension $T_{ce}$ has a zero-moment arm. Thus,

$$(10)(0.707)T_{ce} - (10)(1,000) = 0$$

Hence,

$$T_{ce} = 1,414 \text{ lb}$$

[7]By taking moments in Fig. 5.36 about the line connecting points $a$ and $d$ (see Fig. 5.35), we could get $T_{bc}$ directly using the scalar triple product. We suggest that you try this.

## Example 5.11

A blimp is shown in Fig. 5.38 fixed at the mooring tower $D$ by a ball-joint connection, and held by cables $AB$ and $AC$. The blimp has a mass of 1,500 kg. The simplest resultant force $F$ from air pressure (including the effects of wind) is

$$F = 17{,}500i + 1{,}000j + 1{,}500k \text{ N}$$

at a position shown in the diagram. Compute the tension in the cables as well as the force transmitted to the ball joint at the top of the tower at $D$. Also, what force system is transmitted to the ground at $G$ through the mooring tower? The tower weighs 5,000 N.

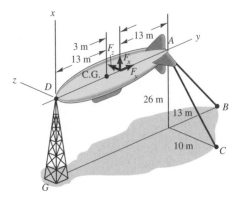

**Figure 5.38.** Tethered blimp.

We shall first consider a free-body diagram of the blimp, as shown in Fig. 5.39. We have five unknown forces here, and we can solve all of them by using equations of equilibrium for this free body.[8] As a first step, we express the cable tensions vectorially. That is,

$$T_{AC} = T_{AC}\left(\frac{-26i - 10k}{\sqrt{26^2 + 10^2}}\right) = T_{AC}(-.933i - .359k)$$

$$T_{AB} = T_{AB}\left(\frac{-26i + 13j}{\sqrt{26^2 + 13^2}}\right) = T_{AB}(-.894i + .447j)$$

We now go back to the basic vector equation of equilibrium. Thus

$$\sum F_i = 0:$$

$$D_x i + D_y j + D_z k - (1{,}500)(9.81)i + 17{,}500i + 1{,}000j + 1{,}500k$$
$$+ T_{AC}(-.933i - .359k) + T_{AB}(-.894i + .447j) = 0$$

**Figure 5.39.** Free-body diagram of blimp.

---

[8]What equation is identically satisfied (thus forming the sixth equation of equilibrium) and thus not helpful in solving for the desired unknowns? What does this say about how the blimp can begin to move with the constraints present? See footnote 6 for the previous example.

## Example 5.11 (Continued)

The scalar equations are

$$D_x + 2{,}785 - .933T_{AC} - .894T_{AB} = 0 \qquad \text{(a)}$$
$$D_y + 1{,}000 + .447T_{AB} = 0 \qquad \text{(b)}$$
$$D_z + 1{,}500 - .359T_{AC} = 0 \qquad \text{(c)}$$

Next take moments about point $D$.

$$\sum (M_i)_D = 0:$$

$$13j \times (9.81)(1{,}500)(-i) + 16j \times (17{,}500i + 1{,}000j + 1{,}500k)$$
$$+ 29j \times T_{AC}(-.933i - .359k) + 29j \times T_{AB}(-.894i + .447j) = 0$$

Carrying out the various cross products, we end up only with $k$ and $i$ components, thus generating two scalar equations.[9] They are

$$-10.41T_{AC} + 24{,}000 = 0 \qquad \text{(d)}$$

$$25.9T_{AB} + 27.1T_{AC} - 88{,}700 = 0 \qquad \text{(e)}$$

We now have five independent equations for five unknowns. We can thus solve these equations simultaneously. From Eq.(d), we have

$$T_{AC} = 2{,}305 \text{ N}$$

From Eq. (e), we have

$$T_{AB} = 1{,}012 \text{ N}$$

From Eq. (c), we have

$$D_z = -672 \text{ N}$$

From Eq. (b), we have

$$D_y = -1{,}452 \text{ N}$$

---

[9]The third equation is $0 = 0$. That is, there are no moments about the $y$ axis, because all forces pass through the $y$ axis.

### Example 5.11 (Continued)

From Eq. (a), we have

$$D_x = 270 \text{ N}$$

Next, consider the mooring tower as a free body (Fig. 5.40). Notice that in showing the forces at the ball joint $D$, we have taken into account *both* of the negative signs shown above for $D_y$ and $D_z$ as well as Newton's third law.

Again, using the basic vector equations of statics, we have on summing forces

$$-270i + 1,452j + 672k + F_x i + F_y j + F_z k - 5,000i = 0$$

Hence,

$$F_x = 5,270 \text{ N}$$
$$F_y = -1,452 \text{ N}$$
$$F_z = -672 \text{ N}$$

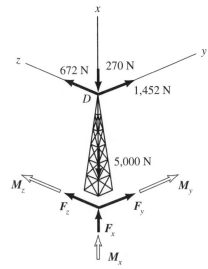

**Figure 5.40.** Free-body diagram of mooring tower.

Now take moments about the base at $F$. We get

$$26i \times (-270i + 1,452j + 672k) + M_x i + M_y j + M_z k = 0$$

From this, we get

$$M_x = 0$$
$$M_y = 17,470 \text{ N-m}$$
$$M_z = -37,800 \text{ N-m}$$

We can conclude that, at the center of the base, the force system from the ground is

$$F = 5,270i - 1,452j - 672k \text{ N}$$

$$C = 17,470j - 37,800k \text{ N-m}$$

The force system acting on the ground at the center of the base is the reaction to the system above. Thus,

$$F_{\text{ground}} = -5,270i + 1,452j + 672k \text{ N}$$
$$C_{\text{ground}} = -17,470j + 37,800k \text{ N-m}$$

**5.58.** The triple pulley sheave and the double pulley sheave weigh 15 lb and 10 lb, respectively. What rope force is necessary to lift a 350-lb engine? What is the force on the ceiling hook?

**Figure P.5.58.**

**5.59.** A multipurpose pry bar can be used to pull nails in the three positions. If a force of 400 lb is required to remove a nail and a carpenter can exert 50 lb, which position(s) must he use?

**Figure P.5.59.**

**5.60.** At what position must the operator of the counterweight crane locate the 50-kN counterweight when he lifts a 10-kN load of steel?

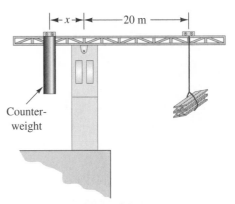

**Figure P.5.60.**

**5.61.** A Jeep winch is used to raise itself by a force of 2 kN. What are the reactions at the Jeep tires with and without the winch load? The driver weighs 800 N, and the Jeep weighs 11 kN. The center of gravity of the Jeep is shown.

**Figure P.5.61.**

**5.62.** A *differential pulley* is shown. Compute $F$ in terms of $W$, $r_1$, and $r_2$.

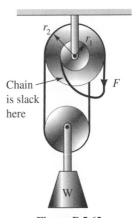

**Figure P.5.62.**

**5.63.** What is the longest portion of pipe weighing 400 lb/ft that can be lifted without tipping the 12,000-lb tractor? Take the center of gravity of the tractor at the geometric center.

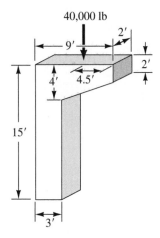

**Figure P.5.63.**

**5.64.** The L-shaped concrete post supports an elevated railroad. The concrete weighs 150 lb/ft³. What are the reactions at the base of the post?

**Figure P.5.64.**

**5.65.** Two hoists are operated on the same overhead track. Hoist A has a 3,000-lb load, and hoist B has a 4,000-lb load. What are the reactions at the ends of the track when the hoists are in the position shown?

**Figure P.5.65.**

**5.66.** An I-beam cantilevered out from a wall weighs 30 lb/ft and supports a 300-lb hoist. Steel (487 lb/ft³) cover plates 1 in. thick are welded on the beam near the wall to increase the moment-carrying capacity of the beam. What are the reactions at the wall when a 400-lb load is hoisted at the outermost position of the hoist?

**Figure P.5.66.**

**5.67.** Find the supporting force system for the cantilever beam shown pinned at C.

**Figure P.5.67.**

**5.68.** Find the supporting force system for the cantilever beams connected to bar AB by pins.

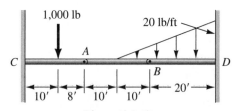

**Figure P.5.68.**

**5.69.** The trailer weighs 50 kN and is loaded with crates weighing 90 kN and 40 kN. What are the reactions at the rear wheel and on the tractor at A?

**Figure P.5.69.**

194

**5.70.** What load $W$ will a pull $P$ of 100 lb lift in the pulley system? Sheaves $A$, $B$, and $C$ weigh 20 lb, 15 lb, and 30 lb, respectively. Assume first that the three sheaves are frictionless and find $W$. Then, calculate $W$ that can be raised at constant speed for the case where the resisting torque in each of sheaves $A$ and $B$ is .01 times the total force at the bearing of each of sheaves $A$ and $B$.

**Figure P.5.70.**

**5.71.** A piece of pop art is being developed. The weight of the body enclosed by the full lines is 2 kN. What is the smallest distance $d$ that the artist can use for cutting a .5-m-diameter hole and still avoid tipping? The body is uniform in thickness.

**Figure P.5.71.**

**5.72.** What is the largest weight $W$ that the crane can lift without tipping? What are the supporting forces when the crane lifts this load? What is the force and couple-moment system transmitted through section $C$ of the beam? Compute the force and couple-moment system transmitted through section $D$. The crane weighs 10 tons, having a center of gravity as shown in the diagram.

**Figure P.5.72.**

**5.73.** A 20-kN block is being raised at constant speed. If there is no friction in the three pulleys, what are forces $F_1$, $F_2$, and $F_3$ needed for the job? The block is not rotating in any way. The line of action of the weight vector passes through point $C$ as shown.

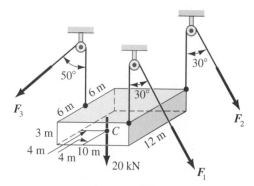

**Figure P.5.73.**

**5.74.** A 10-ton sounding rocket (used for exploring outer space) has a center of gravity shown as C.G.$_1$. It is mounted on a launcher whose weight is 50 tons with a center of gravity at C.G.$_2$. The launcher has three identical legs separated 120° from each other. Leg $AB$ is in the same plane as the rocket and supporting arms $CDE$. What are the supporting forces from the ground? What torque is transmitted from the horizontal arm $CD$ to the ramp $ED$ by the rack and pinion at hinge $D$ to counteract the weight of the rocket?

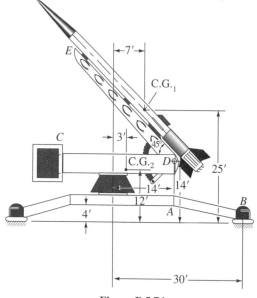

**Figure P.5.74.**

195

**5.75.** A door is hinged at $A$ and $B$ and contains water whose specific weight $\gamma$ is 62.5 lb/ft³. A force $F$ normal to the door keeps the door closed. What are the forces on the hinges $A$ and $B$ and the force $F$ to counteract the water? As noted in Chapter 4, the pressure in the water above atmosphere is given as $\gamma d$, where $d$ is the perpendicular distance from the free surface of the water.

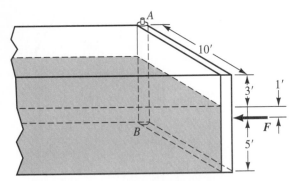

**Figure P.5.75.**

**5.76.** A row of books of length 750 mm and weighing 200 N sits on a three-legged table as shown. The legs are equidistant from each other with one leg $B$ coinciding with the $y$ axis. The other two legs lie along a line parallel to the $x$ axis. If the table weighs 400 N, will it tip? If not, what are the forces on the legs?

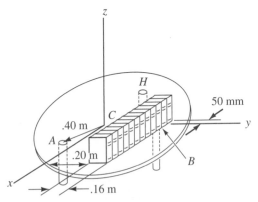

**Figure P.5.76.**

**5.77.** A small helicopter is in a hovering maneuver. The helicopter rotor blades give a lifting force $F_1$ but there results from the air forces on the blades a torque $C_1$. The rear rotor prevents the helicopter from rotating about the $z$ axis but develops a torque $C_2$. Compute the force $F_1$ and couple $C_2$ in terms of the weight $W$. How are $F_3$ and $C_1$ related?

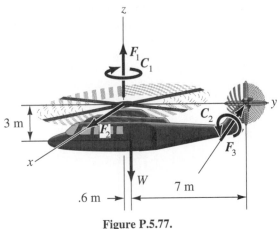

**Figure P.5.77.**

**5.78.** Find the supporting force and couple-moment system for the cantilever beam. What is the force and couple-moment system transmitted through a cross section of the beam at $B$?

**Figure P.5.78.**

**5.79.** A structure is supported by a ball-and-socket joint at $A$, a pin connection at $B$ offering no resistance in the direction $AB$, and a simple roller support at $C$. What are the supporting forces for the loads shown?

**Figure P.5.79.**

**5.80.** Compute the value of $F$ to maintain the 200-lb weight shown. Assume that the bearings are frictionless, and determine the forces from the bearings on the shaft at $A$ and $B$.

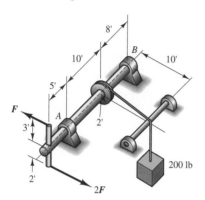

**Figure P.5.80.**

**5.81.** A bar with two right-angle bends supports a force $F$ given as

$$F = 10i + 3j + 100k \text{ N}$$

If the bar has a weight of 10 N/m, what is the supporting force system at $A$?

**Figure P.5.81.**

**5.82.** What is the resultant of the force system transmitted across the section at $A$? The couple is parallel to plane $M$.

**Figure P.5.82.**

**5.83.** Determine the vertical force $F$ that must be applied to the windlass to maintain the 100-lb weight. Also, determine the supporting forces from the bearings onto the shaft. The handle $DE$ on which the force is applied is in the indicated $xz$ plane.

**Figure P.5.83.**

**5.84.** A transport plane has a gross weight of 70,000 lb with a center of gravity as shown. Wheels $A$ and $B$ are locked by the braking system while an engine is being tested under load prior to take off. A thrust $T$ of 3,000 lb is developed by this engine. What are the supporting forces?

**Figure P.5.84.**

197

**5.85.** Two cables *GH* and *KN* support a rod *AB* which connects to a ball-and-socket joint support at *A* and supports a 500-kg body *C* at *B*. What are the tensions in the cable and the supporting forces at *A*?

**Figure P.5.85.**

**5.86.** What change in elevation for the 100-lb weight will a couple of 300 lb-ft support if we neglect friction in the bearings at *A* and *B*? Also, determine the supporting force components at the bearings for this configuration.

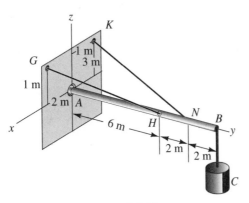

**Figure P.5.86.**

**5.87.** Determine the force *P* required to keep the 150-N door of an airplane open 30° while in flight. The force *P* is exerted in a direction normal to the fuselage. There is a net pressure increase on the outside surface of .02 N/mm². Also, determine the supporting forces at the hinges. Consider that the top hinge supports any vertical force on the door.

**Figure P.5.87.**

**5.88.** What force *P* is needed to hold the door in a horizontal position? The door weighs 50 lb. Determine the supporting forces at *A* and *B*. At *A* there is a pin and at *B* there is a ball-and-socket joint.

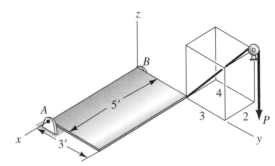

**Figure P.5.88.**

**\*5.89.** A uniform bar of length *l* and weight *W* is connected to the ground by a ball-and-socket joint, and rests on a semicylinder from which it is not allowed to slip down by a wall at *B*. If we consider the wall and cylinder to be frictionless, determine the supporting forces at *A* for the following data:

$$l = 1 \text{ m}$$
$$c = .30 \text{ m}$$
$$r = .20 \text{ m}$$
$$b = .40 \text{ m}$$
$$W = 100 \text{ N}$$

**Figure P.5.89.**

# 5.8  Two Point Equivalent Loadings

We shall now consider a simple case of equilibrium that occurs often and from which simple useful conclusions may readily be drawn.

Consider a rigid body on which the *equivalent* systems of two separate forces are respectively applied at two points *a* and *b* as shown in Fig. 5.41. If the body is in equilibrium, the first basic equation of statics, 5.1(a), stipulates that $F_1 = -F_2$; that is, the forces must be *equal and opposite*. The second fundamental equation of statics, 5.1(b), requires that $C = 0$, indicating that the forces be *collinear* so as not to form a nonzero couple. With points *a* and *b* given as points of application for the two forces in Fig. 5.41, clearly the *common line of action for the forces must coincide with the line segment ab*. Such bodies, where there are only *two points of loading*, are sometimes called "two-force" members. Such members occur often in structural mechanics problems. Furthermore, there is considerable saving of time and labor if the student recognizes them at the outset of any problem.

**Figure 5.42.** Compression and tension members.

**Figure 5.41.** Two-force member.

We often have to deal with pin-connected structural members with loads applied at the pins. If we neglect friction at the pins and also the weight of the members, we can conclude that the equivalent of only two forces act on each member. These forces, then, must be equal and opposite and must have lines of action that are collinear, with the line joining the points of application of the forces. If the member is straight (see Fig. 5.42), the common line of action of the two forces coincides with the centerline of the member. The top member in Fig. 5.42 is a *compression* member, the one below a *tensile* member. Note that the bent member in Fig. 5.43, if weightless, is also a two point loaded member. The line of action of the forces must coincide with the line *ab* connecting the two points of loading. However, the beam in Fig. 5.44 is not a two point loaded member since at the left end there will be a couple moment. And, clearly, such a loading must entail two points of loading to accommodate the two equal and opposite forces comprising the couple. There are then in effect three points of loading for this member. Accordingly, use caution here when dealing with a cantilever beam.

**Figure 5.43.** Line of action of *F* collinear with *ab*.

Before considering an example, it should be emphasized that the forces $F_1$ and $F_2$ in Fig. 5.41 may be the resultants of systems of concurrent forces at *a* and *b*, respectively. Since concurrent forces are always equivalent to their resultant at the point of concurrency, the member in Fig. 5.41 is still a two-force member with the resulting restrictions on the resultants $F_1$ and $F_2$.

**Figure 5.44.** Cantilever beam is not an example of a two point loaded member.

## ■ Example 5.12

A device for crushing rocks is shown in Fig. 5.45. A piston $D$ having an 8-in. diameter is activated by a pressure $p$ of 50 psig (pounds per square inch above that of the atmosphere). Rods $AB$, $BC$, and $BD$ can be considered weightless for this problem. What is the horizontal force transmitted at $A$ to the trapped rock shown in the diagram?

We have here three two-force members coming together at $B$. Accordingly, if we isolate pin $B$ as a free body, we will have three forces acting on the pin. These forces must be collinear with the centerlines of the respective members, as explained earlier (Fig. 5.42).

The force $F_D$ is easily computed by considering the action of the piston. Thus, we get

**Figure 5.45.** Rock crusher.

$$F_D = (50)\frac{\pi 8^2}{4} = 2,510 \text{ lb}$$

Summing forces at pin $B$:

$\underline{\sum F_x = 0}$:

$$F_A \cos 15° - F_C \cos 15° = 0$$

$$F_A = F_C$$

$\underline{\sum F_y = 0}$:

$$2F_A \sin 15° - 2,510 = 0$$

$$F_A = \frac{2,510}{(2)(0.259)} = 4,850 \text{ lb}$$

The force transmitted to the rock in the horizontal direction is then 4,850 $\cos 15° = 4,690$ lb.

∴ Horizontal force transmitted to the rock = 4,690 lb

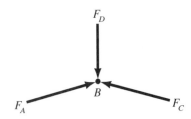

**Figure 5.46.** Free-body diagram of pin $B$.

Of less direct use is the *three-force* theorem. It states that a system of three forces in equilibrium must be *coplanar* and either be *concurrent* or be *parallel*.[10]

## 5.9   Problems Arising From Structures

We will now examine some interesting problems that arise in studies of structural mechanics. Primarily, these problems involve procedures and theory that we have presented in this chapter with some simple additions, (which will be described in the examples to follow), and some use of freshman physics.

[10]To prove this, assume that two of the forces intersect at a point $A$. Show from the basic equations of equilibrium that the forces must be coplanar and concurrent. Now assume the forces do *not* intersect. Setting moments of the system equal to zero about two points along the line of action of one of the forces, show that the system must be coplanar. Now since the forces do not intersect, they must be parallel. The theorem will thus have been proved.

# Example 5.13

Rod $C$ shown in Fig. 5.47 is welded to a rigid drum $A$, which is rotating about its axis at a steady angular speed $\omega$ of 500 RPM. This rod has a mass per unit length $w$, which varies linearly from the base to the tip starting with the value of 20 kg/m at the base to 28 kg/m at the tip. If *normal stress* is defined as the normal force at a section divided by the area of the section (similar to pressure except that the force can be pulling away from the section rather than always pushing against the section), what is the normal stress at any cross section of the cylinder at a distance $r$ from the centerline $B$–$B$ of the drum due only to the motion?

**Figure 5.47.** Rods attached to a rotating rigid drum.

First we expose a section of the rod at a distance $r$ from $B$–$B$ in a *free-body diagram* as shown in Fig. 5.48 in which we denote the normal stress acting on this section as $\tau_{rr}$. The force from this section must restrain the centrifugal force stemming from the angular motion of the portion of the rod beyond position $r$. For this purpose, we consider an infinitesimal slice of the rod (see Fig. 5.48). We shall use the variable $\eta$ to denote the distance to the slice from the centerline $B$–$B$. Furthermore, the thickness of the slice shall be $d\eta$. We shall use $\eta$ to position slices between end position $r$ of the free-body diagram and the tip of the rod. We are using this approach for bookkeeping purposes. Now you will recall from freshman physics that the centrifugal force on this slice is given by

$$df_{cent.} = (dm)(\eta)(\omega^2) = (w\, d\eta)(\eta)(\omega^2) \qquad (a)$$

# ■ Example 5.13 (Continued)

**Figure 5.48.** Free body exposes $\tau_{rr}$ at section $r$. Note that $\eta$ is a dummy variable.

Consequently, the total centrifugal force for the material beyond position $r$ in the free body of Fig. 5.48 will be found by integration to be

$$f_{cent.} = \int_r^{.7} w\eta \, d\eta \, \omega^2 \qquad (b)$$

Next, note that the term $w$ varies linearly with $\eta$ as follows:

$$w = 20 + \frac{\eta - .2}{.5} 8 \text{ kg/m}$$

Clearly from this equation we see that when $\eta = .2$, we get $w = 20$ kg/m and when $\eta = .7$, we get $w = 28$ kg/m. Also, the variation is linear. Now, going back to Eq. (b), we have

$$f_{cent.} = \int_r^{.7} \left( 20 + \frac{\eta - .2}{.5} 8 \right) \eta \left( \frac{(500)(2\pi)}{60} \right)^2 d\eta$$

Integrating    $f_{cent.} = 2.742 \times 10^3 \left[ (20 - 3.2) \frac{\eta^2}{2} + 16 \frac{\eta^3}{3} \right]_r^{.7}$

$$= 2.742 \times 10^3 \left[ 4.116 + 1.829 - 8.40 r^2 - 5.333 r^3 \right]$$

We can give the desired stress $\tau_{rr}$ by dividing by the cross-sectional area $\pi D^2/4 = 7.854 \times 10^{-3} m^2$. We thus have the desired normal stress distribution for sections of the rod as follows:

$$\tau_{rr} = 3.491 \times 10^5 \left[ 5.945 - 8.40 r^2 - 5.333 r^3 \right] \text{ N/m}^2$$

# Example 5.14

Consider a *thin-walled* tank containing air at a pressure of 100 psi above that of the atmosphere [see Fig. 5.49(a)]. The outside diameter $D$ of the tank is 2 ft and the wall thickness $t$ is 1/4 in. We consider as a free body from the tank wall a vanishingly small element such as $ABCE$ in the diagram having the shape of a rectangular parallelepiped. What are the stresses on the cut surfaces of the element? Neglect the weight of the cylinder.

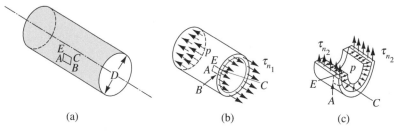

(a)  (b)  (c)

**Figure 5.49.** Thin-walled tank with inside gauge pressure $p$.

We can examine face $BC$ by considering a free body of part of the tank, as shown in Fig. 5.49(b) exposing $BC$. (Note that we have not included the pressure on the inside wall). Because there is a net force from the air pressure only in the *axial* direction of the cylinder, we can expect normal stress $\tau_{n_1}$ (like pressure except that it is tensile rather than compressive) over the cut section of the cylinder, as has been indicated in the diagram. Furthermore, because the wall of the tank is *thin* compared to the diameter, we can assume that the stress $\tau_{n_1}$ is *uniform* across the thickness. Finally, for reasons of *axial symmetry* of geometry and loading, we can expect this stress to be uniform around the entire cross section. Now we may say from considerations of *equilibrium* in the axial direction that

$$\tau_{n_1} \pi \left[ \frac{D^2}{4} - \frac{(D-2t)^2}{4} \right] - p\pi \frac{(D-2t)^2}{4} = 0$$

$$\therefore \tau_{n_1} = \frac{p(D-2t)^2}{D^2 - (D-2t)^2} = \frac{(100)(24 - \frac{1}{2})^2}{24^2 - 23.5^2} = \boxed{2{,}325 \text{ psi}}$$
(a)

Hence, on face $BC$ we have a uniform stress of 2,325 psi. Clearly, this must also be true for face $AE$. This is called an *axial stress*.

To expose face $EC$ next, we consider a half-cylinder of unit length such as is shown in Fig. 5.49(c). Because it is *far* from the ends and because the wall is *thin*, we can assume as an approximation that the stress $\tau_{n_2}$ shown is uniform over the cut section.[11] A pressure $p$ is shown acting

---

[11]Near the ends of the tank the stress distribution *varies* in value because of the proximity of the *complicated geometry* and the contributions toward equilibrium of the end plates.

■ **Example 5.14 (Continued)**

normal to the inside wall surface. (The stress $\tau_{n_1}$ computed earlier is not shown, to avoid cluttering the diagram.) Now we consider an end view of this body in Fig. 5.50 for equilibrium in the vertical direction. We have

$$2\left[(\tau_{n_2})(t)(1)\right] - \int_0^\pi p\left(\frac{D}{2} - t\right)d\theta(1)\sin\theta = 0 \qquad \text{(b)}$$

$$\therefore \tau_{n_2} = \frac{1}{2t}\left[p\left(\frac{D}{2} - t\right)\right](-\cos\theta)\Big|_0^\pi$$

$$\tau_{n_2} = \frac{p\left(\frac{D}{2} - t\right)}{t} = \boxed{4{,}700 \text{ psi}}$$

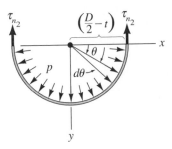

**Figure 5.50.** Free body of part of cylinder.

We point out now that the force in a particular direction from a uniform pressure on a curved surface equals the pressure times the projected area of this surface in the direction of the desired force. (You will learn this in your studies of hydrostatics.) Thus for the case at hand the projected area is that of a rectangle $1 \times (D - 2t)$, so that the second expression of Eq. (b) becomes $p(D - 2t)$. You may readily verify that this gives the same result as above.

The stress $\tau_{n_2}$ is called the *hoop stress*; it is about twice the *axial stress* $\tau_{n_1}$. We show element *ABCE* with the stresses present in Fig. 5.51.

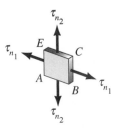

**Figure 5.51.** Free body of an element of the cylinder.

## 5.10    Static Indeterminacy

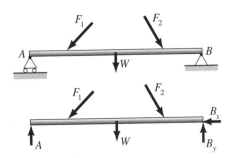

**Figure 5.52.** Statically determinate problem.

Examine the simple beam in Fig. 5.52, with known external loads and weight. If the deformation of the beam is small, and the final positions of the external loads after deformation differ only slightly from their initial positions, we can assume the beam to be rigid and, using the undeformed geometry, we can solve the supporting forces $A$, $B_x$, and $B_y$. This is possible since we have three equations of equilibrium available. Suppose, now, that an additional support is made available to the beam, as indicated in Fig. 5.53. The beam can still be considered a rigid body, since the applied load will shift even less because of deformation. Therefore, the resultant force coming from the ground to counteract the applied loads and weight of the beam must be the *same* as before. In the first case, in which two supports were given, however, a unique set of values for the forces $A$, $B_x$, and $B_y$ gave us the required resultant. In other words, we were able to solve for these forces by statics alone, without further considerations. In the second case, rigid-body statics will give the required *same* resultant supporting force system, but now there are an infinite number of possible combinations of values of the supporting forces that will give us the resultant system demanded by equilibrium of rigid bodies. To decide on the proper combination of supporting forces requires additional computation. Although the deformation properties of the beam were unimportant up to this point, they now become the all-important criterion in apportioning the supporting forces. These problems are termed *statically indeterminate*, in contrast to the statically determinate type, in which statics and the rigid-body assumption suffice. For a given system of loads and masses, two models—the rigid-body model and models taught in other courses involving elastic behavior—are accordingly both employed to achieve a desired end. In summary:

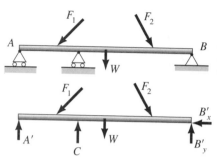

**Figure 5.53.** Statically indeterminate problem.

> *In statically indeterminate problems, we must satisfy both the equations of equilibrium for rigid bodies and the equations that stem from deformation considerations. In statically determinate problems, we need only satisfy the equations of equilibrium.*

In the discussion thus far, we used a beam as the rigid body and discussed the statical determinacy of the supporting system. Clearly, the same conclusions apply to any structure that, without the aid of the external constraints, can be taken as a rigid body. If, for such a structure as a free body, there are as many unknown supporting force and couple-moment components as there are equations of equilibrium, and if these equations can be solved for these unknowns, we say that the structure is *externally statically determinate*.

On the other hand, should we desire to know the forces transmitted *between* internal members of this kind of structure (i.e., one that does not depend on the external constraints for rigidity), we then examine free bodies of these members. When all the unknown force and couple-moment components can be found by the equations of equilibrium for these free bodies, we then say that the structure is *internally statically determinate*.

There are structures that depend on the external constraints for rigidity (see the structure shown in Fig. 5.54). Mathematically speaking, we can say for such structures that the supporting force system always depends on both the internal forces and the external loads. (This is in contrast to the previous case, where the supporting forces could, for the externally statically determinate case, be related directly with the external loads without consideration of the internal forces.) In this case, we do not distinguish between internal and external statical determinacy, since the evaluation of supporting forces will involve free bodies of some or all of the internal members of the structure; hence, some or all of the internal forces and moments will be involved. For such cases, we simply state that the structure is statically determinate if, for all the unknown force and couple-moment components, we have enough equations of equilibrium that can be solved for these unknowns.

**Figure 5.54.** Nonrigid structure.

**5.90.** Draw free-body diagrams for the hoe, arms, and tractor of the backhoe. Consider the weight of each part to act at a central location. The backhoe is not digging at the instant shown. Neglect the weights of the hydraulic systems *CE*, *AB*, and *FH*.

**Figure P.5.90.**

**5.91.** A parking-lot gate arm weighs 150 N. Because of the taper, the weight can be regarded as concentrated at a point 1.25 m from the pivot point. What force must be exerted by the solenoid to lift the gate? What solenoid force is necessary if a 300-N counterweight is placed .25 m to the left of the pivot point?

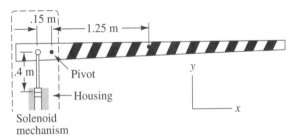

**Figure P.5.91.**

**5.92.** Find the force delivered at *C* in a horizontal direction to crush the rock. Pressure $p_1$ = 100 psig and $p_2$ = 60 psig (pressures measured above atmospheric pressure). The diameters of the pistons are 6 in. each. Neglect the weight of the rods.

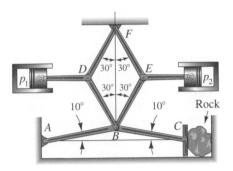

**Figure P.5.92.**

**5.93.** A Broyt X-20 digger carries a 20-kN load as shown. If hydraulic ram *CB* is normal to *BA*, where *A* is the axis of rotation for member *E*, find the force needed by ram *CB*. Do not consider the weights of the members.

**Figure P.5.93.**

**5.94.** What force $F_1$ is needed for equilibrium? Neglect friction. The design is symmetrical about vertical axis. Horizontal members are pinned at *E*.

**Figure P.5.94.**

**5.95.** Find the values of $F$ and $C$ so that members $AB$ and $CD$ fail simultaneously. The maximum load for $AB$ is 15 kN and for $CD$ is 22 kN. Neglect the weight of the members.

**Figure P.5.95.**

**5.96.** The landing carriage of a transport plane supports a stationary total vertical load of 200 kN. There are two wheels on each side of shock strut $AB$. Find the force in member $EC$, and the forces transmitted to the fuselage at $A$, if the brakes are locked and the engines are tested resulting in a thrust of 5 kN, 40% of which is resisted by this landing gear.

**Figure P.5.96.**

**5.97.** Find the magnitudes of the supporting forces for the frame shown. You may only use *two* free-body diagrams for this problem. Set forth a complete system of equations for solving the desired unknowns but do not carry out the algebra for actually solving these equations for the desired unknowns.

**Figure P.5.97.**

**5.98.** (a) Find the supporting forces at $B$. Neglect the weights of the members.
(b) What is the force in the member $CB$?

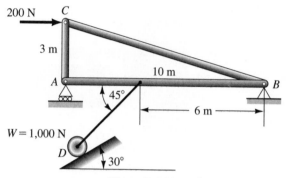

**Figure P.5.98.**

**5.99.** Find the supporting forces at $B$ and $C$. Disc $A$ weighs 200 N. Neglect the weights of the members as well as friction.

**Figure P.5.99.**

**5.100.** Find the supporting force system at $C$. Neglect friction.

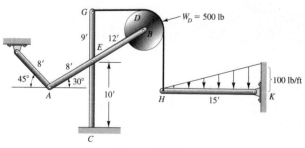

**Figure P.5.100.**

**5.101.** The pavement exerts a force of 1,000 lb on the tire. The tire, brakes, and so on, weigh 100 lb; the center of gravity is taken at the center plane of the tire. Determine the force from the spring and the compression force in *CD*.

**Figure P.5.101.**

**5.102.** A *flyball governor* is shown rotating at a constant speed $\omega$ of 500 rpm. The weights *C* and *D* are each of mass 500 g and are pin-connected to light rods. The centrifugal force on the weights, you will recall from physics, is given as $mr\omega^2$, where *r* is the radial distance to the particle from the axis of rotation and $\omega$ is in rad/sec. Using this centrifugal force, and, imagining that we are rotating with the system, we can consider that we have equilibrium. (This is the D'Alembert principle that you learned in physics.) What is the tension in the rods and the downward force *F* at *B* needed to maintain the configuration shown for the given $\omega$?

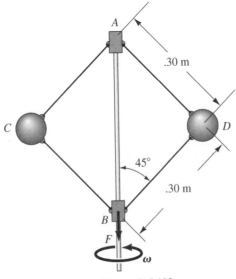

**Figure P.5.102.**

**5.103.** A flyball governor is shown below. If $\omega_1 = 3$ rad/sec, compute the tension in *AG* and *AE*. Neglect the weights of the members but assume that *HC* and *HB* are stiff. The weights *C* and *B* each have a mass of 200 g. What is the force *F* needed to maintain the configuration? (*Hint:* Read the discussion of centrifugal force and D'Alembert's principle in Problem 5.102.)

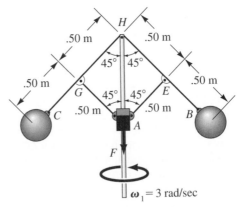

**Figure P.5.103.**

**5.104.** Determine the supporting forces at *A, C, D, G, F,* and *H* for the structure. *C, D, H* and *G* are ball joint connections.

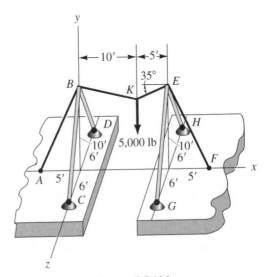

**Figure P.5.104.**

**5.105.** A trap door is kept open by a rod *CD*, whose weight we shall neglect. The door has hinges at *A* and *B* and has a weight of 200 lb. A wind blowing against the outside surface of the door creates a pressure increase of 2 lb/ft². Find the force in the rod, assuming that it cannot slip from the position shown. Also determine the forces transmitted to the hinges. Only hinge *B* can resist motion along direction *AB*.

**Figure P.5.105.**

**5.106.** Find the force *BD*. All connections are ball-and-socket joints. Neglect the weights of all the members. Member *AB* has two 90° bends. Member *BD* is in the *yz* plane.

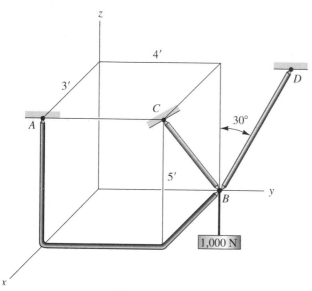

**Figure P.5.106.**

**5.107.** Find the supporting forces at *A* and *C*. You must show and use only one free-body diagram.

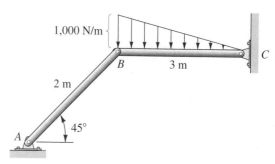

**Figure P.5.107.**

**5.108.** A coupling between two shafts transmits an axial load of 5,000 N. Four bolts having a diameter each of 13 mm, connects the two units. Before loading of the shafts, these bolts have only negligible tensile forces. Assuming that each bolt carries the same load, what is the average normal stress in each bolt stemming from the 5,000 N load?

**Figure P.5.108.**

**5.109.** A circular shaft is suspended from above. The specific weight of the shaft material is $7.22 \times 10^4$ N/m³. What is the tensile stress $\tau_{zz}$ on cross sections of the shaft as a function of *z*?

**Figure P.5.109.**

**5.110.** Do the previous problem for the case where the specific weight $\gamma$ varies with the *square* of *z* starting with the value of $6.50 \times 10^4$ N/m³ at the top and reaching a value of $7.50 \times 10^4$ N/m³ at the bottom.

**5.111.** A cone is suspended from above. The specific weight $\gamma$ of the material is 460 lb/ft³. What is the average normal stress denoted as $\tau_{zz}$ for cross-sections of the cone as a function of $z$?

1.2'

3'

$z$

**Figure P.5.111.**

**5.112.** Solve the preceding problem for the case where the specific weight $\gamma$ varies linearly from the top of the cone ($\gamma = 460$ lb/ft³) the bottom ($\gamma = 385$ lb/ft³). Get the normal stress $\tau_{zz}$ at cross sections of the cone as a function of $z$ but do not bother to collect terms coming out of the integration. Finally, compute from your function the stress $\tau_{zz}$ at a position one foot from the top.

# 5.11  Closure

We have presented in this chapter the *free-body* concept. We hope that the importance of drawing correct free bodies has been made abundantly clear. Do not proceed in your work on a problem until correct, clearly drawn free bodies have been formed. This is the most critical step in problem solving! If you do not master this capability, you will be plagued throughout your future courses with difficulties and frustrations arising from this inadequacy. Also vital, is to know how many *independent equations of equilibrium* you can write for any free-body diagram you draw. Recall, we reached this capability by considering simplest resultants for various categories of force systems and reasoned what was necessary in terms of equations to render the particular simplest resultant system a null vector system and thus to generate the appropriate number of independent equations of equilibrium.

At the end of the chapter, we considered the case of the *statically indeterminate* problem pointing out that in addition to satisfying the rigid-body equations of equilibrium, we also had to consider the deformation of the body, however small the deformation. These considerations are beyond this course and will be studied in the solid mechanics course which you will take soon after the completion of your statics course.[12]

In Chapter 6, we shall consider certain types of bodies that are of great engineering interest. The problems will be statically determinate and will involve nothing that is fundamentally new. We devote a separate chapter to these problems because they contain sign conventions and techniques that are important and complex enough to warrant such a study. We therefore proceed to an introduction of statically determinate structural mechanics problems.

---

[12]For such problems, see I. H. Shames, *Introduction to Solid Mechanics*, 2nd edition, Prentice-Hall, Inc., Englewood Cliffs, N.J., 1989.

**5.113.** Determine the tensions in all the cables. Block $A$ has a mass of 600 kg. Note that $GH$ is in the $yz$ plane.

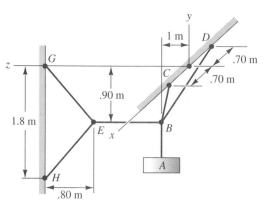

Figure P.5.113.

**5.114.** Determine the force components at $G$. $E$ weighs 300 lb.

Figure P.5.114.

**5.115.** A scenic excursion train with cog wheels for steep inclines weighs 30 tons when fully loaded. If the cog wheels have a mean radius to the contact points of the teeth of 2 ft, what torque must be applied to the driver wheels $A$ if wheels $B$ run free? What force do wheels $B$ transmit to the ground?

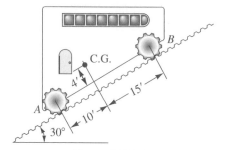

Figure P.5.115.

**5.116.** Find the forces on the block of ice from the hooks at $A$ and $F$.

Figure P.5.116.

**5.117.** Members $AB$ and $BC$ weighing, respectively, 50 N and 200 N are connected to each other by a pin. $BC$ connects to a disc $K$ on which a torque $T_K = 200$ N-m is applied. What torque $T$ is needed on $AB$ to keep the system in equilibrium at the configuration shown?

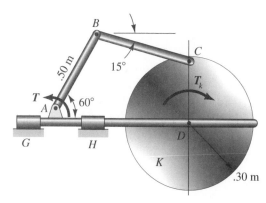

Figure P.5.117.

**5.118.** A transport jet plane has a weight without fuel of 220 kN. If one wing is loaded with 50 kN of fuel, what are the forces in each of the three landing gear?

**Figure P.5.118.**

**5.119.** A rod $AB$ is connected by a ball-and-socket joint to a frictionless sleeve at $A$, and by a ball-and-socket joint to a fixed position at $B$. What are the supporting forces at $B$ and at $A$ if we neglect the weight of $AB$? The 100-lb load is connected to the center of $AB$.

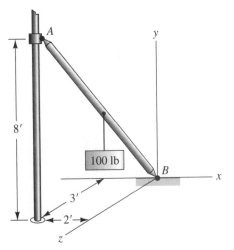

**Figure P.5.119.**

**5.120.** A beam weighing 400 lb is held by a ball-and-socket joint at $A$ and by two cables $CD$ and $EF$. Find the tension in the cables. They are attached to blocks at opposite ends of the beam as shown.

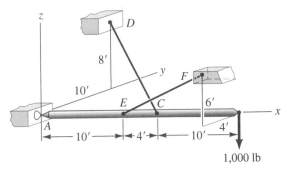

**Figure P.5.120.**

**5.121.** What should the values of $R$ and $M$ be if the supporting rods $AB$ and $CD$ are to fail simultaneously? Rod $AB$ can withstand a 5,000-lb force, and rod $CD$ can withstand an 8,000-lb force. Neglect the weights of the members.

**Figure P.5.121.**

**5.122.** Find $A_x$ and $A_y$ at the bottom support. Do *not* determine any other unknowns. Do this using only *two* free-body diagrams and only *two* equations. Disc $D$ weighs 30 lb and has a diameter of 2 ft. Neglect the weight of the members. Neglect friction everywhere.

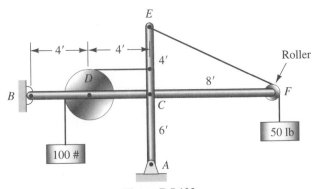

**Figure P.5.122.**

**5.123.** Light rods *BC* and *AC* are pinned together at *C* and support a 300-N load and a 500-N-m couple moment. What are the supporting forces at *A* and *B*?

**5.125.** A bent rod *ADGB* supports two weights—one at the center of *AD* and one at the center of *DG*. There are ball-and-socket joint supports at *A* and *B*. With one scalar equation using the triple scalar product, determine the tension in cable *DC*.

**Figure P.5.123.**

**Figure P.5.125.**

**5.124.** A rod *AB* is held by a ball-and-socket joint at *A* and supports a 100-kg mass *C* at *B*. This rod is in the *zy* plane and is inclined to the *y* axis by an angle of 15°. The rod is 16 m long and *F* is at its midpoint. Find the forces in cables *DF* and *EB*. Cross-hatching indicates part of the *yz* plane.

**5.126.** Find the tension in cable *FH*. The disc *G* weighs 500 lb. Use only one free body.

**Figure P.5.124.**

**Figure P.5.126.**

213

**5.127.** (a) Find the supporting force components at *A* using only one free-body diagram.

(b) Find the force system transmitted through the cross section at *E* of the beam *CD*. Again, use only one free-body diagram for this calculation. *ACD* is a bent member.

**Figure P.5.127.**

**5.128.** A wind is creating a uniform pressure of 0.3 psig on the left side of the door *E* and causes a uniform suction on the right side of the door of −.02 psig. A wire *AB* constrains the door. If hinge *D* allows for movement in the *y* direction, but hinge *C* does not, what are the supporting forces from the hinges? The weight of the door is 200 lb with a center of gravity at the geometric center of the door. The position of end *A* of the supporting wire has coordinates (−6, 5, 15) ft.

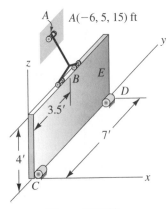

**Figure P.5.128.**

**5.129.** Find the supporting forces. Use no more than two free-body diagrams.

**Figure P.5.129.**

**5.130.** What is the tension in the cables of a 10-ft-wide 12-ft-long 6,000-lb castle drawbridge when the bridge is first raised? When the bridge is at 45°? What are the reactions at the hinge pin?

**Figure P.5.130.**

**5.131.** A small hoist has a lifting capacity of 20 kN. What is the maximum cable tension and the corresponding reactions at *C*? Do not consider weight of beam. At *C* there is a pin connection.

**Figure P.5.131.**

214

**5.132.** A uniform block weighing 500 lb is constrained by three wires. What are the tensions in these wires?

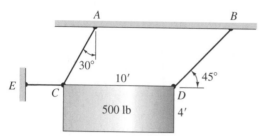

**Figure P.5.132.**

**5.133.** Find the supporting forces for the frame shown.

**Figure P.5.133.**

**5.134.** Four cables support a block of weight 5,000 N. The edges of the block are parallel to the coordinate axes. Point $B$ is at $(7, 7, -15)$. What are the forces in the cables and the direction cosines for cable $CD$?

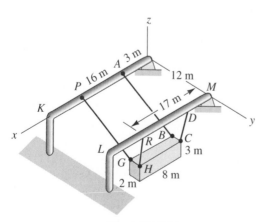

**Figure P.5.134.**

**5.135.** A mechanism consists of two weights $W$ each of weight 50 N, four light linkage rods each of length $a$ equal to 200 mm, and a spring $K$ whose spring constant is 8 N/mm. The spring is unextended when $\theta = 45°$. If held vertically, what is the angle $\theta$ for equilibrium? Neglect friction. The force from the spring equals $K$ times the compression of the spring.

**Figure P.5.135.**

**5.136.** Find the compressive force in pawl $AB$. What is the resultant supporting force system at $E$?

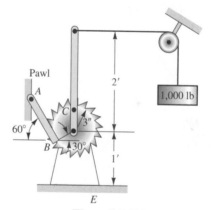

**Figure P.5.136.**

**5.137.** A 10-kN load is lifted in the front loader bucket. What are the forces at the connections to the bucket and to arm $AE$? Hydraulic ram $DF$ is perpendicular to arm $AE$, and $BC$ is horizontal. Points $A$ and $F$ are at the same height above the ground.

**Figure P.5.137.**

215

**5.138.** Determine the supporting forces at $A$ and $B$.

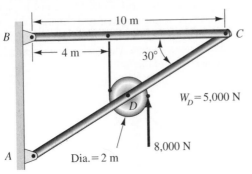

**Figure P.5.138.**

**5.139.** Find the force components on the pin $C$ of the frame shown. Neglect friction everywhere as well as the weights of the members.

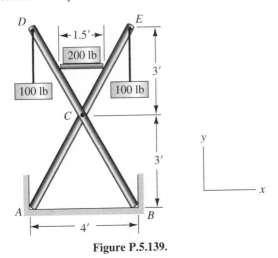

**Figure P.5.139.**

**5.140.** Find the supporting force system at $A$. Use only one free-body diagram.

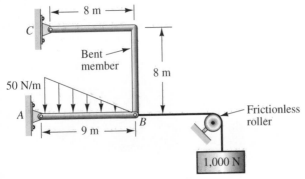

**Figure P.5.140.**

**5.141.** Find the supporting force systems at $A$ and $B$. Note the pin connection at $C$.

**Figure P.5.141.**

**5.142.** A 300-kN tank is climbing up a 30° incline at constant speed. What is the torque developed on the rear drive wheels to accomplish this? Assume that all other wheels are free-turning.

**Figure P.5.142.**

**5.143.** Find the components of the forces acting on pins $A$, $B$, and $C$ connecting and supporting the blocks shown. Block I weighs 10 kN, and block II weighs 30 kN.

**Figure P.5.143.**

**5.144.** What force $F$ do the pliers develop on the pipe section $D$? Neglect friction.

**Figure P.5.144.**

**5.145.** What are the supporting forces for the frame? Neglect all weights except the 10-kN weight.

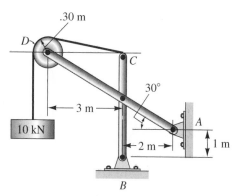

**Figure P.5.145.**

**5.146.** A 20-m circular arch must withstand a wind load given for $0 < \theta < \pi/2$ as

$$f = 5,000\left(1 - \frac{\theta}{\pi/2}\right) \text{N/m}$$

where $\theta$ is measured in radians. Note that for $\theta > \pi/2$, there is no loading. What are the supporting forces? (*Hint:* What is the point for which taking moments is simplest?)

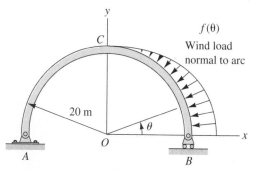

**Figure P.5.146.**

**5.147.** An arch is formed by uniform plates $A$ and $B$. Plate $A$ weighs 5 kN and plate $B$ weighs 2 kN. What are the supporting forces at $C$, $D$, and $E$?

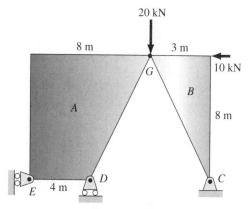

**Figure P.5.147.**

**5.148.** Find the supporting forces at the ball-and-socket connections $A$, $D$, and $C$. Members $AB$ and $DB$ are pinned together through member $EC$ at $B$.

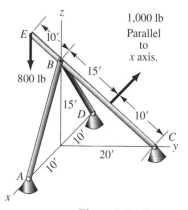

**Figure P.5.148.**

217

**5.149.** A tractor with a bulldozer is used to push an earthmover picking up dirt. If the tractor force on the earthmover is 150 kN in a horizontal direction, what are the reactions of the bulldozer on the tractor at $B$ and $A$?

**Figure P.5.149.**

**5.151.** A Bucyrus–Erie Dynahoe digger is partially shown. To develop the indicated forces in the bucket, what forces must hydraulic cylinders $HB$ and $CD$ develop? Consider only the 3-kN and 5-kN loads and not the weights of the members.

**Figure P.5.151.**

**5.150.** A hydraulic-lift platform for loading trucks supports a weight $W$ of 5,000 lb. Only one side of the system has been shown; the other side is identical. If the diameter of the pistons in the cylinders is 4 in., what pressure $p$ is needed to support $W$ when $\theta = 60°$? The following data apply.

$$l = 24 \text{ in.}, \quad d = 60 \text{ in.}, \quad e = 10 \text{ in.}$$

Neglect friction everywhere. (Hint: Only two free-body diagrams need be drawn.)

**5.152.** A block of material weighing 200-lb is supported by members $KC$ and $HB$, whose weight we neglect, a ball-and-socket-joint support at $A$, and a smooth, frictionless support at $E$. Members $KC$ and $HB$ have directions collinear with diagonals of the block as shown. What are the supporting forces for this block?

**Figure P.5.150.**

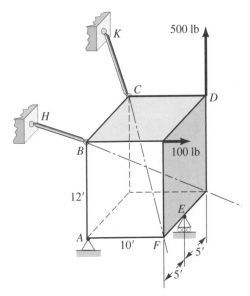

**Figure P.5.152.**

**\*5.153.** A bar can rotate parallel to plane $A$ about an axis of rotation normal to the plane at $O$. A weight $W$ is held by a cord that is attached to the bar over a small pulley that can rotate freely as the bar rotates. Find the value of $C$ for equilibrium if $h = 300$ mm, $W = 30$ N, $\phi = 30°$, $l = 700$ mm, and $d = 500$ mm.

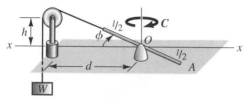

**Figure P.5.153.**

**5.154.** Find the supporting forces at $A$ and $B$ in the frame. Neglect weights of members.

**Figure P.5.154.**

**5.155.** A bolt cutter has a force of 130 N applied at each handle. What is the force on the bolt from the cutter edge?

**Figure P.5.155.**

**5.156.** A steam locomotive is developing a pressure of .20 N/mm² gage. If the train is stationary, what is the total traction force from the two wheels shown? Neglect the weight of the various connecting rods. Neglect friction in piston system and connecting rod pins.

**Figure P.5.156.**

**5.157.** Find the supporting forces at $A$, $B$, and $C$. Neglect the weight of the rod. Use only one free-body diagram.

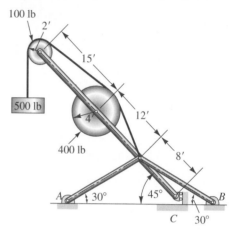

**Figure P.5.157.**

**5.158.** The 5,000 lb van $A$ of an airline food catering truck rises straight up until its floor is level with the airplane floor. A hydraulic ram pulls on the right bottom support of the lift mechanism at which we have rollers to prevent friction. The two members of the lift mechanism are pinned at their center. The center of gravity of the van is its geometric center. What is the ram force for this position?

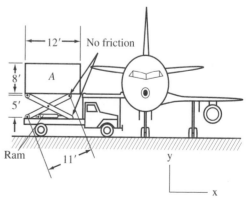

**Figure P.5.158.**

219

**5.159.** (a) Using only one free-body diagram, find the supporting forces from the ground at $A$ and at $B$.

(b) At a position 2.5 ft to the left of $D$, find the force system acting on the cross-section of the member $CD$.

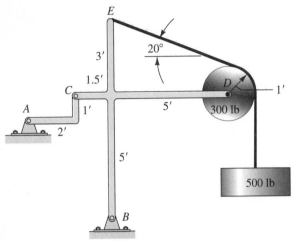

**Figure P.5.159.**

**5.160.** For the structure shown, determine the force in the cable $EF$.

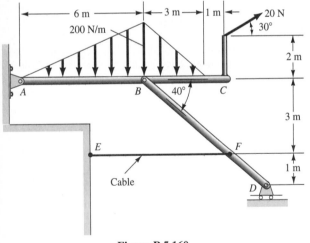

**Figure P.5.160.**

**5.161.** Find the supporting forces at $A$ and $B$. The cylinder can turn freely.

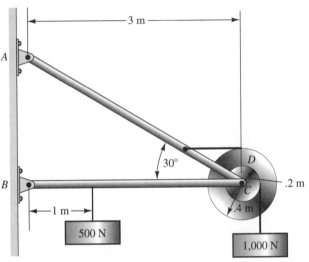

**Figure P.5.161.**

**5.162.** Find the supporting force system at $A$ and $B$. Cylinder $C$ weighs 4,000 N and is free to rotate without friction.

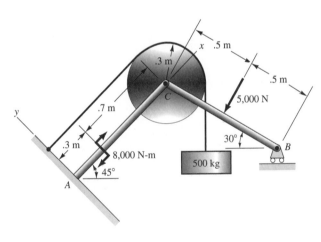

**Figure P.5.162.**

# Introduction to Structural Mechanics

## Part A: Trusses

## 6.1  The Structural Model

A *truss* is a system of members that are fastened together at their ends to support stationary and moving loads.[1] Everyday examples of trusses are shown in Figs. 6.1 and 6.2. Each member of a truss is usually of uniform cross-section along its length; however, the various members typically have different cross-sectional areas because they must transmit different forces. Our purpose in Part A of this chapter is to set forth methods for determining forces in members of an elementary class of trusses.

As a first step, we shall divide trusses into two main categories according to geometry. A truss consisting of a coplanar system of members is called a *plane truss*. Examples of plane trusses are the sides of a bridge (see Fig. 6.1) and a roof truss (see Fig. 6.2). A three-dimensional system of members, on the other hand, is called a *space truss*. A common example of a space truss is the tower from an electric power transmission system (see Fig. 6.3). Both plane trusses and space trusses consist of members having cross-sections resembling the letters H, I, and L. Such members are commonly used in many structural applications. These members are fastened together to form a truss by being welded, riveted, or bolted to intermediate structural elements called *gusset plates* such as has been shown in Fig. 6.4(a) for the case of a plane truss. The analysis of forces and moments in such connections is clearly quite complicated. Fortunately, there is a way of simplifying these connections

---

[1]A *truss* is different than a *frame* (see the footnote on p. 155) in that the members of a truss are always connected together at the ends of the members, as will soon become evident, whereas a frame has some members with connections not at the ends of the member.

**Figure 6.1.** Foot bridge near author's former home. Sides of structure are plane trusses.

**Figure 6.2.** Roof trusses that are plane trusses.

**Figure 6.3.** Space trusses supporting transmission lines sending power into the northeast grid of the United States.

such as to incur very little loss in accuracy in determining forces in the members. Specifically, if the centerlines of the members are *concurrent* at the connections, such as is shown in Fig. 6.4(a) for the coplanar case, then we can replace the complex connection at the points of concurrency by a simple pin connection in the coplanar truss and a simple ball-and-socket connection for the space truss. Such a replacement is called an *idealization* of the system. This is illustrated for a plane truss in Fig. 6.4, where the actual connector or joint is shown in (a) and the idealization as a pinned joint is shown in (b).

In order to maximize the load-carrying capacity of a truss, the external loads must be applied at the joints. The prime reason for this rule is the fact that the members of a truss are long and slender, thus rendering compression members less able to carry loads transverse to their centerlines away from the ends.[2] If the weights of the members are neglected, as is sometimes the case, it should be apparent that each member is a *two-force member*, and accordingly is either a tensile member or a compression member. If the weight is not negligible, the common practice as an approximation is to apply half the weight of a member to each of its two joints. Thus, the idealization of a member as a two-force member is still valid.

[2]You will understand these limitations more clearly when you study buckling in your strength of materials course.

(a) (b)

**Figure 6.4.** (a) Gusset plate; (b) idealization.

**Figure 6.5.** (a) Simple space truss.

**Figure 6.6.** Simple plane truss.

## 6.2 The Simple Truss

An idealized truss as described in Section 6.1 is termed *just-rigid* if the removal of any of its members destroys its rigidity. If removing a member does not destroy rigidity, the structure is said to be *over-rigid*. We shall be concerned with just-rigid trusses in Part A of this chapter.[3]

The most elementary just-rigid truss is one with three members connected to form a triangle. Just-rigid space trusses may be built up from this triangle by adding for each new joint three new members, as is shown *in Fig. 6.5*. Trusses constructed in this manner are called *simple space trusses*. The *simple plane truss* is built up from an elementary triangle by adding two new members for each new pin as shown in Fig. 6.6. Clearly, the simple plane truss is just-rigid.

A simple relationship exists between the number of joints $j$ and the number of members $m$ in a simple truss. You can directly verify by examining the simple space truss in Fig. 6.5 that $m$ is related to $j$ as follows:

$$m = 3j - 6 \tag{6.1a}$$

Similarly, for the simple plane truss in Fig. 6.6 you can verify that

$$m = 2j - 3 \tag{6.1b}$$

You will learn in more advanced structures courses that Eqs. 6.1 (a) and 6.1 (b) hold generally for just-rigid space trusses and for just-rigid plane trusses, respectively.

We now show that if the supporting force system is statically determinate, we can compute the forces in all the members of simple trusses. Specifically, in examining the ball joints of simple space trusses, we can see that in the general three-dimensional case, a ball joint with only three unknown

[3]Over-rigid structures are studied in courses of strength of materials and structural mechanics. They are internally statically indeterminate and deformation must be taken into account when computing forces in the members.

forces acting on it from the members can always be found. (One such joint is the last joint formed.) Each unknown force from a member onto this joint must have a direction collinear with that member, and hence has a known direction. There are, then, only three unknown scalars, and since we have a concurrent force system they can be determined by statics alone. We then find another joint with only three unknowns and so carry on the computations until the forces in the entire structure have been evaluated. For the simple plane truss, a similar procedure can be followed. The free body of at least one joint has only two unknown forces. We have a concurrent, coplanar force system, and we accordingly can solve the corresponding two equilibrium equations in two unknowns at that joint. We then proceed to the other joints, thereby evaluating all member forces by the use of statics alone.

## 6.3   Solution of Simple Trusses

Generally, the first step in a truss analysis is to compute the supporting forces in the overall truss. This calculation of the external forces or reactions that must exist to keep the truss in equilibrium is independent of whether the truss, internally, is statically determinate or statically indeterminate. Simply regard the truss as a rigid body to which forces are applied, some known (given applied forces) and some unknown (reactions),[4] and solve for the reactions as we did in Chapter 5. We have shown a simple plane truss in Fig. 6.7(a) and have shown the features of the truss in Fig. 6.7(b) that are essential for the calculations of the reactions. Note that members such as *CB, DB,* and *DE* need not be shown in the free body since they provide *internal* forces for the body.

Once the free-body diagram has been carefully drawn, use three equations of equilibrium to determine the reactions of a plane truss (six equations for a space truss). It is highly advisable to then check your results by using another (dependent) equation of equilibrium. You will be using the computed reactions for many subsequent calculations involving forces in internal members. Accordingly, with much work at stake, it is important to start off with a correct set of reactions.

We shall present two methods for determining the forces in the members of the truss. One is called the *method of joints* and the other is called the *method of sections*. As will be seen in the following sections, the prime difference between these methods lies in the choice of the free bodies to be used.

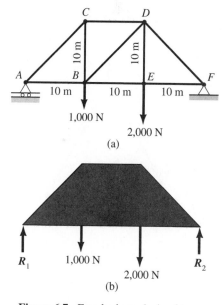

Figure 6.7. Free-body analysis of truss.

## 6.4   Method of Joints

In the method of joints, the free-body diagrams to be used, once the reactions are determined, are the pins or ball joints and the forces applied to them by the attached members and external loads. Note that we have already alluded

---

[4]Supporting forces are often called *reactions* in structural mechanics.

to this method in Section 6.2. Consider first the triangular plane truss shown in Fig. 6.8(a). Notice we have already determined the reactions.

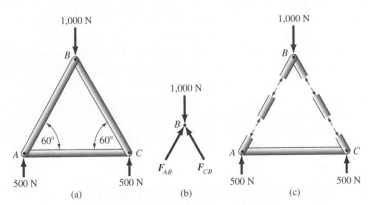

**Figure 6.8.** Method of joints—joint B.

Next, consider the free body of pin $B$ [Fig. 6.8(b)]. The unknown forces from the members are shown collinear with the centerlines of these members since they are two-force members. We can solve for these forces by setting the sum of forces equal to zero in the horizontal and vertical directions, to get

$$F_{AB} = F_{CB} = 577 \text{ N}$$

Because both forces are *pushing* against pin $B$, the corresponding members are *compressive* rather than tension members. We can most readily see this fact by considering Fig. 6.8(c), where members $AB$ and $CB$ have been cut at various places. Notice that $AB$ is also pushing against pin $A$ as does $CB$ against pin $C$. Thus, once having decided that the members are compressive members as a result of considerations at a pin at one end of the member, we can conclude that the member is pushing with equal force against the pin at the other end. To make for speed and accuracy as we go from one joint to another, we recommend that, once the nature of the loading in a member has been established by considerations at a pin, we mark down this value using a $T$ for tension or a $C$ for compression after it on the truss diagram, as shown in Fig. 6.9(a). Note also that appropriate arrows are drawn in the members. These arrows represent forces developed by the *members on the pins*. Hence, for *compression* the arrows point *toward* the pins, and for *tension* they point *away* from the pins. Accordingly, if we now consider the free body of pin $A$ as shown in Fig. 6.9(b), we know the direction and value of the force on $A$ from member $AB$.

If a negative value is found for a force at a pin, the sense of the force has been taken incorrectly at the outset. With this in mind, we decide whether the member associated with the force is a tension or compression member and we label the member accordingly, as shown in Fig. 6.9(a) for use later in examining the pin at the other end of the member as a free body.

We now consider the solution of a plane truss problem by the method of joints in greater detail.

**Figure 6.9.** Procedure for method of joints. (a) Notation for members $AB$ and $CB$; (b) free-body diagram of $A$.

# Example 6.1

A simple plane truss is shown in Fig. 6.10. Two 1,000-lb loads are shown acting on pins $C$ and $E$. We are to determine the force transmitted by each member. Neglect the weight of the members.

In this simple loading, we see by inspection that there are 1,000-lb vertical forces at each support. We shall begin, then, by studying pin $A$, for which there are only two unknowns.

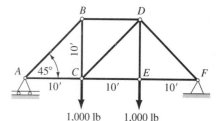

**Figure 6.10.** Plane truss.

***Pin A.*** The forces on pin $A$ are the known 1,000-lb supporting force and two unknown forces from the members $AB$ and $AC$. The orientation of these forces is known from the geometry of the truss, but the magnitude and sense must be determined. To help in interpreting the results, put the forces in the same position as the corresponding members in the truss diagram as is shown in Fig. 6.11. That is, avoid the force diagram in Fig. 6.12, which is equivalent to the one in Fig. 6.11 but which may lead to errors in interpretation. There are two unknowns for the concurrent coplanar force system in Fig. 6.11 and thus, if we use the scalar equations of equilibrium, we may evaluate $F_{AB}$ and $F_{AC}$:

$$\sum F_x = 0:$$
$$F_{AC} - 0.707 F_{AB} = 0$$

$$\sum F_y = 0:$$
$$-0.707 F_{AB} + 1,000 = 0$$

Therefore,

$$F_{AB} = 1,414 \text{ lb} \; ; \qquad F_{AC} = 1,000 \text{ lb}$$

**Figure 6.11.** Pin $A$.

Since both results are positive, we have chosen the proper senses for the forces. We can then conclude on examining Fig. 6.11 that $AB$ is a compression member, whereas $AC$ is a tension member.[5] In Fig. 6.13, we have labeled the members accordingly.

If we next examine pin $C$, clearly since there are three unknowns involved for this pin, we cannot solve for the forces by equilibrium equations at this time. However, pin $B$ can be handled, and once $F_{BC}$ is known, the forces on pin $C$ can be determined.

**Figure 6.12.** Pin $A$—avoid this diagram.

***Pin B.*** Since $AB$ is a compression member (see Fig. 6.13) we know that it exerts a force of 1,414 lb directed against pin $B$ as has been shown in Fig. 6.14. As for members $BC$ and $BD$, we assign senses as shown.

[5]Had we used Fig. 6.12 as a free body, the state of loading in the members (i.e., tension or compression) would not be clear. Therefore, we strongly recommend putting forces representing members in positions coinciding with the members.

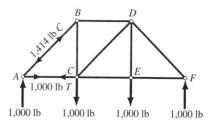

**Figure 6.13.** Notation for members $AB$ and $AC$.

## Example 6.1 (Continued)

**Figure 6.14.** Pin $B$.

Summing forces on pin $B$ (Fig. 6.14), we get

$$\sum F_x = 0:$$

$$(1{,}414)(0.707) + F_{BD} = 0$$

$$\boxed{F_{BD} = -1{,}000 \text{ lb}}$$

$$\sum F_y = 0:$$

$$(1{,}414)(0.707) + F_{BC} = 0$$

$$\boxed{F_{BC} = -1{,}000 \text{ lb}}$$

Here we have obtained two negative quantities, indicating that we have made incorrect choices of sense. Keeping this in mind, we can conclude that member $BD$ is a compression member, whereas member $BC$ is a tension member. Notice that we have shown these forces properly in Fig. 6.15.

We can proceed in this manner from joint to joint. At the last joint all the forces will have been computed without using it as a free body. Thus, it is available to be used as a check on the solution. That is, the sum of the known forces for the last joint in the $x$ and $y$ directions should be zero or close to zero, depending on the accuracy of your calculations. We urge you to take advantage of this check. The final solution is shown in Fig. 6.15. Notice that member $CD$ has zero load. This does not mean that we can get rid of this member. Other loadings expected for the truss will result in nonzero force for $CD$. Furthermore, without $CD$ the truss will not be rigid.

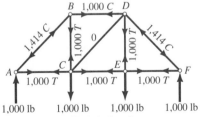

**Figure 6.15.** Solution for truss.

# Example 6.2

A bridge truss is shown in Fig. 6.16 supporting at its pins half of a roadbed weighing 1,000 lb per foot. A truck is shown on the bridge having esti-mated loads on pins $E$, $G$, and $I$ of the truss equaling, respectively, 1 ton, 1.5 tons, and 2.5 tons. The members weigh 45 lb/ft. Include the weight of the members by putting half the weight of a member at each of its two sup-porting pins. Find the supporting forces.

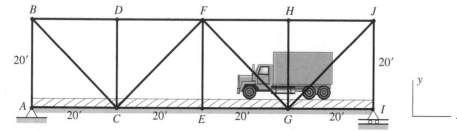

**Figure 6.16.** A bridge truss supporting a roadbed and a truck.

We first determine the forces on the pins from the weight of the members denoting them as $(W_1)_i$. Thus we have

***Loads 1: Weights of members***

$$(W_1)_A = (W_1)_I = 2\left[\frac{1}{2}(20)(45)\right] = 900 \text{ lb}$$

$$(W_1)_B = (W_1)_J = 2\left[\frac{1}{2}(20)(45)\right] + \frac{1}{2}\left(\frac{20}{.707}\right)(45) = 1{,}536.5 \text{ lb}$$

$$(W_1)_C = (W_1)_F = (W_1)_G = 3\left[\frac{1}{2}(20)(45)\right] + 2\left[\frac{1}{2}\left(\frac{20}{.707}\right)(45)\right] = 2{,}623 \text{ lb}$$

$$(W_1)_D = (W_1)_H = (W_1)_E = 3\left[\frac{1}{2}(20)(45)\right] = 1{,}350 \text{ lb}$$

Next we get to the roadbed.

***Loads 2: Weights from the roadbed***

$$(W_2)_A = (W_2)_I = \frac{1}{2}(500)(20) = 5{,}000 \text{ lb}$$
$$(W_2)_C = (W_2)_E = (W_2)_G = (500)(20) = 10{,}000 \text{ lb}$$

Finally, we list the loads from the truck.

***Loads 3: Weights from the truck***

$$(W_3)_E = 2{,}000 \text{ lb}$$
$$(W_3)_G = 3{,}000 \text{ lb}$$
$$(W_3)_I = 5{,}000 \text{ lb}$$

## Example 6.2 (Continued)

We are now ready to determine the supporting forces for the truss. Taking the entire truss as a free body we first take moments about joint $I$ (see Fig. 6.17).

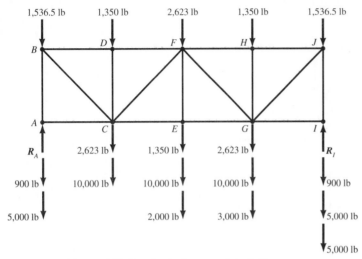

**Figure 6.17.** Free-body diagram of the bridge truss.

$$\sum M_I = 0:$$

$$-R_A(80) + (1,536.5 + 900 + 5,000)(80) + (1,350 + 2,623$$
$$+10,000)(60) + (2,623 + 1,350 + 10,000 + 2,000)(40)$$
$$+(1,350 + 2,623 + 10,000 + 3,000)(20) = 0$$

$$\boxed{R_A = 30,146 \text{ lb}}$$

$$\sum M_A = 0:$$

$$R_I(80) - (1,536.5 + 900 + 5,000 + 5,000)(80) - (1,350$$
$$+ 2,623 + 10,000 + 3,000)(60) - (2,623 + 1,350$$
$$+ 10,000 + 2,000)(40) - (1,350 + 2,623 + 10,000)(20) = 0$$

$$\boxed{R_I = 36,646 \text{ lb}}$$

Check:

$$\sum F_y = 0:$$

$$30,146 + 36,646 - (2)(900) - (2)(1,536.5) - (3)(2,623) - (3)(1,350)$$
$$-(2)(5,000) - (3)(10,000) - 2,000 - 3,000 - 5,000 = 0$$
$$0 = 0$$

## Example 6.3

Ascertain the forces transmitted by each member of the three-dimensional truss [Fig. 6.18(a)].

We can readily find the supporting forces for this simple structure by considering the whole structure as a free body and by making use of the symmetry of the loading and geometry. The results are shown in Fig. 6.18(b).

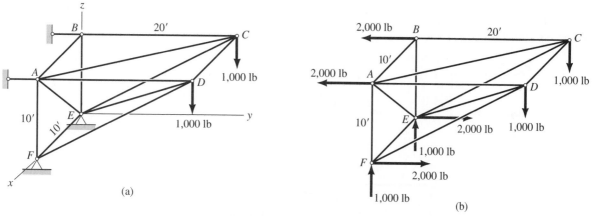

**Figure 6.18.** (a) Space truss and (b) free-body diagram.

***Joint F.*** It is clear, on an inspection of the forces in the *x* direction acting on joint *F*, that $F_{FE} = 0$, since all other forces are in a plane at right angles to it. These other forces are shown in Fig. 6.19. Summing forces in the *y* and *z* directions, we get

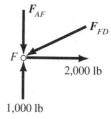

**Figure 6.19.** Free-body diagram of joint *F*.

$$\underline{\sum F_y = 0:}$$

$$-F_{FD} \frac{20}{\sqrt{20^2 + 10^2}} + 2,000 = 0$$

Therefore,

$$F_{FD} = 2,240 \text{ lb compression}$$

$$\underline{\sum F_z = 0:}$$

$$-F_{AF} + 1,000 - 2,240 \frac{10}{\sqrt{500}} = 0$$

Therefore,

$$F_{AF} = 1,000 - 1,000 = 0$$

$$F_{AF} = 0$$

## Example 6.3 (Continued)

***Joint B.*** Going to joint *B,* we see [Fig. 6.18(b)] that $F_{AB} = 0$ and $F_{BE} = 0$, since there are no other force components on pin *B* in the directions of these members. Finally, $F_{BC} = 2,000 \text{ lb tension}$.

***Joint A.*** Let us next consider joint *A* (Fig. 6.20). We can express force $\boldsymbol{F}_{AC}$ and $\boldsymbol{F}_{AE}$ vectorially. Thus,

$$\boldsymbol{F}_{AC} = F_{AC} \frac{-10\boldsymbol{i} + 20\boldsymbol{j}}{\sqrt{10^2 + 20^2}} = F_{AC}(-.447\boldsymbol{i} + .894\boldsymbol{j}) \text{ lb}$$

$$\boldsymbol{F}_{AE} = F_{AE} \frac{-10\boldsymbol{i} - 10\boldsymbol{k}}{\sqrt{10^2 + 10^2}} = F_{AE}(-.707\boldsymbol{i} - .704\boldsymbol{k}) \text{ lb}$$

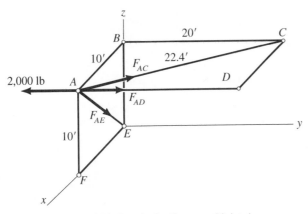

**Figure 6.20.** Free-body diagram of joint *A.*

Summing forces, we have

$$-2,000\boldsymbol{j} + F_{AD}\boldsymbol{j} + F_{AC}(-.447\boldsymbol{i} + .894\boldsymbol{j}) + F_{AE}(-.707\boldsymbol{i} - .707\boldsymbol{k}) = \boldsymbol{0}$$

Hence,

$$.894F_{AC} + F_{AD} = 2,000 \tag{a}$$
$$-.447F_{AC} - .707F_{AE} = 0 \tag{b}$$
$$-.707F_{AE} = 0 \tag{c}$$

We see that $F_{AE} = F_{AC} = 0$ and $F_{AD} = 2,000 \text{ lb tension}$.

***Joint D.*** We now consider joint *D* (Fig. 6.21). Forces $\boldsymbol{F}_{FD}$ and $\boldsymbol{F}_{ED}$ are expressed as follows:

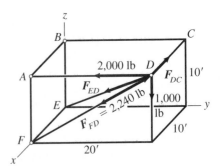

**Figure 6.21.** Free-body diagram of joint *D.*

# Example 6.3 (Continued)

$$F_{ED} = F_{ED} \frac{-10i - 20j - 10k}{\sqrt{10^2 + 20^2 + 10^2}} = F_{ED}(-.408i - .816j - .408k) \text{ lb}$$

$$F_{FD} = F_{FD} \frac{20j + 10k}{\sqrt{20^2 + 10^2}} = 2,240(.894i + .447k) \text{ lb}$$

Hence, summing forces, we get

$$-2,000j - 1,000k - F_{DC}i + 2,240(.894j + .447k)$$
$$+ F_{ED}(-.408i - .816j - .408k) = 0 \tag{d}$$

Thus,

$$-2,000 + 2,000 - .816F_{ED} = 0 \tag{e}$$

$$F_{DC} + .408F_{ED} = 0 \tag{f}$$

$$-1,000 + 1,000 - .408F_{ED} = 0 \tag{g}$$

We see here that $\boxed{F_{ED} = 0}$ and $\boxed{F_{DC} = 0}$ .

*Joint E.* The only nonzero forces on joint $E$ are the supporting forces and $F_{CE}$, as shown in Fig. 6.22(a). We solve $\boxed{F_{CE} = 2,240 \text{ lb compression}}$ .

*Joint C.* As a check on our problem, we can examine joint $C$. The only nonzero forces are shown on the joint [Fig. 6.22(b)]. The reader may readily verify that the solution checks.

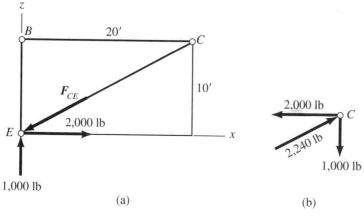

**Figure 6.22.** (a) Free-body diagram of joint $E$ and (b) joint $C$.

Before proceeding with the problems, it will be well to comment on the loading of plane roof trusses. Usually there will be a series of separated parallel trusses supporting the loading from the roof such as is shown in Fig. 6.23, where a wind pressure is shown on a roof as *p*. Now the *inside* truss can be considered to support the loading over a region extending halfway to each neighboring truss (shown as distance *d*). Furthermore, pins *A* and *B* support the force exerted on area *lhmk* while pins *B* and *C* support the forces exerted on area *lrvh*. When dealing with the entire inside truss as a free body, you can use the resultant force from pressure over *krvm*. However, when dealing with the pins as a free body you must use the forces coming on to each pin as described above and *not* the total resultant, which was used for the free body of the entire internal truss. Clearly, the outside trusses support half the loads described above.

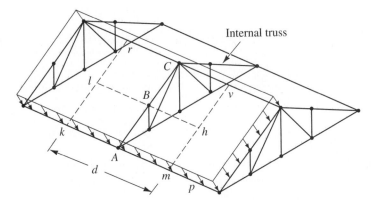

**Figure 6.23.** Roof trusses supporting a wind load.

Finally, we wish to remind you that a curved member in a truss, such as appears in Problems 6.5 and 6.8, is a two-force member with forces coming only from the pins. Recall that, for such members, the force transmitted to the pins must be collinear with the line connecting the pins, such as is shown in Fig. 6.24.

**Figure 6.24.** Curved two-force member.

# PROBLEMS

**6.1.** State which of the trusses shown are simple trusses and which are not.

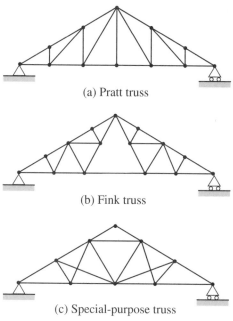

(a) Pratt truss

(b) Fink truss

(c) Special-purpose truss

**Figure P.6.1.**

**6.2.** Find the forces transmitted by each member of the truss.

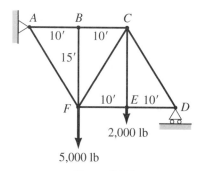

**Figure P.6.2.**

**6.3.** The simple country-road bridge has floor beams to carry vehicle loads to the truss joints. Find the forces in all members for a truck-loaded weight of 160 kN. Floor beams 1 are supported by pins $A$ and $B$, while floor beams 2 are supported by pins $B$ and $C$.

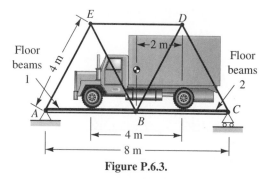

**Figure P.6.3.**

**6.4.** A rooftop pond is filled with cooling water from an air conditioner and is supported by a series of parallel plane trusses. What are the forces in each member of an inside truss? The roof trusses are spaced at 10 ft apart. Water weights 62.4 lb/ft.[3]

**Figure P.6.4.**

**6.5.** Find the forces transmitted by the straight members of the truss. $DC$ is circular.

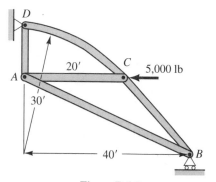

**Figure P.6.5.**

**6.6.** Roof trusses such as the one shown are spaced 6 m apart in a long, rectangular building. During the winter, snow loads of up to 1 kN/m² (or 1 kPa) accumulate on the central portion of the roof. Find the force in each member for a truss not at the ends of the building.

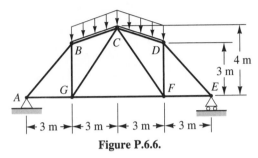

**Figure P.6.6.**

**6.7.** The bridge supports a roadway load of 1,000 lb/ft for each of the two trusses. Each member weighs 30 lb/ft. Compute the forces in the members, accounting approximately for the weight of the members.

**Figure P.6.7.**

**6.8.** Find the forces in the straight members of the truss.

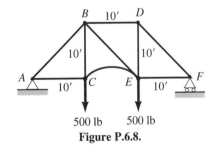

**Figure P.6.8.**

**6.9.** Find the forces in the members of the truss.

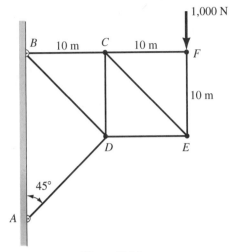

**Figure P.6.9.**

**6.10.** In Example 6.1, include the weights of the members approximately. The members each weigh 100 lb/ft.

**6.11.** Determine the forces in the members. The pulleys at $C$ and $F$ each weigh 300 N. Neglect all other weights. Be sure you have a check on your solution.

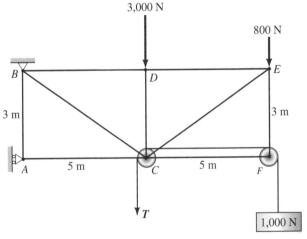

**Figure P.6.11.**

236

**6.12**  Roadway and vehicle loads are transmitted to the highway bridge truss as the idealized forces shown. Each load is 100 kN. What are the forces in the members?

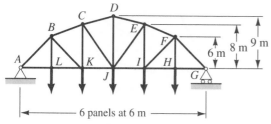

Figure P.6.12.

**6.13.**  A hoist weighing 5 kN lifts railroad cars for truck repair. The hoist has a 150-kN capacity and hangs from a truss with an L-shaped member to clear boxcars. What are the forces in the straight members for full capacity of the hoist?

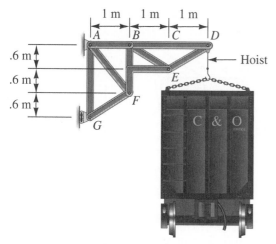

Figure P.6.13.

**6.14.**  A 5-kN traveling hoist has a 50-kN capacity and is suspended from a beam weighing 1 kN/m, which, in turn, is fastened to the roof truss at *I* and *G* as shown. In addition, wind pressures of up to 2 kN/m² (or 2 kPA) act on the side of the roof. The resulting force is transmitted to pins *A* and *J*. If the trusses are spaced 5 m apart, what are the forces in each member of the truss when the hoist is in the middle of the span?

Figure P.6.14.

**6.15.**  Find the forces in the members of the truss. The 1,000-lb force is parallel to the *y* axis, and the 500-lb force is parallel to the *z* axis.

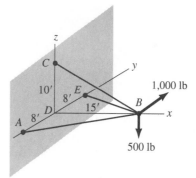

Figure P.6.15.

**6.16.**  Find the forces in the members and the supporting forces for the space truss *ABCD*. Note that *BDC* is in the *xz* plane.

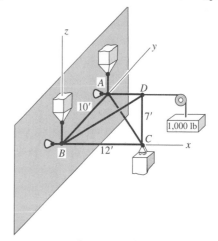

Figure P.6.16.

237

**6.17.** A steel space truss *ABCDE* with members having identical cross-sections supports a 50-kN vertical load as well as a 10-kN horizontal load and rests on smooth, mutually perpendicular surfaces. Assume that the contact between the space truss and the smooth surface is at the ball joints. What are the forces in the members at *D*?

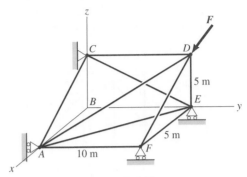

**Figure P.6.18.**

**6.19.** The plane of ball-and-socket joints *CDHE* of the space truss is in the *zy* plane, while the plane of *FGDE* is parallel to the *xz* plane. Note that this is *not* a simple space truss. Nevertheless, the forces in the members can be ascertained by choosing a desirable starting joint and proceeding by statics from joint to joint. Determine the forces in all the members and then determine the supporting forces. At *A* and *H* we have ball and socket joints.

**Figure P.6.17.**

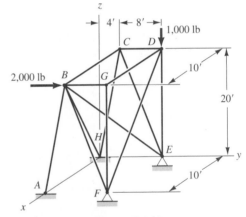

**Figure P.6.19.**

**6.18.** Find the forces in the members of the space truss under the action of a force *F* given as

$$F = 10i - 6j - 12k \text{ kN}$$

Note that *C* is a ball-and-socket joint while *A*, *F*, and *E* are on rollers.

**6.20.** Find the forces in all the members of Example 6.2.

# 6.5 Method of Sections

In the method of sections that we shall use for plane trusses, we employ free-body diagrams that are generally different than that of the method of joints, as was pointed out earlier. *A free body in this method is formed by cutting away a portion of a truss and including at the cut sections the forces that are transmitted across these sections.* We then use the equations of equilibrium for these free bodies. In this way, we can expose for calculation individual members well inside a truss and avoid the laborious process of proceeding joint by joint until reaching a joint on which the desired unknown force acts.

Generally a free body is created by passing a section (or cut) through the truss such as section *A–A* or section *B–B* in Fig 6.25(a). Note that the section can be straight or curved. The corresponding right hand free-body diagrams [see Fig. 6.25(b) for cut *A–A* and Fig. 6.25(c) for cut *B–B*] involve

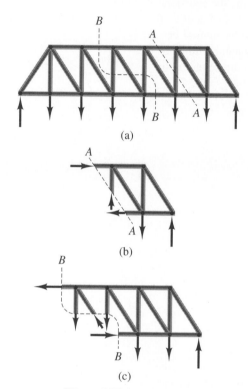

**Figure 6.25.** Section cuts.

coplanar force systems. We have, accordingly, three equations of equilibrium available for each free body. One can also choose to use left-hand free-body diagrams for these cuts. Note that in contrast to the method of joints, one or more equilibrium equations can most profitably be moment equations. The choice of the section (or sections) to find the desired unknowns inside a truss involves ingenuity on the part of the engineer. He/she will want the fewest and simplest sections to find desired forces for one or more members inside the truss. The method of sections is used for efficiently finding limited information. The method of joints for such problems is by contrast one of "brute force."

We now illustrate the method of sections in the following examples.

## Example 6.4

In Example 6.1 suppose that we wish to know the force in member $CE$ only.

To avoid the laborious joint-by-joint procedure, we employ a portion of the truss to the left of cut $K–K$, as shown in Fig. 6.26. Notice that the forces from the other part of the truss acting on this part through the cut members have been included, and in this way the desired force has been exposed. The sense of these exposed forces is not known, but we do know the orientations, as explained in our earlier discussions. Using the equations of equilibrium and taking advantage of the fact that the lines of action of some of the exposed unknown forces are concurrent at the location of certain joints, we may readily solve for the unknowns if they number three or less. To determine $F_{CE}$, we take moments about a point corresponding to joint $D$ through which the lines of action of forces $F_{BD}$ and $F_{CD}$ pass:

$$\sum M_D = 0:$$
$$-(1,000)(20) + (1,000)(10) + 10F_{CE} = 0$$

**Figure 6.26.** Cut $K–K$.

Therefore,

$$F_{CE} = 1,000 \text{ lb}$$

Our ingenuity here has led us to one equation with only one unknown, the desired force $F_{CE}$. By observing the free-body diagram in Fig. 6.26, we can clearly see that $CE$ is a tension member.

If we desire $F_{BD}$ also, we can take moments about point $C$ through which the lines of action of $F_{CE}$ and $F_{CD}$ pass. However, $F_{BD}$ now comes out negative, indicating that we have made an incorrect choice of sense. With this in mind, we can conclude that $BD$ is in compression.

Perhaps a suitable single section with sufficient unknowns for a solution cannot be found. We may then have to take several sections before we can expose the desired force in a free body with enough simultaneous equations to effect a solution. These problems are no different from the ones we studied in Chapter 5, where several free-body diagrams were needed to generate a complete set of equations containing the unknown quantity. We now consider such a problem.

# Example 6.5

A plane truss is shown in Fig. 6.27 for which only the force in member $AB$ is desired. The supporting forces have been determined and are shown in the diagram.

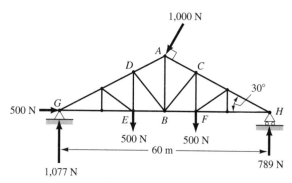

**Figure 6.27.** Plane truss.

In Fig. 6.28(a) we have shown a cut $J$–$J$ of the truss exposing force $F_{AB}$. (This is the same force diagram as that which results from the free-body diagram of pin $A$.) We have here three unknown forces for which only two equations of equilibrium are available. We must use an additional free body.

Thus, in Fig. 6.28(b) we have shown a second cut $K$–$K$. Note that by taking moments about joint $B$, we can solve for $F_{AC}$ directly. With this information we can then return to the first cut to get the desired unknown $F_{AB}$. Accordingly, we have, for free body II:

$$\underline{\sum M_B = 0} :$$

$$-(10)(500) + (30)(789) - (F_{AC})(\sin 30°)(30) = 0$$

(Note we have transmitted $F_{AC}$ to joint $H$ in evaluating its moment contribution.) Solving for $F_{AC}$ we get

$$F_{AC} = 1{,}245 \ \text{N}$$

(a) Free body I from cut $J$–$J$          (a) Free body II from cut $K$–$K$

**Figure 6.28.** Free bodies needed for the computation of force $F_{AB}$.

## ■ Example 6.5 (Continued)

Summing forces for free body I, we have[6]

$$\underline{\sum F_x = 0:}$$

$$F_{DA}\cos 30° - F_{AC} \cos 30° - 1{,}000 \sin 30° = 0$$

Therefore,

$$F_{DA} = 1{,}822 \text{N}$$

$$\underline{\sum F_y = 0:}$$

$$F_{DA} \sin 30° + F_{AC} \sin 30° + F_{AB} - 1{,}000 \cos 30° = 0$$

Therefore,

$$F_{AB} = -667 \text{ N}$$

We see that member $AB$ is a tension member rather than a compression member as was our initial guess in drawing the free-body diagrams.

[6]What now follows is the same as the method of joints applied to pin $A$.

In retrospect, you will note that, in the method of joints, errors made early will of necessity propagate through the calculations. There is, on the other hand, much less likelihood of this occurring in the method of sections. However, for simple trusses with many members, we may profitably use the method of joints in conjunction with a computer for which the brute-force approach of the method of joints is ideally suited. It is to be pointed out that there is computer software available which makes this kind of computation routine and quick.

## *6.6    Looking Ahead—Deflection of a Simple, Linearly Elastic Truss

In the solids or structures courses you will soon be taking, you will be given instruction for determining the movement of the pins of a linearly elastic, simple truss stemming from the external loads. It is true that you can now determine the forces in a simple truss and from this you can determine the change in length of each member. However, using this data as well as the accompanying changes of orientation of the members, you will find it next to impossible to get the movements of the pins for all but the most trivial trusses.

You will learn later of a neat method of doing this. In this method, each movable (unconstrained) pin is imagined to be given first a hypothetical movement[7] $\delta_1$ in one of two orthogonal directions, say the $x$ direction here, while no movement is allowed for that pin in the $y$ direction. All other pins are held fixed for the preceding action. We now evaluate via geometry the changes in length stemming from $\delta_1$ of all members affected by this hypothetical displacement. Also, the energy of deformation[8] for these members is computed. The aforementioned displacement is shown in Fig. 6.29 for $\delta_1$. In addition to the energy of deformation of the affected members, we compute the work done by external forces that undergo movement from $\delta_1$, keeping the forces constant during this movement. Here the work is $F(\cos \alpha)\delta_1$. This procedure is carried out for each such $x$ and $y$ displacement for all of the movable pins. We then add up all the energy terms and all the work terms, multiplying the latter by $-1$. We then have a function of all of the $n$ $\delta$'s comprising the $x$ and $y$ components of all movable pins. This function, which is commonly denoted as $\pi$, is then extremized with respect to the $n$ $\delta$'s. We then form $n$ simultaneous equations from

$$\left(\frac{\partial \pi}{\partial \delta}\right)_i = 0 \qquad i = 1, 2, \dots n \qquad (6.2)$$

**Figure 6.29.** A simple truss has its free pin $H$ given a hypothetical displacement $\delta_1$ while keeping all other displacement components fixed. Note that only members $CH$, $EH$, and $HB$ are involved. Work done is $F(\cos \alpha)\delta_1$.

Solving for the $\delta$'s gives the deflection of each pin. This formulation is a form of the method of *total potential energy*. You will later realize that Eqs. (6.2) are possibly the most powerful equations in solid mechanics.

What is particularly interesting here is that Eqs. (6.2), set up properly, are equivalent to equations of *equilibrium*. That is, we have here the strange situation whereby the extremization of a function is, among other things, equivalent to a most important equation.

The same shift can take place in other engineering sciences where we can sometimes shift from a key differential equation with certain boundary conditions to the process of extremizing a function (or some related quantity called a *functional* which, simply speaking, is a function of a function). Notice that for our simple truss we are going from physically obvious equations of equilibrium to a more obscure, more mathematical approach of extremizing a function. Yet (note this carefully) the latter approach yields more useful information of a very practical nature despite its more obscure, more mathematical nature.

In Chapter 10, we shall see a limited form of the total potential energy principle which is applicable for rigid bodies. In the present chapter for the truss, the members were deformable and we mentioned that we needed the energy of deformation for the use of the total potential energy principle.

---

[7]This is called a *virtual displacement* and will be discussed further in Chapter 10 when we discuss the method of virtual work.

[8]You will learn to calculate the energy of deformation in your solid mechanics or your structures course. The formula is $\frac{EA\delta^2}{2L}$ where $\delta$ is the axial change in length of the member, $A$ is the cross-sectional area, $E$ is the modulus of elasticity, and $L$ is the length of the member.

**6.21.** Find the forces in members *CB* and *BE* of the plane truss.

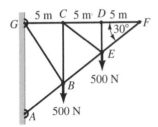

**Figure P.6.21.**

**6.22.** For the roof truss: (a) Find the forces transmitted by member *DC*. (b) What is the force transmitted by *DE*?

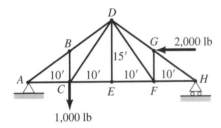

**Figure P.6.22.**

**6.23.** Determine the force transmitted by member *KU* in the plane truss. The vertical members are 9m apart.

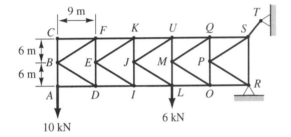

**Figure P.6.23.**

**6.24.** In the roof truss of Problem 6.6, find the force in member *GF*. Remember loads are applied to the pins.

**6.25.** Find the forces in members *CD*, *DG*, and *HG* in the plane truss.

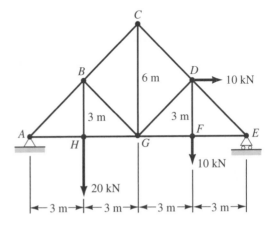

**Figure P.6.25.**

**6.26.** In Problem 6.13 find the force in members *BF* and *AB*.

**6.27.** The roof is subjected to a wind loading of 20 lb/ft². Find the forces in members *LK* and *KJ* of an interior truss if the trusses are spaced 10 ft apart.

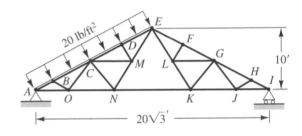

**Figure P.6.27.**

**6.28.** The guideways for a large overhanging crane are suspended from certain joints of the truss (*M, K, J,* and *G*). Find the forces in members *BC, BK, DE, DI,* and *EF*. Neglect the truss and guideway weights. Guideways only transmit supported loads to pins and are not considered part of the truss structurally.

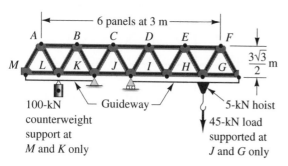

100-kN counterweight support at *M* and *K* only

Guideway

5-kN hoist
45-kN load supported at *J* and *G* only

**Figure P.6.28.**

**6.29.** In Example 6.2, determine the forces in members *FG* and *CE*.

**6.30.** (a) Find the forces in members *DG* and *DF* by the method of *sections*. State whether the members are tension or compression members.

(b) Determine the forces in members *AC, AB, CB,* and *CD* by the method of *joints*. Label the diagram in an appropriate manner and indicate whether members are in tension or compression.

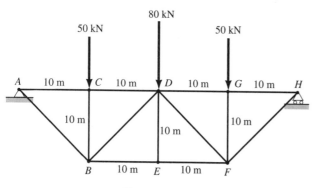

**Figure P.6.30.**

**6.31.** A pair of trusses supports a roadway weighing 500 lb/ft. By method of sections, find the forces in *DF* and *DE*. The roadway is supported at pins *A, D, F,* and *H* on the two trusses.

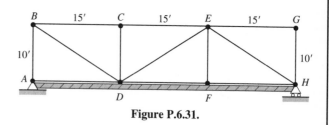

**Figure P.6.31.**

**6.32.** Find the force in members *HE, FH, FE,* and *FC* of the truss.

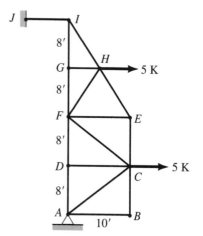

**Figure P.6.32.**

**6.33.** Find the force in member *JF* in the truss.

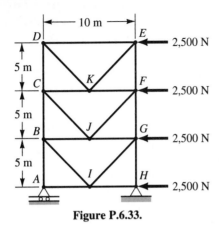

**Figure P.6.33.**

**6.34.** Find the force in members *FI*, *EF*, and *DH* in the truss. Neglect the weight of the pulleys.

**Figure P.6.34.**

**6.35.** A railroad engine is starting to cross the deck-type truss bridge shown. If the weight of the engine is idealized by the four 50-kip loads,[9] find the forces in members *AB*, *BL*, *CK*, *CL*, *LK*, *DK*, *KJ*, and *DJ*.

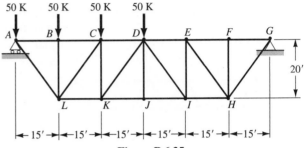

**Figure P.6.35.**

**6.36.** A diesel train engine is moving along a deck-type truss which is part of a bridge. If the loads on positions *A*, *B*, *C*, and *D* are estimated to be 50 kips each,[9] find the forces in members *LC*, *KL*, *FG*, and *HG*.

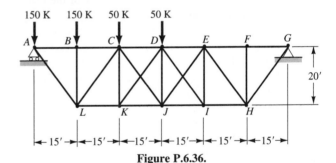

**Figure P.6.36.**

[9]A kip is a kilopound, or 1,000 lb.

# Part B: Section Forces in Beams[10]

## 6.7   Introduction

In Part A, we considered a number of problems involving members loaded axially along the axes of the members. The resultant force at any section was established easily as a single axial force. We shall now consider thin prismatic members that are loaded *transversely* as well as axially. Generally, when such members are loaded transversely, we call them *beams*. Of considerable use will be certain components of the resultant force system acting on *cross-sections* of the beams. We shall set forth methods in this section for computing these quantities. We consider beams with a vertical plane of symmetry along the axis of the beam.

## 6.8   Shear Force, Axial Force, and Bending Moment

Consider first a beam with an arbitrary intensity of loading $w(x)$ in the plane of symmetry and a load $P$ along the direction of the beam applied at the end $A$ as shown in Fig. 6.30(a). It will be assumed that the supporting forces have been determined. To find the force transmitted across the cross-sectional interface at position $x$, we take a portion of the beam as a free body so as to "expose" the section of the beam at $x$ as shown in Fig. 6.30(b). Since we have a coplanar-loading distribution, we know from rigid-body mechanics that, depending on the problem, we can replace the distribution at section $x$ most simply by a single force or a single couple in the plane of the external loads. If the resultant is most simply a single force, we know that it must have a particular line of action. This line of action does not usually go through the center of the cross-section. Since the actual position of the intersection of this force with the cross-section is of little interest in beam theory, we deliberately take the position of the resultant force to be at the center of the cross-section at all times and include the proper couple moment $M_z$ to accompany the force. Furthermore, we decompose the force in orthogonal components—in this case, a vertical force $V_y$, and a horizontal force $H$. These quantities are shown in Fig. 6.30(b). Since these quantities are used to such a great extent in structural work, we have associated names with them. They are $V_y \equiv$ *shear-force* component, $H \equiv$ *axial-force* component, $M_z \equiv$ *bending-moment* component.[11] If we had a three-dimensional load, there would have been one

---

[10]In many engineering programs, this topic is considered in the solid mechanics or strength of materials course which follows the statics course. There, the results are used abundantly throughout the course. It is entirely feasible then to bypass Part B without penalty in this book. For a more detailed treatment along the same lines as in this chapter, see I.H. Shames, *Introduction to Solid Mechanics,* 2nd Ed., Prentice-Hall Inc., Englewood Cliffs, NJ, 1985.

[11]For curved beams, shear forces $V$ are always tangent to the cross-section, whereas axial force $H$ is always normal to the cross-section.

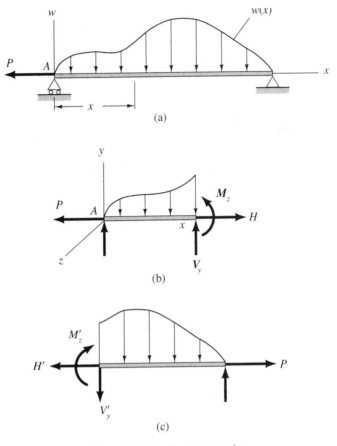

**Figure 6.30.** Resultant at a section.

additional shear component $V_z$ (see Fig. 6.31), one additional bending-moment component $M_y$, and a couple moment along the axis of the beam $M_x$, which we shall call the *twisting moment*.

Notice in Fig. 6.30(c) that a second free-body diagram has been drawn which exposes the "other side" of the cross-section at position $x$. The shear force, axial force, and bending moment for this section have been primed in the diagram. We know from Newton's third law that they should be equal and opposite to the corresponding unprimed quantities in part (b) of the diagram. We can thus choose for our computations either a left-hand or a right-hand free-body diagram. But this poses somewhat of a problem for us when we come to reporting the signs of the transmitted forces and couple moments at a section. We cannot use the direction of a force or couple moment at the section. Clearly, this would be inadequate since the sense of the force or couple moment at a section would depend on whether a left-hand or a right-hand free-body diagram was used. To associate an unambiguous sign for shear force, axial force, and bending moment at a section, we adopt the following convention:

*A force component for a section is positive if the area vector of the cross-section and the force component both have senses either in the positive or in the negative directions of any one or two of the references axes.*[12]

The same is true for the bending moment.

Thus, consider Fig. 6.30. For the left-hand free-body diagram, the area vector for section $x$ points in the positive $x$ direction. Note also that $H$, $V_y$, and the vectorial representation of $M_z$ also point in positive directions of the $xyz$ axes. Hence, according to our convention we have drawn a positive shear force, a positive axial force, and a positive bending moment at the section at $x$. For the right-hand free-body diagram, the cross-sectional area vector points in the negative $x$ direction. And, since $H'$, $V_y'$, and $M_z'$ point in negative directions of the $x$, $y$, and $z$ axes, these components are again positive for the section at $x$ according to our convention. Clearly, by employing this convention, we can easily and effectively specify the force system at a section without the danger of ambiguity.

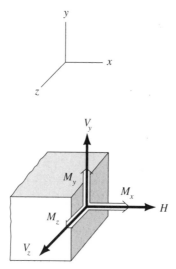

**Figure 6.31.** Section resultant for three-dimensional loading.

As pointed out earlier, we can solve for $V_y$, $H$, and $M_z$ at section $x$ using rigid-body mechanics for either a left-hand or a right-hand free-body diagram provided that we know all the external forces. The quantities $V_y$, $H$, and $M_z$ will depend on $x$, and for this reason, it is the practice to sketch shear-force and bending-moment diagrams to convey this information for the entire beam.

We now illustrate the computation of $V$ and $M$.

[12]Some authors employ the reverse convention for shear force from the one that we have proposed. Our convention is consistent with the usual convention used in the theory of elasticity for the sign of stress at a point, and it is for this reason that we have employed this convention rather than the other one.

■ **Example 6.6**

We shall express the shear-force and bending-moment equations for the simply supported beam shown in Fig. 6.32(a), whose weight we shall neglect. The support forces obtained from equilibrium are 500 N each.

To get the shear force at a section $x$, we isolate either the left or right side of the beam at $x$ and employ the equations of equilibrium on the resulting free body. If $x$ lies between $A$ and $C$ of the beam, the only noninternal force present for a left-hand free body is the left supporting force [see Fig. 6.32(b)]. Notice that we have used directions for $V$ and $M$ (there is no need for subscripts in the simple problem) corresponding to the *positive* states from the point of view of our convention. Clearly, the *algebraic* sign we get for these quantities from equilibrium calculations will then correspond to the *convention* sign. If $x$ is between $C$ and $B$ for such a free body, two external forces appear [see Fig. 6.32(c)]. Therefore, if the shear force is to be expressed as a function of $x$, clearly separate equations covering the two ranges, $0 < x < l/2$ and $l/2 < x < l$, are necessary. Summing forces we then get

$0 < x < l/2$:
$$500 + V = 0; \quad \text{therefore,} \quad V = -500 \text{ N} \quad\quad\text{(a)}$$

$l/2 < x < l$:
$$500 - 1{,}000 + V = 0; \quad \text{therefore,} \quad V = 500 \text{ N} \quad\quad\text{(b)}$$

Notice from the above results that there is a sudden change of the shear force from −500 N to +500 N as we pass the position of the concentrated 1,000-N load. Clearly then, the value of shear must perforce be *indeterminate* at the position of this concentrated load. It is for this very reason that, in the ranges of applicability of Eqs. (a) and (b), we have excluded the positions of the three concentrated loads of the problem. Note further that if there were only a distributed load starting at point $C$, there would not be discontinuity in shear and so we would not have to delete the position of point $C$ in the range applicability of the shear equations.

Now let us turn to the bending-moment equations. Again, we must consider two discrete regions. Taking moments about position $x$, we get

$0 \le x \le l/2$:
$$-500x + M = 0; \quad \text{therefore,} \quad M = 500x \text{ N-m} \quad\quad\text{(c)}$$

$l/2 \le x \le l$:
$$-500x + 1{,}000\left(x - \tfrac{l}{2}\right) + M = 0;$$
$$\text{therefore,} \quad M = 500(l - x) \text{ N-m} \quad\quad\text{(d)}$$

# Example 6.6 (Continued)

If there were a point couple present in the loading, there obviously would be a discontinuity in the bending-moment equation at the position of the point couple exactly as there were discontinuities at the point loads in the shear equations above. We would then not include such points in the bending-moment equations. In the present problem there are no point couples and so, in Eqs. (c) and (d), we include the entire beam in the combined ranges of these equations.

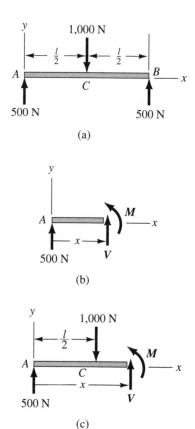

**Figure 6.32.** Simply supported beam.

It is generally the practice to express the shear and bending-moment equations successively under common ranges of applicability. In such circumstances, we shall adopt the practice of excluding from any range of applicability *any* points of discontinuity for *either* the shear or the bending-moment equations.

## Example 6.7

Determine the shear-force and bending-moment equations for the simply supported beam shown in Fig. 6.33. Neglect the weight of the beam.

We must first find the supporting forces for the beam. Hence, we have using the right-hand rule

$$\sum M_B = 0:$$

$$-R_1(22) + (50)(8)(14) + (1,000)(14) - 500 = 0$$

Therefore,

$$R_1 = 868 \text{ lb}$$

**Figure 6.33.** Simply supported beam.

$$\sum M_A = 0:$$

$$R_2(22) - 500 - (50)(8)(8) - (1,000)(8) = 0$$

Therefore,

$$R_2 = 532 \text{ lb}$$

In Fig. 6.34(a) we have shown a free-body diagram exposing sections between the left support and the uniform load. Summing forces and taking moments about a point in the section, where we have drawn $V$ and $M$ as positive according to our convention, we get

$\underline{0 < x \le 4:}$

$$868 + V = 0; \quad \text{therefore,} \quad V = -868 \text{ lb}$$
$$-868x + M = 0; \quad \text{therefore,} \quad M = 868x \text{ ft-lb}$$

The next interval is between the beginning of the uniform load and the point force. Thus, observing Fig. 6.34(b):

$\underline{4 \le x < 8:}$

for $V$,

$$868 - 50(x - 4) + V = 0$$
$$\therefore V = 50x - 1,068 \text{ lb}$$

and for $M$,

$$-868x + \frac{50(x - 4)^2}{2} + M = 0$$
$$\therefore M = -25x^2 + 1,068x - 400 \text{ ft-lb}$$

We now consider the interval between the point force and the end of the uniform load. Thus, observing Fig. 6.34(c):

(b)

**Figure 6.34.** Free-body diagrams for various ranges.

$\underline{8 < x \le 12:}$

for $V$,

$$868 - 50(x - 4) - 1,000 + V = 0$$
$$\therefore V = 50x - 68 \text{ lb}$$

## Example 6.7 (Continued)

and for $M$,

$$-868x + \frac{50(x-4)^2}{2} + 1{,}000(x-8) + M = 0$$
$$\therefore M = -25x^2 + 68x + 7{,}600 \text{ ft-lb}$$

The next interval is between the end of uniform loading and the point couple. We can now replace the uniform loading by its resultant of 400 lb, as shown in Fig. 6.34(d). Thus,

$\underline{12 \leq x < 18:}$

for $V$,

$$868 - 400 - 1{,}000 + V = 0$$
$$\therefore \quad V = 532 \text{ lb}$$

and for $M$,

$$-868x + 1{,}400(x-8) + M = 0$$
$$\therefore M = -532x + 11{,}200 \text{ ft-lb}$$

The last interval goes from the point couple to the right support. It is to be pointed out that the point couple *does not* contribute directly to the shear force and we could have used the above formulation for $V$ for interval $18 < x < 22$. However, the couple *does* contribute directly to the bending moment, thus requiring the additional interval. Accordingly, using Fig. 6.34(e), we get

$\underline{18 < x < 22:}$

$$V = 532 \text{ lb} \quad \text{(as in previous interval)}$$

Whereas for $M$ we have

$$-868x + 1{,}400(x-8) - 500 + M = 0$$
$$\therefore M = -532x + 11{,}700 \text{ ft-lb}$$

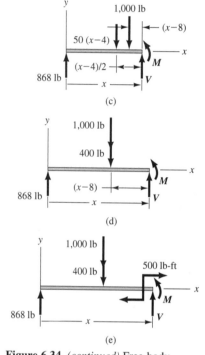

**Figure 6.34.** *(continued)* Free-body diagrams for various ranges.

We wish to point out now that we can determine shear-force and bending-moment equations in a less formal manner than what has been shown thus far. In this connection, it will be useful to note that from a consideration of equilibrium a downward force $P$, as shown in Fig. 6.35(a), induces on sections to the right a positive shear force (see insert) of value $+P$, whereas an upward force of $P$ induces on sections to the right of it a negative shear force $-P$ [see Fig. 6.35(b)}. Also, an upward force $P$ induces on sections at a distance $\xi$ to the right of it a positive bending moment $P\xi$ [see Fig. 6.36(a)], whereas a downward force $P$ induces on sections $\xi$ to the right of it a negative bending moment $-P\xi$ [Fig. 6.36(b)]. Finally, as can be seen in Fig. 6.37(a), a clockwise couple moment $C$ requires from equilibrium a positive bending moment $+C$ on the sections to the right of it (it does not require from equilibrium a shear force), whereas a counterclockwise couple moment $C$ [Fig. 6.37(b)] requires a negative bending moment $-C$ on sections to the right of it. In the following example, we shall show how by this reasoning we may more directly formulate the shear-force and bending-moment equations.

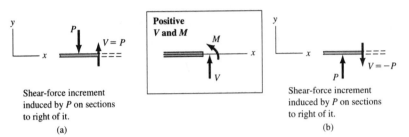

Shear-force increment induced by $P$ on sections to right of it.

(a)

Shear-force increment induced by $P$ on sections to right of it.

(b)

**Figure 6.35.** Shear induced by $P$.

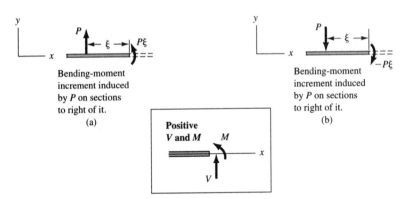

Bending-moment increment induced by $P$ on sections to right of it.

(a)

Bending-moment increment induced by $P$ on sections to right of it.

(b)

**Figure 6.36.** Bending moment induced by $P$.

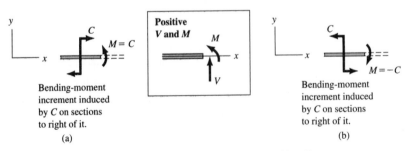

Bending-moment increment induced by $C$ on sections to right of it.

(a)

Bending-moment increment induced by $C$ on sections to right of it.

(b)

**Figure 6.37.** Bending moment induced by $C$.

# Example 6.8

Evaluate the shear-force and bending-moment equations for the beam shown in Fig. 6.38.

A free-body diagram of the beam is shown in Fig. 6.39. We can immediately compute the supporting forces as follows, remembering to use the right-hand rule.

$$\sum M_2 = 0:$$

$$-R_1(26) + (500)(21) - 800 + (500)(5) = 0$$

Therefore,

$$R_1 = 469 \text{ lb}$$

$$\sum M_1 = 0:$$

$$R_2(26) - (500)(21) - 800 - (500)(5) = 0$$

Therefore,

$$R_2 = 531 \text{ lb}$$

We shall now directly give the shear force $V$ and bending moment $M$ while viewing Fig. 6.39. Thus,

$\underline{0 < x < 5:}$

$$V = -469 \text{ lb}$$

$$M = 469x \text{ ft-lb}$$

$\underline{5 < x < 13:}$

$$V = -469 + 500 = 31 \text{ lb}$$

$$M = 469x - 500(x - 5) = -31x + 2,500 \text{ ft-lb}$$

$\underline{13 < x \le 16:}$

$$V = 31 \text{ lb} \quad \text{(same as previous interval)}$$

$$M = 469x - 500(x - 5) + 800 = -31x + 3,300 \text{ ft-lb}$$

$\underline{16 \le x < 26:}$

$$V = -469 + 500 + 50(x - 16) = -769 + 50x \text{ lb}$$

$$M = 469x - 500(x - 5) + 800 - \frac{50(x - 16)^2}{2}$$

$$= -25x^2 + 769x - 3,100 \text{ ft-lb}$$

We shall present effective methods of sketching the shear-force and bending-moment diagrams in Section 6.9.

**Figure 6.38.** Simply supported beam.

**Figure 6.39.** Free-body diagram of beam.

Before we proceed further, it must be carefully pointed out that the replacement of a distributed load by a single resultant force is only meaningful for the particular free body on which the force distribution acts. Thus, to compute the reactions for the entire beam taken as a free body (Fig. 6.40), we can

**Figure 6.40.** Uniform loading.

replace the weight distribution $w_0$ by the total weight at position $L/2$ (Fig. 6.41). For the bending moment at $x$, the resultant of the loading for the free

**Figure 6.41.** Resultant for $w_0$ for entire beam.

body shown in Fig. 6.42 becomes $w_0 x$ and is midway at position $x/2$. In other words, *in making shear-force and bending-moment equations and diagrams, we cannot replace loading distributions over the entire beam by a resultant and then proceed*; there is inherent in these equations an infinite number of free bodies, each shorter than the beam itself, which makes the above-mentioned replacements invalid for shear-force and bending-moment considerations.

**Figure 6.42.** Resultant for $w_0$ for portion $x$ of beam.

# PROBLEMS

*In Problems 6.37 through 6.48 make use of free-body diagrams.*

**6.37.** Formulate the shear-force and bending-moment equations for the simply supported beam. Do not include the weight of the beam.

**Figure P.6.37.**

**6.38.** Formulate the shear-force and bending-moment equations for the cantilever beam. Do not include the weight of the beam.

**Figure P.6.38.**

**6.39.** Determine the shear-force and bending-moment equations for the simply supported beam.

**Figure P.6.39.**

**6.40.** For the beam shown, what is the shear force and bending-moment at the following positions?

(a) 5 ft from the left end

(b) 12 ft from the left end

(c) 5 ft from the right end

**Figure P.6.40.**

**6.41.** Formulate the shear-force and bending-moment equations for the simply supported beam.

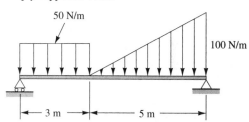

**Figure P.6.41.**

**6.42.** Compute shear force and bending moments for the bent beam as functions of $s$ along the centerline of the beam.

**Figure P.6.42.**

**6.43.** A simply supported beam is loaded in two planes. This means there will be shear-force components $V_y$ and $V_z$ and bending-moment components $M_z$ and $M_y$. Compute these as functions of $x$. The beam is 40 ft in length.

**Figure P.6.43.**

257

**6.44.** What are the shear force, bending moment, and axial force for the three-dimensional cantilever beam? Give your results separately for the three portions *AB*, *BC*, and *CD*. Neglect the weight of the member. Use *s* as the distance along the centerline from *D*.

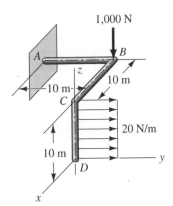

**Figure P.6.44.**

**6.45.** Oil flows from a tank through a pipe *AB*. The oil weighs 40 lb/ft³ and, in flowing, develops a drag on the pipe of 1 lb/ft. The pipe has an inside diameter of 3 in. and a length of 20 ft. Flow conditions are assumed to be the same along the entire length of the pipe. What are the shear force, bending moment, and axial force along the pipe? The pipe weighs 10 lb/ft.

**Figure P.6.45.**

**6.46.** Determine the shear force, bending moment, and axial force as functions of *θ* for the circular beam.

**Figure P.6.46.**

**6.47.** A hoist can move along a beam while supporting a 10,000-lb load. If the hoist starts at the left and moves from $\bar{x} = 3$ to $\bar{x} = 12$, determine the shear force and bending moment at *A* in terms of $\bar{x}$. At what position $\bar{x}$ do we get the maximum shear force at *A* and the maximum bending moment at *A*? What are their values?

**Figure P.6.47.**

**6.48.** A pipe weighs 10 lb/ft and has an inside diameter of 2 in. If it is full of water and the pressure of the water is that of the atmosphere at the entrance *A*, compute the shear force, axial force, and bending moment of the pipe from *A* to *D*. Use coordinate *s* measured from *A* along the centerline of the pipe.

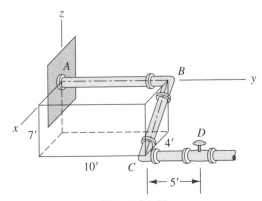

**Figure P.6.48.**

**6.49.** After finding the supporting forces, determine for Problem 6.37 the shear-force and bending-moment equations without the further aid of free-body diagrams.

**6.50.** Determine the shear-force and bending-moment equations for Problem 6.38 without the aid of free-body diagrams.

**6.51.** In Problem 6.39, after determining the supporting forces, determine the shear-force and bending-moment equations without the aid of free-body diagrams.

**6.52.** In Problem 6.40, after finding the supporting forces, write the shear force and bending moment as a function of $x$ for the beam without the aid of free-body diagrams.

**6.53.** Give the shear-force and bending-moment equations for the cantilever beam. Except for determining the supporting forces, do not use free-body diagrams.

**6.54.** Formulate the shear-force and bending-moment equations for the simply supported beam. [*Suggestion:* For the domain $5 < x < 15$, it is simplest to replace the indicated downward triangular load, going from 400 N/m to zero, by a uniform 400-N/m uniform downward load from $x = 5$ to $x = 15$ plus a triangular upward load going from zero to 400 N/m in the interval.]

**Figure P.6.53.**

**Figure P.6.54.**

## 6.9 Differential Relations for Equilibrium

In Section 6.8, we considered free bodies of *finite size* comprising variable portions of a beam in order to ascertain the resultant force system at sections along the beam. We shall now proceed in a different manner by examining an *infinitesimal slice* of the beam. Equations of equilibrium for this slice will then yield *differential equations* rather than algebraic equations for the variables $V$ and $M$.

Consider a slice $\Delta x$ of the beam shown in Fig. 6.43. We adopt the convention that intensity of loading $w$ in the positive coordinate direction is positive. We shall assume here that the weight of the beam has been included in the intensity of loading so that all forces acting on the element have been shown on the free-body diagram of the element in Fig. 6.44. Note we have employed positive shear-force and bending-moment convention as presented in Section 6.9. We now apply the equations of equilibrium. Thus, summing forces:

$$\underline{\sum F_y = 0:}$$

$$-V + (V + \Delta V) + w\,\Delta x = 0$$

Taking moments about corner $a$ of the element, we get

$$\underline{\sum M_a = 0:}$$

$$-M + V\Delta x - (w\Delta x)(\beta\Delta x) + (M + \Delta M) = 0$$

**Figure 6.43.** Element $\Delta x$ of beam.

**Figure 6.44.** Free-body diagram of element.

259

where $\beta$ is some fraction which, when multiplied by $\Delta x$, gives the proper moment arm of the force $w \, \Delta x$ about corner $a$. These equations can be written in the following manner after we cancel terms and divide through by $\Delta x$:

$$\frac{\Delta V}{\Delta x} = -w$$

$$\frac{\Delta M}{\Delta x} = -V + w\beta \, \Delta x$$

In the limit as $\Delta x \to 0$, we get the following differential equations:

$$\frac{dV}{dx} = -w \tag{6.3a}$$

$$\frac{dM}{dx} = -V \tag{6.3b}$$

We may next integrate Eqs. 6.3(a) and 6.3(b) from position 1 along the beam to position 2. Thus, we have

$$(V)_2 - (V)_1 = -\int_1^2 w \, dx$$

Therefore,

$$(V)_2 = (V)_1 - \int_1^2 w \, dx \tag{6.4}$$

and

$$(M)_2 - (M)_1 = -\int_1^2 V \, dx$$

Therefore,

$$(M)_2 = (M)_1 - \int_1^2 V \, dx \tag{6.5}$$

Equation (6.4) means that the change in the shear force between two points on a beam equals minus the area under the loading curve between these points provided that there is no point force present in the interval.[13] Note that, if $w(x)$ is positive in an interval, the area under this curve is positive in this interval; if $w(x)$ is negative in an interval, the area under this curve is negative in this interval. Similarly, Eq. 6.5 indicates that the change in bending moment between two points on a beam equals minus the area of the shear-force diagram between these points provided that there are no point couple moments applied in the interval. If $V(x)$ is positive in an interval, the area under this curve is positive in this interval; if $V(x)$ is negative in an interval, the area under the curve is negative for this interval. In sketching the diagram, we shall make use of Eq. 6.4 and 6.5 as well as the differential equations 6.3.

---

[13]The differential equation 6.3(a) is only meaningful with a continuous loading present, while Eq. 6.3(b) is only valid in the absence of point couple moments.

# ■ Example 6.9 ■

Sketch the shear-force and bending-moment distributions for the simply supported beam shown in Fig. 6.45 and label the key points.

**Figure 6.45.** Simply supported beam.

The supporting forces $R_1$ and $R_2$ are found by rules of statics. Thus,

$$\sum M_B = 0:$$
$$-R_1(20) + (500)(14) + (50)(10)(10/2) - 100 = 0$$

Therefore,

$$R_1 = 470 \text{ lb}$$

$$\sum M_A = 0:$$
$$R_2(20) - (500)(6) - (50)(10)(15) - 100 = 0$$

Therefore,

$$R_2 = 530 \text{ lb}$$

In sketching the diagrams, we shall employ Eqs. 6.3, 6.4, and 6.5—i.e., the differential equations of equilibrium and their integrals. Accordingly, we first draw the loading diagram in Fig. 6.46(a), and we shall then sketch the shear-force and bending-moment diagrams directly below without the aid of the shear-force and bending-moment equations, evaluating key points as we go.

Note as we start on the shear diagram that the 470-lb supporting force induces a negative shear of −470 lb just to the right of the support. Now from $A$ to $C$, the area under the loading curve is zero and so, in accordance with Eq. 6.4, there is no change in the value of shear between $A$ and $C$. Hence, $V_C = -470$ lb, as shown in Fig 6.46(b). Also, since $w = 0$ between $A$ and $C$, the slope of the shear curve should be zero, in accordance with Eq. 6.3(a). And so we have a horizontal line for $V$ between $A$ and $C$. Now as we cross $C$, the 500-lb downward force will induce a positive increment of shear of value 500 on sections to the right of it. Accordingly, $V$ *jumps* from −470 lb to +30 lb as we cross $C$. Between $C$ and $D$ there is no loading $w$, so $V_D = V_C$ and we have a 30-lb shear force at point $D$. Again, since $w = 0$ in this interval, the slope of the shear curve is zero and we have a horizontal line for the shear curve between $C$ and $D$. Since there is no concentrated

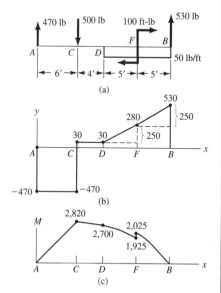

**Figure 6.46.** Shear-force and bending-moment diagrams.

## Example 6.9 (Continued)

load at $D$, there is no sudden change in shear as we cross this point. Next, the change in shear between $D$ and $B$ is minus the area of the loading curve[14] in this interval in accordance with Eq. 6.4. But this area is $(-50)(10) = -500$. Hence, from Eq. 6.4 the value of $V_B$ (just to the left of the support) is $V_D - (-500) = 530$ lb. Also, since $w$ is negative and constant between $D$ and $B$, the slope of the shear curve should be positive and constant, in accordance with Eq. 6.3(a). Hence, we can draw a straight line between $V_D = 30$ lb and $V_B = 530$ lb. As we now cross the right support force, we see that it induces a negative shear of 530 lb on sections to the right of the support, and so at $B$ the shear curve comes back to zero.

We now proceed with the bending-moment curve. With no point couple moment present at $A$, the value of $M_A$ must be zero. The change in moment between $A$ and $C$ is then minus the area underneath the shear curve in this interval. We can then say from Eq. 6.5 that $M_C = M_A - (-470)(6) = 0 + 2,820 = 2,820$ ft-lb, and we denote this in the moment diagram. Furthermore the value of $V$ is a negative constant in the interval and, accordingly [see Eq. 6.3(b)], the slope of the moment curve is positive and constant. We can then draw a straight line between $M_A$ and $M_C$. Between $C$ and $D$ the area for the shear diagram is 120 lb-ft, and so we can say that $M_D = M_C - (120) = 2,820 - 120 = 2,700$ ft-lb. Again, with $V$ constant and positive in the interval, the slope of the moment curve must be negative and constant in the interval and has been so drawn. Between $D$ and $F$ the area under the shear curve is readily seen to be $(30)(5) + \frac{1}{2}(5)(250) = 775$ ft-lb. Hence, the bending moment goes from 2,700 ft-lb at $D$ to 1,925 ft-lb at $F$. Now the shear curve is positive and *increasing* in value as we go from $D$ to $F$. This means that the slope of the bending-moment curve is negative and becoming *steeper* as we go from $D$ to $F$. As we go by $F$ we encounter the 100-ft-lb point couple moment and we can say that this point couple moment induces a positive 100-ft-lb moment on sections to the right of point $F$. Accordingly, there is a sudden increase in bending moment of 100 ft-lb at $F$, as has been shown in the diagram. The area of the shear diagram between $F$ and $B$ is readily seen from Fig. 6.46(b) to be $(280)(5) + \frac{1}{2}(5)(250) = 2,025$ ft-lb. We see then that the bending moment goes to zero at $B$. Since the shear force is positive and *increasing* between $F$ and $B$, we conclude that the slope of the bending-moment curve is negative and becoming *steeper* as we approach $B$. We have thus drawn the shear-force and bending-moment diagram and have labeled all key points.

Note that to be correct, both the shear-force and bending-moment curves must go to zero at the end of the beam to the right of the right support. This serves as a check on the correctness of the calculations.

---

[14]Note that the point couple moment has a zero net force and so need not be of concern in the interval from $D$ to $B$ as far as shear is concerned. However, it will be a point where sudden change occurs in the bending-moment diagram.

In Example 6.9, we can get equations and diagrams of shear force and bending moment independently of each other. With simple loadings such as point forces, point couples, and uniform distributions, this can readily be done. Indeed, this covers many problems that occur in practice. Usually, all that is needed are the labeled diagrams of the kind that we set forth in the previous problem. In problems with more complex loadings, we usually set forth the equations in the customary manner and then sketch the curves using the *equations* to give key values of $V$ and $M$ (the areas for the various curves are no longer the simple familiar ones, thus precluding advantageous use of Eqs. 6.4 and 6.5); the key points are then connected by curves sketched by making use of the slope relations as in Example 6.9.

It will be helpful to remember that if a curve has *increasing* magnitude (absolute value), the subsequent curve must have a *steepening* slope over the corresponding range. On the other hand, if a curve has *decreasing* magnitude (absolutely value), the subsequent curve must have a *flattening* slope over the corresponding range.

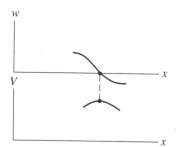

**Figure 6.47.** At $w = 0$, possible maximum for $V$.

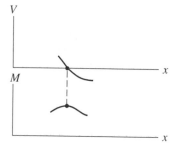

**Figure 6.48.** At $V = 0$, possible maximum for $M$.

You will note in the preceding examples that the key points of the shear-force and bending-moment diagrams were evaluated and marked. The maximum value of both the shear force and bending moment were easily depicted from these diagrams. We wish to note, in this regard, that at points on shear-force and bending-moment curves where there is zero value of slope, there may be possible maximum values of shear force and bending moment, respectively, for the beam. This is illustrated for shear force in Fig. 6.47 and for bending moment in Fig. 6.48. Note that where the loading curve $w$ crosses the $x$ axis, we accordingly have the position of a possible maximum shear force; similarly, where the shear curve $V$ crosses the $x$ axis, we accordingly have the position of a possible maximum bending moment. These respective positions and corresponding values of shear force and bending moment should be evaluated and marked in the diagram. Note however, that these are *local maximums* and that elsewhere there may be shears or bending moments exceeding these local maxima.

# PROBLEMS

**6.55.** After finding the supporting forces of the cantilever beam, sketch the shear-force and bending-moment diagrams labeling key points.

**Figure P.6.55.**

**6.56.** What is the maximum negative bending moment in the region between the supports for the simply supported beam?

**Figure P.6.56.**

**6.57.** Find the supporting forces for the simply supported beam. Then sketch the shear-force and bending-moment diagrams, labeling key points.

**Figure P.6.57.**

**6.58.** Sketch the shear-force and bending-moment diagrams and compute key points for the overhanging beam.

**Figure P.6.58.**

**6.59.** A simply supported beam *AB* is shown. A bar *CD* is welded to the beam. After determining the supporting forces, sketch the shear-force and bending-moment diagram and determine the maximum bending moment. [*Hint:* Find the position for *V* = 0 using similar triangles.]

**Figure P.6.59.**

**6.60.** For the cantilever beam, sketch the shear-force and bending-moment diagrams evaluating key points only.

**Figure P.6.60.**

**6.61.** Sketch the shear-force and bending-moment diagrams for the sinusoidally loaded beam. What is the maximum bending moment?

Figure P.6.61.

**6.62.** Formulate the shear-force and bending-moment equations for the beam. Sketch the shear and moment diagrams.

Figure P.6.62.

**6.63.** A simply supported I-beam is shown. A hole must be cut through the web to allow passage of a pipe that runs horizontally at right angles to the beam.

(a) Where, within the marked 24-ft section, would the hole least affect the moment-carrying capacity of the beam?

(b) In the same marked section, where should the hole go to least affect the shear-carrying capacity of the beam?

Figure P.6.63.

**6.64.** A cantilever beam supports a parabolic and a triangular load. What are the shear-force and bending-moment equations? Sketch the shear-force and bending-moment diagrams. See the suggestion in Problem 6.54 regarding the triangular load.

Figure P.6.64.

**6.65.** Determine the shear-force and bending-moment equations for the beam. Then sketch the diagrams using the aforementioned equations if necessary to ascertain key points in the diagrams, such as the position between the supports where $V = 0$. What is the bending moment there?

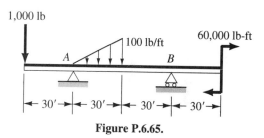

Figure P.6.65.

265

# Part C: Chains and Cables

## 6.10    Introduction

We often encounter relatively flexible cables or chains that are used to support loads. In suspension bridges, for example, we find a coplanar arrangement in which a cable supports a large load. The weight of the cable itself in such cases may often be considered negligible. In transmission lines, on the other hand, the principal force is the weight of the cable itself. In Part C, we shall evaluate the shape of and the tension in the cables for both these cases.

To facilitate computations, the model of the structural system will be assumed to be perfectly flexible and inextensible. The flexibility assumption means that at the center of any cross section of the cable only a tensile force is transmitted and there can be no bending moment there. The force transmitted through the cable must, under these conditions, be tangent to the cable at all positions along the cable. The inextensibility assumption means that the length of the cable is constant.

## 6.11    Coplanar Cables; Loading is a Function of $x$

We shall now consider the case of a cable suspended between two rigid supports $A$ and $B$ under the action of a loading function $w(x)$ given per unit length as measured in the *horizontal* direction. This loading will be considered to be coplanar with the cable and directed vertically, as shown in Fig. 6.49. Consider an element of the cable of length $\Delta s$ as a free body (Fig. 6.50).

**Figure 6.49.** Coplanar cable; $w = w(x)$.

**Figure 6.50.** Element of cable.

Summing forces in the $x$ and $y$ directions, respectively, we get

$$-T \cos \theta + (T + \Delta T) \cos(\theta + \Delta \theta) = 0 \qquad (6.6a)$$

$$-T \sin \theta + (T + \Delta T) \sin (\theta + \Delta \theta) - w_{av} \Delta x = 0 \qquad (6.6b)$$

where $w_{av}$ is the average loading over the interval $\Delta x$. Dividing by $\Delta x$ and taking the limit as $\Delta x \to 0$, we have

$$\lim_{\Delta x \to 0}\left[ \frac{(T + \Delta T)\cos(\theta + \Delta\theta) - T\cos\theta}{\Delta x} \right] = 0$$

$$\lim_{\Delta x \to 0}\left[ \frac{(T + \Delta T)\sin(\theta + \Delta\theta) - T\sin\theta}{\Delta x} \right] = w$$

The term $w$ is now the loading at position $x$. The left sides of the equations above are derivatives in accordance with elementary calculus, and so we can say for these equations:

$$\frac{d(T\cos\theta)}{dx} = 0 \tag{6.7a}$$

$$\frac{d(T\sin\theta)}{dx} = w \tag{6.7b}$$

From Eq. 6.7(a), we conclude that

$$T\cos\theta = \text{constant} = H \tag{6.8}$$

where clearly the constant $H$ represents the *horizontal* component of the tensile force anywhere along the cable. Integrating Eq. 6.7(b), we get

$$T\sin\theta = \int w(x)\,dx + C_1' \tag{6.9}$$

where $C_1'$ is a constant of integration. Solving for $T$ in Eq. 6.8 and substituting into Eq. 6.9, we get

$$\frac{\sin\theta}{\cos\theta} = \frac{1}{H}\int w(x)\,dx + C_1$$

Noting that $\sin\theta/\cos\theta = \tan\theta = dy/dx$, we have, on carrying out a second integration:

$$y = \frac{1}{H}\int\left[\int w(x)\,dx\right]dx + C_1 x + C_2 \tag{6.10}$$

Equation 6.10 is the deflection curve for the cable in terms of $H$, $w(x)$, and the constants of integration. The constants of integration must be determined by the boundary conditions at the supports $A$ and $B$.

## Example 6.10

A cable is shown in Fig. 6.51 terminating at points at the same elevation. The loading distribution is uniform, given by constant $w$. Other known data are the span, $l$, and the sag, $h$. The maximum force in the cable, the shape of the cable, and the length of the cable are desired. Neglect the weight of the cable itself.

We have placed the reference at the center of the cable for simplicity as shown in the diagram. Noting that $w(x) = w = $ constant for this problem, we can proceed directly with the integrations in Eq. 6.10. Thus, we have

$$y = \frac{1}{H}\int\left[\int w\, dx\right]dx + C_1 x + C_2 = \frac{1}{H}\int wx\, dx + C_1 x + C_2$$

**Figure 6.51.** Cable with sag $h$.

Therefore,

$$y = \frac{1}{H}\frac{wx^2}{2} + C_1 x + C_2 \tag{a}$$

The deflection curve is thus a *parabola*. We now require that $y = dy/dx = 0$, when $x = 0$. Thus, the constants $C_1 = C_2 = 0$. The deflection curve then is simply

$$y = \frac{w}{2H}x^2 \tag{b}$$

To get the constant $H$, we set $y = h$ for $x = l/2$. Thus,

$$h = \frac{w}{2H}\frac{l^2}{4}$$

Therefore,

$$H = \frac{wl^2}{8h} \tag{c}$$

The deflection curve is now fully established in terms of the data of the problem in the form

$$y = \frac{w}{2(wl^2/8h)}x^2 = 4\frac{hx^2}{l^2} \tag{d}$$

$$\therefore\quad y = 4\frac{hx^2}{l^2}$$

We next compute the *maximum tension* in the cable. Equation 6.8 can be used for this purpose. Solving for $T$, we get

$$T = \frac{H}{\cos\theta} \tag{e}$$

from which it is apparent that the maximum value of $T$ occurs where $\theta$ is greatest. Examining the slope of the deflection curve using Eq. (b),

$$\frac{dy}{dx} = \frac{w}{H}x \tag{f}$$

## Example 6.10 (Continued)

it is apparent that the largest $\theta$ occurs at $x = l/2$ (i.e., at the supports). Hence, from above we have, for $\theta_{max}$:

$$\theta_{max} = \tan^{-1}\left(\frac{dy}{dx}\right)_{x=l/2} = \tan^{-1}\left(\frac{w}{H}\frac{l}{2}\right) \qquad (g)$$

Consequently, we get for $T_{max}$:

$$T_{max} = \frac{H}{\cos\left[\tan^{-1}(wl/2H)\right]} \qquad (h)$$

From trigonometric consideration of the denominator,

$$T_{max} = \frac{H(4H^2 + w^2l^2)^{1/2}}{2H} = H\left[1 + \left(\frac{wl}{2H}\right)^2\right]^{1/2} \qquad (i)$$

Substituting for $H$ using Eq. (c), we then get, on rearranging the terms,

$$T_{max} = \frac{wl}{2}\sqrt{1 + \left(\frac{l}{4h}\right)^2} \qquad (j)$$

Finally, to determine the *length of the cable* for the given conditions, we must perform the following integration:

$$L = 2\int_0^{s_{max}} ds = 2\int_0^{s_{max}} \sqrt{dx^2 + dy^2} = 2\int_0^{l/2}\sqrt{1 + \left(\frac{dy}{dx}\right)^2}\,dx \qquad (k)$$

Now the slope, $dy/dx$, equals $wx/H$ [see Eq. (f)], which on substituting for $H$ [see Eq. (c)] becomes $8hx/l^2$. Therefore,

$$L = 2\int_0^{l/2}\sqrt{1 + \left(\frac{8hx}{l^2}\right)^2}\,dx$$

This may be integrated using a formula to be found in Appendix I to give

$$L = \left[x\sqrt{1 + \left(\frac{8hx}{l^2}\right)^2} + \frac{l^2}{8h}\sinh^{-1}\frac{8hx}{l^2}\right]_0^{l/2}$$

Substituting limits, we have

$$L = \left[\frac{l}{2}\sqrt{1 + \left(\frac{4h}{l}\right)^2} + \frac{l^2}{8h}\sinh^{-1}\frac{4h}{l}\right]$$

Rearranging so that the result is given as a function of the sag ratio $h/l$ and the span $l$, we get finally

### ▪ Example 6.10 (Continued)

$$L = \frac{l}{2}\left[\sqrt{1 + 16\left(\frac{h}{l}\right)^2} + \frac{1}{4h/l}\sinh^{-1}\frac{4h}{l}\right] \tag{l}$$

Another possible approach to determining the length of the cable is to expand the integrand in Eq. (k) as a power series using the *binomial theorem*. Thus we have

$$L = 2\int_0^{l/2}\left[1 + \frac{1}{2}\left(\frac{dy}{dx}\right)^2 - \frac{1}{8}\left(\frac{dy}{dx}\right)^4 + \dots\right]dx \tag{m}$$

provided that $|dy/dx| < 1$ at all positions along the interval.[15] Now, employ Eq. (f) and Eq. (c) to replace $dy/dx$ in Eq. (m) to get

$$L = 2\int_0^{l/2}\left(1 + \frac{1}{2}\left(\frac{8h}{l^2}\right)^2 x^2 - \frac{1}{8}\left(\frac{8h}{l^2}\right)^4 x^4 + \dots\right)dx \tag{n}$$

We can integrate a power series term by term and so we have, for $L$:

$$L = l\left[1 + \frac{8}{3}\left(\frac{h}{l}\right)^2 - \frac{32}{5}\left(\frac{h}{l}\right)^4 + \dots\right] \tag{o}$$

For cables having small slopes (i.e., small sag ratio h/l), the series converges rapidly and only the first few terms need generally be employed.

---

[15]Otherwise, the series diverges. Hence, this approach is limited to cases where the slope of the cable is less than 45°.

In Example 6.10, the supports are at the same level and consequently the position of zero slope is known (i.e., it is at the midpoint). We found it simplest to set our reference $xy$ at this point. In problems at the end of this section, the supports may not be at the same level. For such cases, the reference is best taken at one of the supports. Also, the slope of the cable is often known at some point, and the problem may then be solved in much the same way as Example 6.10.

## 6.12  Coplanar Cables; Loading is the Weight of the Cable Itself

In the previous development, the loading was given as a function of $x$. Let us now consider the case of a cable loaded only by its own weight. The loading function is now most easily expressed as a function of $s$, the position along the cable. Equations 6.6 apply to this case provided that we replace $\Delta x$ by $\Delta s$.

Dividing through by $\Delta s$ and taking the limit as $\Delta s \to$ zero, we get equations analogous to Eqs. 6.7.

$$\frac{d(T\cos\theta)}{ds} = 0$$

$$\frac{d(T\sin\theta)}{ds} = w(s)$$

Integrating, we have

$$T\cos\theta = H \tag{6.11a}$$

$$T\sin\theta = \int w(s)\,ds + C_1' \tag{6.11b}$$

Eliminating $T$ from Eqs. 6.11, we get, as in the previous development:

$$\frac{dy}{dx} = \frac{1}{H}\int w(s)\,ds + C_1$$

The right side of the equation is a function of $s$. Thus, we cannot directly integrate as a next step. Accordingly, note that

$$dy = (ds^2 - dx^2)^{1/2}$$

Hence, from this equation,

$$\frac{dy}{dx} = \left[\left(\frac{ds}{dx}\right)^2 - 1\right]^{1/2} \tag{6.13}$$

Substituting for $dy/dx$ in Eq. 6.12 using the preceding result, we get

$$\left[\left(\frac{ds}{dx}\right)^2 - 1\right]^{1/2} = \frac{1}{H}\int w(s)\,ds + C_1$$

Solving for $ds/dx$, we have

$$\frac{ds}{dx} = \left\{1 + \left[\frac{1}{H}\int w(s)\,ds + C_1\right]^2\right\}^{1/2}$$

Separating variables and integrating, we get

$$x = \int \frac{ds}{\left\{1 + \left[\frac{1}{H}\int w(s)\,ds + C_1\right]^2\right\}^{1/2}} + C_2 \tag{6.14}$$

As a first step, if possible determine the constant $C_1$ by applying a slope-boundary condition to Eq. 6.12. With this $C_1$ in Eq. 6.14, solve for $s$ as a function of $x$. Next, substitute for $s$ in Eq. 6.12 using this relation. Finally, integrate Eq. 6.12 with respect to $x$ to get $y$ as a function of $x$. Boundary conditions must then be used to determine $H$ as well as the remaining constant of the integration. The following examples will illustrate how these steps are carried out.

# Example 6.11

Consider a uniform cable having a span $l$ and a sag $h$ as shown in Fig. 6.52. The weight per unit length $w$ of the cable is a constant. Determine the shape of the cable when it is loaded only by its own weight.

**Figure 6.52.** Uniform cable loaded by its own weight.

For simplicity, we have placed a reference at the center of the span where the slope of the cable is zero. Accordingly, consider Eq. 6.12 for this case:

$$\frac{dy}{dx} = \frac{1}{H}\int w(s)\,ds + C_1 = \frac{w}{H}s + C_1 \tag{a}$$

When $s = 0$ we require that $dy/dx = 0$, whereupon $C_1$ is zero. Now consider Eq. 6.14:

$$x = \int \frac{ds}{\left\{1 + \left[(1/H)\int w\,ds + C_1\right]^2\right\}^{1/2}} + C_2$$

$$= \int \frac{ds}{\left\{1 + \left[(w/H)s\right]^2\right\}^{1/2}} + C_2 \tag{b}$$

Integrating the right side of the equation using integration formula 10 from Appendix I, we get

$$x = \frac{H}{w}\sinh^{-1}\frac{sw}{H} + C_2 \tag{c}$$

## Example 6.11 (Continued)

The constant $C_2$ must also be zero, since $x = 0$ at $s = 0$. Solving for $s$ from Eq. (c), we get

$$s = \frac{H}{w} \sinh \frac{xw}{H} \qquad (d)$$

Substituting for $s$ in Eq. (a) using the preceding result, we have

$$\frac{dy}{dx} = \sinh \frac{w}{H} x \qquad (e)$$

Integrating, we get

$$y = \frac{H}{w} \cosh \frac{w}{H} x + C_3$$

Since $y = 0$ at $x = 0$, the constant $C_3$ becomes $-H/w$. We then have the deflection curve:

$$y = \frac{H}{w} \left( \cosh \frac{w}{H} x - 1 \right) \qquad (f)$$

This curve is called a *catenary curve*.[16]

To determine $H$, we set $y = h$ when $x = l/2$. Thus,

$$h = \frac{H}{w} \left( \cosh \frac{wl}{2H} - 1 \right) \qquad (g)$$

This equation can be solved by trial and error by the student or by a computer. We may then proceed to determine the maximum force in the cable as well as the length of the cable in the manner followed in Example 6.10.

---

[16]The Latin for chain is *catena*.

# Example 6.12

A water skier is shown in Fig. 6.53 dangling from a kite that is towed by a powerboat at a speed of 30 mph. The boat develops a thrust of 200 lb. The drag on the boat from the water is estimated as 100 lb. At the support $A$, the rope has a tangent of 30°. If the man weighs 150 lb, find the height and the lift of the kite as well as the maximum tension in the rope. The kite weighs 25 lb. The uniform rope is 50 ft long and weighs .5 lb/ft. Neglect aerodynamic effects on the rope.

**Figure 6.53.** Analyze tow rope $AB$.

We start with Eq. 6.12, which becomes for this case:

$$\frac{dy}{dx} = \frac{w}{H}s + C_1 \tag{a}$$

Using a reference at $A$ as shown in the diagram, we know that $dy/dx = \tan 30° = .577$ when $s = 0$. Thus, we get for $C_1$,

$$C_1 = .577$$

Equation 6.14 is considered next. We have

$$x = \int \frac{ds}{\left\{1 + \left[(w/H)s + .577\right]^2\right\}^{1/2}} + C_2$$

Integrating by making a change in variable to replace $[(w/H)s + .577]$ and using the integration formula 10 in Appendix I, we get

$$x = \frac{H}{w}\sinh^{-1}\left(\frac{w}{H}s + .577\right) + C_2 \tag{b}$$

Solving for $s$, we have

$$s = \frac{H}{w}\left\{\sinh\left[(x - C_2)\frac{w}{H}\right] - .577\right\} \tag{c}$$

Substituting for $s$ in Eq. (a) using Eq. (c), we get

$$\frac{dy}{dx} = \sinh\left[(x - C_2)\frac{w}{H}\right] \tag{d}$$

Integrating again, we have

$$y = \frac{H}{w}\cosh\left[(x - C_2)\frac{w}{H}\right] + C_3 \tag{e}$$

We must now evaluate the unknown constants $C_2$, $C_3$, and $H$ using the boundary conditions and data of the problem. First, since $H$ is the horizontal component of force transmitted by the rope, we know that $H$ is the thrust of the boat minus the drag of the water. Thus,

$$H = 100 \text{ lb}$$

Also, $x = 0$ when $s = 0$, so that from Eq. (c), we get $C_2$ as follows:

$$\sinh\left(-\frac{.5}{100}C_2\right) = .577$$

## Example 6.12 (Continued)

Therefore,

$$-\frac{.5}{100} C_2 = \sinh^{-1} .577 = .549$$

Hence,

$$C_2 = -109.8$$

Finally, note that $x = 0$ when $y = 0$. From Eq. (e), we can then get constant $C_3$ in the following manner:

$$C_3 = -\frac{100}{.5} \cosh\left[\frac{-.5}{100}(-109.8)\right]$$

$$= -200 \cosh .548 = -231$$

We may now evaluate the position $x'$, $y'$ of point $B$ of the kite. To get $x'$, we insert for $s$ in Eq. (b) the value of 50 ft. Thus,

$$x' = \frac{100}{.5} \sinh^{-1}\left[\frac{.5}{100} 50 + .577\right] - 109.8 = 40.9 \text{ ft}$$

Now from Eq. (e) we can get $y'$ and consequently the desired height.

$$y' = \frac{100}{.5} \cosh\left[(40.9 + 109.8)\frac{.5}{100}\right] - 231$$

$$\boxed{y' = 28.6 \text{ ft}} \tag{f}$$

The maximum tension in the rope occurs at point $B$, where $\theta$ is greatest. To get $\theta_{max}$, we go back to Eq. (a). Thus,

$$\left(\frac{dy}{dx}\right)_{max} = \tan \theta_{max} = \frac{.5}{100}(50) + .577 = .827$$

Therefore,

$$\theta_{max} = 39.6° \tag{g}$$

Hence, from Eq. 6.11(a) we have for $T_{max}$:

$$T_{max} = \frac{100}{\cos 39.6°} = \boxed{130 \text{ lb}} \tag{h}$$

To get the lifting force of the kite, we draw a free-body diagram of point $B$ of the kite as shown in Fig. 6.54. Note that $F_y$ and $F_x$ are, respectively, the aerodynamic lift and drag forces on the kite. The lift force $F_y$ of the kite then becomes

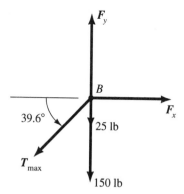

**Figure 6.54.** Free-body diagram of kite support.

$$F_y = 175 + T_{max} \sin 39.6° = \boxed{258 \text{ lb}} \tag{i}$$

**6.66.** Find the length of a cable stretched between two supports at the same elevation with span $l = 200$ ft and sag $h = 50$ ft, if it is subjected to a vertical load of 4 lb/ft uniformly distributed in the horizontal direction. (Assume that the weight of the cable is either negligible or included in the 4-lb/ft distribution.) Find the maximum tension.

**6.67.** A cable supports an 8000-kg uniform bar. What is the equation describing the shape of the cable and what is the maximum tension in the cable?

**Figure P.6.67.**

**6.68.** A cable supports a uniform loading of 100 lb/ft. If the lowest point of the cable occurs 20 ft from point $A$ as shown, what is the maximum tension in the cable and its length? Use $A$ as the origin of reference.

**Figure P.6.68.**

**6.69.** A uniform cable is shown whose weight we shall neglect. If a loading given as $5x$ N/m is imposed on the cable, what is the deflection curve of the cable if there is a zero slope of the curve at point $A$? What is the maximum tension?

**Figure P.6.69.**

**6.70.** The left side of a cable is mounted at an elevation 7 m below the right side. The sag, measured from the left support, is 7 m. Find the maximum tension if the cable has a *uniform* loading in the vertical direction of 1500 N/m. [*Suggestion:* Place reference at position of zero slope and determine the location of this point from the boundary conditions.]

**Figure P.6.70.**

**6.71.** A blimp is dragging a chain of length 400 ft and weight 10 lb/ft. A thrust of 300 lb is developed by the blimp as it moves against an air resistance of 200 lb. How much chain is on the ground and how high is the blimp? The vertical lift of the blimp on the cable is taken as 1,000 lb.

**Figure P.6.71.**

**6.72.** A large balloon has a buoyant force of 100 lb. It is held by a 150-ft cable whose weight is .5 lb/ft. What is the height $h$ of the balloon above the ground when a steady wind causes it to assume the position shown? What is the maximum tension in the cable?

**Figure P.6.72.**

**6.73.** What is the deflection curve for the uniform cable shown weighing 30 N/m? Find the maximum tension. Compute the height $h$ of the support $B$.

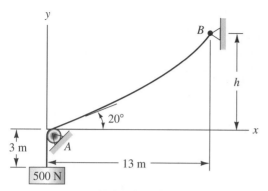

**Figure P.6.73.**

**6.74.** A search boat is dragging the lake floor for stolen merchandise using a 100-m chain weighing 100 N/m. The tension of the chain at support point $B$ is 5,000 N and the chain makes an angle of 50° there. What is the height of point $B$ above the lake bed? Also, what length of chain is dragging along on the bottom? Do not consider buoyant effects.

**Figure P.6.74.**

**6.75.** A cable weighing 3 lb/ft is stretched between two points on the same level. If the length of the cable is 450 ft and the tension at the points of support is 1,500 lb, find the sag and the distance between the points of support. Put reference at left support.

**6.76.** A flexible, inextensible cable is loaded by concentrated forces. If we neglect the weight of the cable, what are the supporting forces at $A$ and $B$? What are the tensions in the chord $AC$ and the angle $\alpha$? [*Hint:* Proceed by using finite free bodies and working from first principles.]

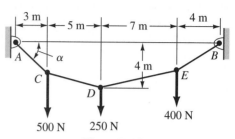

**Figure P.6.76.**

**6.77.** A system of two inextensible, flexible cables is shown supporting a 2,000-lb platform in a horizontal position. What are the inclinations of the cable segments $AB$, $BC$, and $DE$ to accomplish this and what lengths should they be? Neglect the weight of the segments and note the hint in Problem 6.76.

**Figure P.6.77.**

# 6.13  Closure

Essentially what we have done in this chapter is to apply previously developed material to situations of singular importance in engineering. Further information on structures can be found in books on strength of materials and structural mechanics. We turn again to new material in Chapter 7, where we will discuss the Coulomb laws of friction.

**6.78.** A 3-kN traveling hoist has a 27-kN capacity and is suspended from a beam weighing .5 kN/m. The beam is fastened to several trusses spaced 4 m apart. What are the forces in each truss member when the fully loaded hoist is located at point $C$ directly under the truss shown? Assume that the hoist acts on pin $C$ and that pin $C$ also supports half of the I-beam $G$ between each of the adjacent trusses.

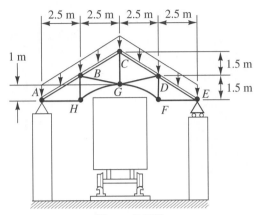

**Figure P.6.79.**

**6.80.** Find forces in all the members of the space truss. Note that $ACE$ is in the $xz$ plane (shaded).

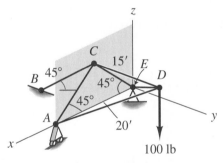

**Figure P.6.80.**

**6.81.** Determine the forces in members $BG$, $BF$, and $CE$ for the plane truss.

**Figure P.6.78.**

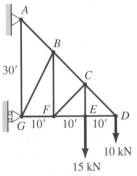

**Figure P.6.81.**

**6.79.** The truss is used to support the roof of a low-clearance train-car repair shed (hence the curved members). The roof is subjected to a snow load of 1 kN/m.² What are the forces in the straight members if the trusses are 10 m apart? Consider an inner truss.

**6.82.** Express the shear-force and bending-moment equations with the aid of free-body diagrams. Then express $V$ and $M$ without the diagrams.

Figure P.6.82.

**6.83.** Express the shear-force and bending-moment equations without the aid of free-body diagrams.

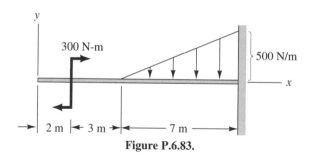

Figure P.6.83.

**6.84.** Give the shear-force and bending-moment equations for the beam, and sketch shear-force and bending-moment diagrams. At what position between supports is the bending moment equal to zero?

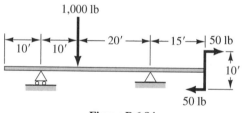

Figure P.6.84.

**6.85.** Sketch the shear-force and bending-moment diagrams labeling key points.

Figure P.6.85.

**6.86.** Find the shape of a cable stretched between two points on the same level, $l$ units apart with sag $h$, and subjected to a vertical loading of

$$w(x) = 5 \cos \frac{\pi x}{l} \ \text{N/m}$$

distributed in the horizontal direction. The coordinate $x$ is measured from the zero slope position of the cable.

**6.87.** (a) By inspection, which members in the truss shown have a zero force for the given loads?

(b) For the Fink truss in Fig. P.6.1, with vertical loads on the bottom pins, which members will carry a zero force?

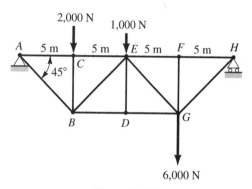

Figure P.6.87.

**6.88.** A 4,000-N load is being raised at constant speed. What are the forces in the truss?

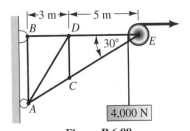

**Figure P.6.88.**

**6.89.** A truss supports a roadway load of 800 lb/ft per truss. Concentrated loads have been shown representing approximations of vehicle loading for each truss at some instant of time. The bridge has six 20-ft panels. Determine the forces in members *EG*, *FH*, and *IJ*.

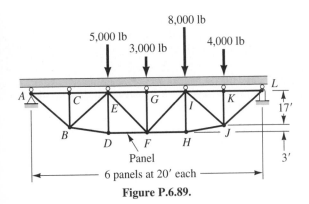

**Figure P.6.89.**

**6.90.** After finding the supporting forces for the simply supported beam *AB*, express the shear-force and bending-moment equations without the aid of free-body diagrams. The 10-kN load is applied to a bracket welded to the beam *AB*.

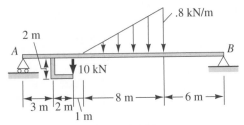

**Figure P.6.90.**

**6.91.** A uniform cable weighs 1 lb/ft. It is connected to a uniform rod at *B*. This rod is free to swing about hinge *C*. If a force component $F_x$ of 200 lb is exerted at *A* as shown, what is the resulting angle of inclination $\alpha$? The cable is 50 ft long and the rod is 20 ft long. What is the weight of the rod?

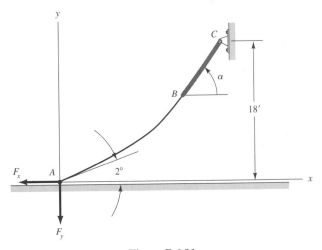

**Figure P.6.91.**

# CHAPTER 7

# Friction Forces

## 7.1 Introduction

*Friction* is the force distribution at the surface of contact between two bodies that prevents or impedes sliding motion of one body relative to the other. This force distribution is tangent to the contact surface and has, for the body under consideration, a direction at every point in the contact surface that is in opposition to the possible or existing slipping motion of the body at that point.

Frictional effects are associated with energy dissipation and are therefore sometimes considered undesirable. At other times, however, this means of changing mechanical energy to heat is a beneficial one, as for example in brakes, where the kinetic energy of a vehicle is dissipated into heat. In statics applications, frictional forces are often necessary to maintain equilibrium.

*Coulomb friction* is that friction which occurs between bodies having dry contact surfaces, and is not to be confused with the action of one body on another separated by a film of fluid such as oil. These latter problems are termed *lubrication problems* and are studied in the fluid mechanics courses. Coulomb, or *dry*, friction is a complicated phenomenon, and actually not much is known about its true nature.[1] The major cause of dry friction is believed to be the microscopic roughness of the surfaces of contact. Interlocking microscopic protuberances oppose the relative motion between the surfaces. When sliding is present between the surfaces, some of these protuberances either are sheared off or are melted by high local temperatures. This is the reason for the high rate of "wear" for dry-body contact and indicates why it is desirable to separate the surfaces by a film of fluid.

---

[1]For a more complete discussion of friction, see F. P. Bowden and D. Tabor, *The Friction and Lubrication of Solids*, Oxford University Press, New York, 1950.

We have previously employed the terms "smooth" and "rough" surfaces of contact. A "smooth" surface can only support a normal force. On the other hand, a "rough" surface in addition can support a force tangent to the contact surface (i.e., a friction force). In this chapter, we shall consider certain situations whereby the friction force can be directly related to the normal force at a surface of contact. Other than including this new relationship, we use only the usual static equilibrium equations.

## 7.2   Laws of Coulomb Friction

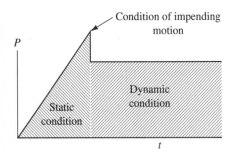

**Figure 7.1.** Idealized plot of applied force $P$.

Everybody has gone through the experience of sliding furniture along a floor. We exert a continuously increasing force which is completely resisted by friction until the object begins to move—usually with a lurch. The lurch occurs because once the object begins to move, there is a decrease in frictional force from the maximum force attained under static conditions. An idealized plot of this force as a function of time is shown in Fig. 7.1 where the force $P$ applied to the furniture, idealized as a block in Fig. 7.2, is shown to drop from the highest or limiting value to a lower value which is constant with time. This latter constant value is independent of the velocity of the object. The condition corresponding to the maximum value is termed the condition of *impending motion* or *impending slippage*.

By carrying out experiments on blocks tending to move without rotation or actually moving without rotation on flat surfaces, Coulomb in 1781 presented certain conclusions which are applicable at the condition of *impending slippage* or once *slippage has begun.* These have since become known as Coulomb's laws of friction. For block problems, he reported that:

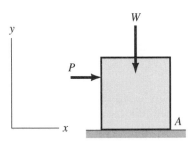

**Figure 7.2.** Idealization of furniture.

1. The total force of friction that can be developed is independent of the magnitude of the area of contact.
2. For low relative velocities between sliding objects, the frictional force is practically independent of velocity. However, the sliding frictional force is less than the frictional force corresponding to impending slippage.
3. The total frictional force that can be developed is proportional to the normal force transmitted across the surface of contact.

Conclusions 1 and 2 may come as a surprise to most of you and be contrary to your "intuition." Nevertheless, they are accurate enough statements for many engineering applications. More precise studies of friction, as was pointed out earlier, are complicated and involved. We can express conclusion 3 mathematically as:

$$f \propto N$$

Therefore,

$$f = \mu N \qquad (7.1)$$

where $\mu$ is called the *coefficient of friction.*

Equation 7.1 is valid *only at conditions of impending slippage or while the body is slipping.* Since the limiting static friction force exceeds the dynamic friction force, we differentiate between coefficients of friction for those conditions. Thus, we have coefficients of *static* friction and coefficients of *dynamic* friction, $\mu_s$ and $\mu_d$, respectively. The accompanying table is a small list of static coefficients that are commonly used. The corresponding coefficients of friction for dynamic conditions are about 25% less.

## Static Coefficients of Friction[2]

| | |
|---|---|
| Steel on cast iron | .40 |
| Copper on steel | .36 |
| Hard steel on hard steel | .42 |
| Mild steel on mild steel | .57 |
| Rope on wood | .70 |
| Wood on wood | .20 –.75 |

Let us consider carefully the simple block problem used to develop the laws of Coulomb. Note that we have:

1. A plane surface of contact.
2. An impending or actual motion which is in the same direction for all area elements of the contact surface. Thus, there is no impending or actual rotation between the bodies in contact.
3. The further implication that the properties of the respective bodies are uniform at the contact surface. Thus, the coefficient of friction $\mu$ is constant for all area elements of the contact surface.

What do we do if any of these conditions is violated? We can always choose an *infinitesimal* part of the area of contact between the bodies. Such an infinitesimal area can be considered plane even though the general surface of contact of which it is an infinitesimal part is not. Furthermore, the relative motion at this infinitesimal contact surface may be considered as along a straight line even though the finite surface of which it is a part may not have such a simple straight motion. Finally, for the infinitesimal area of contact, we may consider the materials to be uniform even though the properties of the material vary over the finite area of contact. In short, when conditions 1 through 3 do not prevail, we can still use Coulomb's law *in the small* (i.e., at infinitesimal contact areas) and then integrate the results. We shall call such problems *complex* surface contact problems and we shall examine a series of such problems in Section 7.4.

[2]F. P. Bowden and D. Tabor, *The Friction and Lubrication of Solids*, Oxford University Press, New York, 1950.

## *7.3   A Comment Concerning the Use of Coulomb's Law

As a simple illustration of why curvature of the surface of contact is second order and hence negligible, consider yourself at a location on the earth where it is perfectly round as a planet. As you look around you over a small area compared to a significant portion of the earth's surface, there is no evidence to your observation or in what you normally do that indicates the presence of curvature of the earth. In the same way, an infinitesimal area can be considered flat even if is part of a finite curved surface when considering Coulomb's law.

Furthermore, a stationary, nonrotating observer in inertial space looking at a similar small area on the earth's surface at the equator sees the velocity of the area essentially as given by $R\omega$ where $R$ is the radius of the earth and $\omega$ is its angular velocity. All parts of this small area will have this same velocity up to only second-order variation as seen by this observer. And, so from this viewpoint, all points within the area are essentially translating with the aforementioned speed. To explain further, we will now say that this small area has a maximum dimension given by the length $r$. The *rotational* speed of any point seen by an inertial and hence nonrotating observer at the area and translating *with* the area must be no greater than $r\omega$. Clearly then, this rotational speed is *negligibly small* when compared with $R\omega$. Hence, an infinitesimal area at a finite distance from the axis of rotation of a rotating surface can be considered as having primarily a translational velocity and so permits the use of Coulomb's law "in the small."

## 7.4   Simple Contact Friction Problems

We now examine simple contact problems where Coulomb's laws apply to the contact surface *as a whole* without requiring integration procedures. We shall thus consider uniform blocklike bodies akin to those used by Coulomb. Also, we shall consider bodies which however complex have very *small* contact surfaces, such as in Fig. 7.3(a). Clearly, the whole contact surface can then be considered an infinitesimal plane area and for reasons set forth earlier, we shall directly use Coulomb's laws when appropriate as has been shown in Fig. 7.3(b).

Before proceeding to the examples, we have one additional point to make. For a finite simple surface of contact, such as the block shown in Fig. 7.2, we must note that we do not generally know the line of action for the simplest resultant supporting force $N$, since we do not generally know the normal force distribution between the two bodies. Hence, we cannot take moments for such free-body diagrams without introducing additional unknown distances in the equation. Consequently, for such problems we limit ourselves to summing forces only. This is not true, however, when we have a *point* contact such as in Fig. 7.3(a). The line of action of the supporting force must be at the point of contact, and we can thus take moments without introducing additional unknown distances.

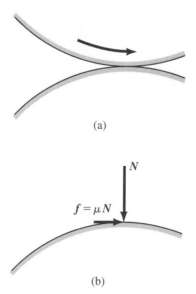

(a)

(b)

**Figure 7.3.** (a) Small contact surface;
(b) Coulomb's laws applied.

Two common classes of statics problems involve dry friction. In one class, we know that motion is impending, or has been established and is uniform, and we desire information about certain forces that are present. We can then express friction forces at surfaces of contact where there is impending or actual slippage as $\mu N$ according to Coulomb's law and, using $f_i$ for other friction forces, proceed by methods of statics. However, the proper direction must be given to *all* friction forces. That is, *they must oppose possible, impending, or actual relative motion at the contact surfaces.* In the second class of problems, external loads on a body are given, and we desire to determine whether the friction forces present are sufficient to maintain equilibrium. One way to attack this latter type of problem is to assume that impending motion exists in the various possible directions, and to solve for the external forces required for such conditions. By comparing the actual external forces present with those required for the various impending motions, we can then deduce whether the body can be restrained by frictional forces from sliding.

The following examples are used to illustrate the two classes of problems.

## Example 7.1

An automobile is shown in Fig. 7.4(a) on a roadway inclined at an angle $\theta$ with the horizontal. If the coefficients of static and dynamic friction between the tires and the road are taken as 0.6 and 0.5, respectively, what is the maximum inclination $\theta_{max}$ that the car can climb at uniform speed? It has rear-wheel drive and has a total loaded weight of 3,600 lb. The center of gravity for this loaded condition has been shown in the diagram.

Let us assume that the drive wheels do not "spin"; that is, there is zero relative velocity between the tire surface and the road surface at the point of contact. Then, clearly, the maximum friction force possible is $\mu_s$ times the normal force at this contact surface, as has been indicated in Fig. 7.4(b).[3]

We can consider this to be a *coplanar* problem with three unknowns, $N_1$, $N_2$, and $\theta_{max}$. Accordingly, since the friction force is restricted to a point, three equations of equilibrium are available. Using the reference $xy$ shown in the diagram, we have:

$$\sum F_x = 0:$$
$$.6N_1 - 3,600 \sin \theta_{max} = 0 \qquad \text{(a)}$$

$$\sum F_y = 0:$$
$$N_1 + N_2 - 3,600 \cos \theta_{max} = 0 \qquad \text{(b)}$$

$$\sum M_A = 0:$$
$$10N_2 - \left(3,600 \cos \theta_{max}\right)(5) + \left(3,600 \sin \theta_{max}\right)(1) = 0 \qquad \text{(c)}$$

To solve for $\theta_{max}$, we eliminate $N_1$ from Eqs. (a) and (b), getting as a result the equation

$$N_2 = 3,600 \cos \theta_{max} - 6,000 \sin \theta_{max} \qquad \text{(d)}$$

Now, eliminating $N_2$ from Eq. (c) using Eq. (d), we get

$$18,000 \cos \theta_{max} - 56,400 \sin \theta_{max} = 0$$

Therefore,

$$\tan \theta_{max} = .320 \qquad \text{(e)}$$

Hence,

$$\boxed{\theta_{max} = 17.7°} \qquad \text{(f)}$$

If the drive wheels were caused to spin, we would have to use $\mu_d$ in place of $\mu_s$ for this problem. We would then arrive at a smaller $\theta_{max}$, which for this problem would be 14.7°.

[3]You will notice that there is no friction force on the front wheels. This is so because there is no torque coming from the automobile's transmission onto these wheels, while at the same time the wheels are rotating at constant speed. Note we are neglecting "rolling" resistance stemming from the deformation of the road surface and the tire, a small force to be considered in Section 7.8, a starred section.

(a)

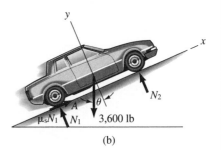

(b)

**Figure 7.4.** (a) Find maximum $\theta$; (b) free-body diagram using Coulomb's law.

# Example 7.2

Using the data of Example 7.1, compute the torque needed by the drive wheels to move the car at a uniform speed up an incline where $\theta = 15°$. Also, assume that the brakes have "locked" while the car is in a parked position on the incline. What force is then needed to tow the car either up the incline or down the incline with the brakes in this condition? The diameter of the tire is 25 in.

A free-body diagram for the first part of the problem is shown in Fig. 7.5(a). Note that the friction force $f$ will now be determined by Newton's law and not by Coulomb's law, since we do not have impending slippage between the wheel and the road for this case. Accordingly, we have, for $f$:

$$\sum F_x = 0:$$
$$f - 3{,}600 \sin 15° = 0$$

Therefore,

$$f = 932 \text{ lb}$$

The torque needed is then computed using the rear wheels as a free body [see Fig. 7.5(a)]. Taking moments about $A$, we have

$$\text{torque} = (f)(r) = (932)\left(\frac{25/2}{12}\right) = \boxed{971 \text{ ft-lb}}$$

For the second part of the problem, we have shown the required free body in Fig. 7.5(b). Note that we have used Coulomb's law for the friction forces with the *dynamic* friction coefficient $\mu_d$ on all wheels. We now write the equations of equilibrium for this free body.

$$\sum F_x = 0:$$
$$T_{up} - .5\left(N_1 + N_2\right) - 3{,}600 \sin 15° = 0 \tag{a}$$
$$\sum F_y = 0:$$
$$\left(N_1 + N_2\right) - 3{,}600 \cos 15° = 0 \tag{b}$$

Solving for $N_1 + N_2$ from Eq. (b), and substituting into Eq. (a), we can now solve for $T_{up}$. Hence,

$$T_{up} = (.5)(3{,}600)(.966) + 932 = \boxed{2{,}670 \text{ lb}} \tag{c}$$

### Example 7.2 (Continued)

For towing the car down the incline we must reverse the direction of the friction forces as shown in Fig. 7.5(c). Solving for $T_{down}$ as in the previous calculation, we get

$$T_{down} = (.5)(3,600)(.966) - 932 = \boxed{807 \text{ lb}} \qquad (d)$$

**Figure 7.5.** Free-body diagrams: (a) climbing at uniform speed; (b) under tow upward; (c) under tow downward.

# Example 7.3

In Fig. 7.6 a strongbox of mass 75 kg rests on a floor. The static coefficient of friction for the contact surface is .20. What is the largest force $P$ and what is the highest position $h$ for applying this force that will not allow the strongbox to either slip on the floor or to tip ?

The free-body diagram for the strongbox is shown in Fig. 7.7. The condition of *impending* motion has been recognized by the use of Coulomb's law. Furthermore, by concentrating the supporting and friction forces at the left corner, we are stipulating *impending tipping* for the problem. These two impending conditions impose the largest possible values of $P$ and $h$ that we are seeking.

The pertinent forces constitute a coplanar system of forces at the midplane of the strongbox. We proceed with the scalar equations of equilibrium:

$$\sum F_y = 0:$$
$$N = 75g = 736\ N$$

$$\sum F_x = 0:$$
$$P = .2N = 147.15\ N$$

$$\sum M_a = 0:$$
$$-(75g)(.3) + (147.15)h = 0$$

Therefore, we get for the largest $P$ and the largest $h$ the following results:

$$P_{max} = 147.15\ N \qquad h_{max} = 1.50\ m$$

Thus, the height of the applied load must be less than or equal to 1.50 m in order to avoid tipping.

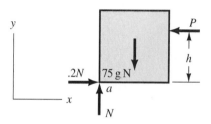

**Figure 7.6.** Strongbox being pushed.

**Figure 7.7.** Impending tipping and slipping.

The three examples presented illustrated the *first* type of friction problem wherein we know the nature of the motion or impending motion present in the system and we determine certain forces or positions of certain forces. In the last example of this series, we illustrate the *second* type of friction problem set forth earlier—namely the problem of deciding whether bodies will move or not move under prescribed external forces.

# Example 7.4

The coefficient of static friction for all contact surfaces in Fig. 7.8 is .2. Does the 50-lb force move the block $A$ up, hold it in equilibrium, or is it too small to prevent $A$ from coming down and $B$ from moving out? The 50-lb force is exerted at the midplane of the blocks so that we can consider this a coplanar problem.

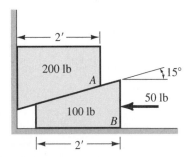

**Figure 7.8.** Do blocks move?

We can compute a force $P$ in place of the 50-lb force to cause impending motion of block $B$ to the left, and a force $P$ for impending motion of block $B$ to the right. In this way, we can judge by comparison the action that the 50-lb force will cause.

The free-body diagrams for impending motion of block $B$ to the left have been shown in Fig. 7.9, which contains the unknown force $P$ mentioned above. We need not be concerned about the correct location of the centers of gravity of the blocks, since we shall only add forces in the analysis. (We do not know the line of action of the normal forces at the contact surfaces and therefore cannot take moments.) Summing forces on block $A$, we get

$$N_2 \cos 15°\mathbf{j} - N_2 \sin 15°\mathbf{i} - .2N_1\mathbf{j} - 200\mathbf{j}$$
$$+ N_1\mathbf{i} - .2N_2 \cos 15°\mathbf{i} - .2N_2 \sin 15°\mathbf{j} = \mathbf{0}$$

The scalar equations are:

$$N_1 - .259N_2 - .1932N_2 = 0$$
$$.966N_2 - .2N_1 - 200 - .0518N_2 = 0$$

Solving simultaneously, we get

$$N_2 = 243 \text{ lb}, \qquad N_1 = 109.8 \text{ lb}$$

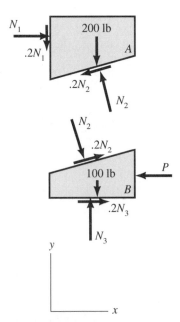

**Figure 7.9.** Impending motion of $B$ to the left.

## Example 7.4 (Continued)

For the free-body diagram of $B$, we have, on summing forces,

$$-N_2 \cos 15°j + N_2 \sin 15°i - Pi + .2N_3i + N_3j$$
$$- 100j + .2N_2 \cos 15°i + .2N_2 \sin 15°j = 0$$

This yields the following scalar equations:

$$-P + 62.9 + .2N_3 + 46.9 = 0$$
$$-235 + N_3 - 100 + 12.6 = 0$$

Solving simultaneously, we have

$$P = 174 \text{ lb}$$

Clearly, the stipulated force of 50 lb is insufficient to induce a motion of block $B$ to the left, so further computation is necessary.

Next, we reverse the direction of force $P$ and compute what its value must be to move the block $B$ to the right. The frictional forces in Fig. 7.9 are all reversed, and the vector equation of equilibrium for block $A$ becomes

$$N_2 \cos 15°j - N_2 \sin 15°i + .2N_1j - 200j$$
$$+ N_1i + .2N_2 \cos 15°i + .2N_2 \sin 15°j = 0$$

The scalar equations are:

$$N_1 - .259N_2 + .1932N_2 = 0$$
$$.966N_2 + .2N_1 - 200 + .0518N_2 = 0$$

Solving simultaneously, we get

$$N_2 = 194.1 \text{ lb}, \qquad N_1 = 12.80 \text{ lb}$$

For free body $B$ we have, on summing forces,

$$-N_2 \cos 15°j + N_2 \sin 15°i + Pi - .2N_3i + N_3j$$
$$- 100j - .2N_2 \cos 15°i - .2N_2 \sin 15°j = 0$$

The following are the scalar equations:

$$P - .2N_3 + 50.3 - 37.5 = 0$$
$$-100 + N_3 - 187.5 - 10.05 = 0$$

Solving, we get $P = 46.7$ lb. Thus, we would have to *pull to the right* to get block $B$ to move in this direction. We can now conclude from this study that the blocks are in equilibrium.

# PROBLEMS

**7.1.** A block has a force $F$ applied to it. If this force has a time variation as shown in the diagram, draw a simple sketch showing the friction force variation with time. Take $\mu_s = .3$ and $\mu_d = .2$ for the problem.

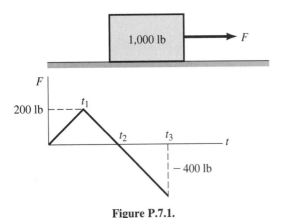

Figure P.7.1.

**7.2.** Show by increasing the inclination $\phi$ on an inclined surface until there is impending slippage of supported bodies, we reach the *angle of repose* $\phi_s$ so that $\tan \phi_s = \mu_s$.

**7.3.** To what minimum angle must the driver elevate the dump bed of the truck to cause the wooden crate of weight $W$ to slide out? For wood on steel, $\mu_s = .6$ and $\mu_d = .4$.

Figure P.7.3.

**7.4.** A platform is suspended by two ropes which are attached to blocks that can slide horizontally. At what value of $W$ does the platform begin to descend? Will $W$ start tipping?

Figure P.7.4.

**7.5.** Explain how a violin bow, when drawn over a string, maintains the vibration of the string. Put this in terms of friction forces and the difference in static and dynamic coefficients of friction.

**7.6.** What is the value of the force $F$, inclined at 30° to the horizontal, needed to get the block just started up the incline? What is the force $F$ needed to keep it just moving up at a constant speed? The coefficients of static and dynamic friction are .3 and .275, respectively.

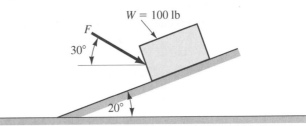

Figure P.7.6.

**7.7.** Bodies $A$ and $B$ weigh 500 N and 300 N, respectively. The platform on which they are placed is raised from the horizontal position to an angle $\theta$. What is the *maximum* angle that can be reached before the bodies slip down the incline? Take $\mu_s$ for body $B$ and the plane as .2 and $\mu_s$ for body $A$ and the plane as .3.

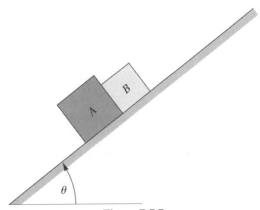

Figure P.7.7.

**7.8.** What is the minimum value of $\mu_s$ that will allow the rod $AB$ to remain in place? The rod has a length of 3.3 m and it has a weight of 200 N.

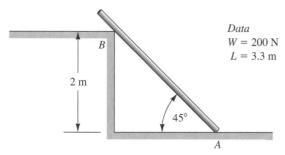

Data
$W = 200$ N
$L = 3.3$ m

**Figure P.7.8.**

**7.9.** Find the minimum force $P$ to get block $A$ moving.

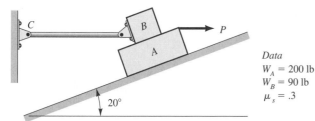

Data
$W_A = 200$ lb
$W_B = 90$ lb
$\mu_s = .3$

**Figure P.7.9.**

**7.10.** What minimum force $F$ is needed to start body $A$ moving to the right if $\mu_s = .25$ for all surfaces? The following weights are given:

$$W_A = 125 \text{ N} \qquad W_B = 50 \text{ N} \qquad W_{AB} = 100 \text{ N}$$

The length of $AB$ is 2.5 m.

**Figure P.7.10.**

**7.11.** A 30-ton tank is moving up a 30° incline. If $\mu_s = .6$ for the contact surface between tread and ground, what *maximum* torque can be developed at the rear drive sprocket with no slipping? What maximum towing force $F$ can the tank develop? Take the mean diameter of the rear sprocket as 2 ft.

**Figure P.7.11.**

**7.12.** A 500-lb crate $A$ rests on a 1,000-lb crate $B$. The centers of gravity of the crates are at the geometric centers. The coefficients of static friction between contact surfaces are shown in the diagram. The force $T$ is increased from zero. What is the first action to occur?

**Figure P.7.12.**

**7.13.** What force $F$ is needed to get the 300-kg block moving to the right? The coefficient of static friction for all surfaces is .3.

**Figure P.7.13.**

**7.14.** A chute is shown having sides that are at right angles to each other. The chute is 30 ft in length with end $A$ 10 ft higher than end $B$. Cylinders weighing 200 lb are to slide down the chute. What is the *maximum* allowable static coefficient of friction so there cannot be sticking of the cylinders along the chute?

$D = 4'$

**Figure P.7.14.**

**7.15.** (a) Can the tractor move the 60-kN weight? (Prove this in some way)
(b) What is the tension in the connecting chain if the tractor moves the weight at constant speed?
(c) What is the torque for this action from the engine on one of the .8-m-diameter drive sprockets?

$W_{TOTAL} = 200$ kN

$\mu_s = .6$

Drive sprocket

60 kN

$\mu_d = .5$

$15°$

**Figure P.7.15.**

**7.16.** In Example 5.12, the wheels at $A$ have rusted so as not to rotate at all, and so there can only be sliding on the floor. The coefficients of friction are $\mu_s = .4$ and $\mu_d = .2$. Using the same data as in the Example, compute the horizontal force on the rock for impending crushing action and then compute the horizontal force at the instant crushing starts. Assume that the action is slow so that we do not need to consider inertial effects. The plate $A$ weighs 100 lb.

**7.17.** In Problem 5.92, the rollers on the block $C$ have rusted and can no longer roll at all along the floor. There is then a dry friction at the contact surface between the block $C$ and the floor. For this surface, the static coefficient of friction $\mu_s = .4$. If $p_1 = \frac{1}{2} p_2$, what pressures in the cylinders must be reached for the impending crushing action on the rock, which requires from the block at $C$ a force of 7,000 lb in the horizontal direction? Block $C$ weighs 200 lb.

**7.18.** A frame can slide up or down along the vertical guide rods or can remain stationary depending on the loads. What force $P$ will cause impending slippage downward for the given applied loads? Member $GA$ has a loose fit with circular rod $EC$, and member $BH$ has a loose fit with circular rod $JD$. The coefficient of friction everywhere is $\mu_s = .3$. Neglect the weights of the members.

$E$

.3 m

100 N-m

$A$

$P$

$G$

500 N

$J$

80 mm

20 mm

$B$

$H$

Guide rods

.15 m

20 mm

$C$

$D$

**Figure P.7.18.**

**7.19.** Given that $\mu_s = .2$ for all surfaces, find the force $P$ needed to start the block $A$ to the right.

1,000 N

$20°$

$B$

$A$

2,000 N

$P$

**Figure P.7.19.**

**7.20.** The cylinder shown weighs 200 N and is at rest. What is the friction force at $A$? If there is impending slippage, what is the static coefficient of friction? The supporting plane is inclined at $60°$ to the horizontal.

**Figure P.7.20.**

**Figure P.7.22.**

**7.21.** Armature $B$ is stationary while rotor $A$ rotates with angular speed $\omega$. In armature $B$ there is a braking system. If $\mu_d = .4$, what is the braking torque on $A$ for a force $F$ of 300 N? Note that the rod on which $F$ is applied is pinned at $C$ to the armature $B$. Neglect friction between $B$ and the brake pads $G$ and $H$.

**7.23.** An insect tries to climb out of a hemispherical bowl of radius 600 mm. If the coefficient of static friction between insect and bowl is .4, how high up does the insect go? If the bowl is spun about a vertical axis, the bug gets pushed out in a radial direction by the force $mr\omega^2$, as you learned in physics. At what speed $\omega$ will the bug just be able to get out of the bowl?

**Figure P.7.21.**

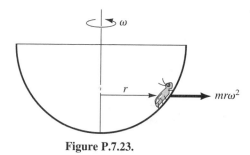

**Figure P.7.23.**

**7.24.** A block $A$ of mass 500 kg rests on a stationary support $B$ where the static coefficient of friction $\mu_s = .4$. On the right side, support $C$ is on rollers. The dynamic coefficient of friction $\mu_d$ of the support $C$ with body $A$ is .2. If $C$ is moved at constant speed to the left, how far does it move before body $A$ begins to move?

**7.22.** A 200-lb load is placed on the luggage rack of the 4,500-lb station wagon. Will the station wagon climb the hill more easily or with greater difficulty with the luggage than without? Explain. The static coefficient of friction is .55.

**Figure P.7.24.**

**7.25.** In Problem 5.94, find the minimum force $F_1$ acting on the block needed to maintain equilibrium. Block $B$ weighs 100 lb and the static coefficient of friction between the block and the floor is 0.276. Neglect the weights of the other members of the system.

**7.26.** If the static coefficient of friction at $C$ is .4, what is the minimum torque $T$ needed to start a counterclockwise rotation of $BA$ about hinge $A$? The following are the weights of the bodies involved:

$$AB = 3 \text{ lb} \qquad BC = 4 \text{ lb} \qquad C = 2 \text{ lb}$$

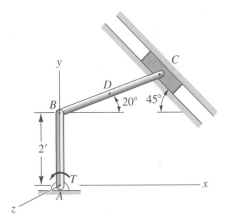

**Figure P.7.26.**

**7.27.** What will ensue if the weights are released from a state of rest? Examine all possibilities to justify your result.

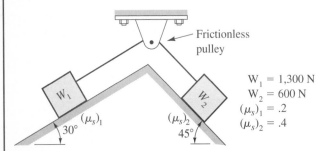

**Figure P.7.27.**

**7.28.** In a preliminary grinding operation for a 1,500-N car engine block, the grinding wheel is pushed against the block with a 500-N force. What force must be exerted by the hydraulic ram to move the block to the right if (a) the wheel rotates clockwise and (b) the wheel rotates counterclockwise? The static coefficient of friction between the grinding wheel and the block is .7 and between the table and the block is .2.

**Figure P.7.28.**

**7.29.** An 8,000-lb tow truck with four-wheel drive and 36 in. diameter tires develops a torque of 750 lb-ft at each axle. What is the *heaviest* car that can be towed up a 10° slope if $\mu_s = .3$?

**7.30.** A 7-m ladder weighing 250 N is being pushed by force $F$. What is the *minimum* force needed to get the ladder to move? The static coefficient of friction for all contact surfaces is .4.

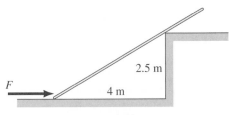

**Figure P.7.30.**

**7.31.** In Problem 7.30 if $F$ is released will the ladder begin to slide down?

**7.32.** Can a force $P$ roll the 50-lb cylinder over the step? The static coefficient of friction is .4. What is the value of $P$ if this can be done?

**Figure P.7.32.**

296

**7.33.** The block of weight $W$ is to be moved up an inclined plane. A rod of length $c$ with negligible weight is attached to the block and the force $F$ is applied to the top of this rod. If the coefficient of static friction is $\mu_s$, determine in terms of $a$, $d$, and $\mu_s$ the *maximum* length $c$ for which the block will begin to slide rather than tip.

**Figure P.7.33.**

**7.34.** Determine the range of values of $W_1$ for which the block will either slide up the plane or slide down the plane. At what value of $W_1$ is the friction force zero? $W_2 = 100$ lb.

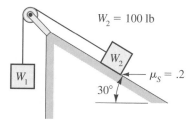

**Figure P.7.34.**

**7.35.** A 200-kN tractor is to push a 60-kN concrete beam up a 15° incline at a construction site. If $\mu_d = .5$ between beam and dirt and if $\mu_s$ is .6 between tractor tread and dirt, can the tractor do the job? If so, what torque must be developed on the tractor drive sprocket which is .8 m in diameter? What force $P$ is then developed to push the beam?

**Figure P.7.35.**

**7.36.** What is the *minimum* static coefficient of friction required to maintain the bracket and its 500-lb load in a static position? (Assume point contacts at the horizontal centerlines of the arms.) The center of gravity is 7 in. from the shaft centerline. Hint: Note that there is clearance between the vertical shaft and the horizontal arms.

**Figure P.7.36.**

**7.37.** If the static coefficient of friction in Problem 7.36 is .2, at what *minimum* distance from the centerline of the vertical shaft can we support the 500-lb load without slipping?

**7.38.** A rod is held by a cord at one end. If the force $F = 200$ N, and if the rod weighs 450 N, what is the *maximum* angle $\alpha$ that the rod can be placed for $\mu_s$ between the rod and the floor equal to .4? The rod is 1 m in length.

**Figure P.7.38.**

297

**7.39.** Suppose that an ice lifter is used to support a hard block of material by friction only. What is the *minimum* coefficient of static friction, $\mu_s$, to accomplish this for any weight $W$ and for the geometry shown in the diagram?

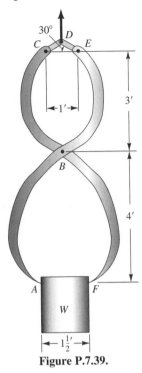

**Figure P.7.39.**

**7.40.** A rectangular case is loaded with uniform vertical thin rods such that when it is full, as shown in (a), the case has a total weight of 1,000 lb. The case weighs 100 lb when empty and has a coefficient of static friction of .3 with the floor as shown in the diagram. A force $T$ of 200 lb is maintained on the case. If the rods are unloaded as shown in (b), what is the limiting value of $x$ for equilibrium to be maintained?

**Figure P.7.40.**

**7.41.** A beam supports load $C$ weighing 500 N. At supports $A$ and $B$, the static coefficient of friction is .2. At the contact surface between load $C$ and the beam, the dynamic coefficient of friction is .75. If force $F$ moves $C$ steadily to the left, how far does it move before the beam begins to move? The beam weighs 200 N. Neglect the height $t$ of the beam in your calculations.

**Figure P.7.41.**

**7.42.** Do Problem 7.41 for the case where the height $t$ is taken into account. Take $t = 120$ mm.

**7.43.** A rod is supported by two wheels spinning in opposite directions. If the wheels were horizontal, the rod would be placed centrally over the wheels for equilibrium. However, the wheels have an inclination of 20° as shown, and the rod must be placed at a position off center for equilibrium. If the coefficient of friction is $\mu_d = .8$, how many feet off center must the rod be placed?

**Figure P.7.43.**

**P.7.44.** How much force $F$ must be applied to the wedge to begin to raise the crate? Neglect changes in geometry. What force must the stopper block provide to prevent the crate from moving to the left? The static coefficient of friction between all surfaces is .3.

**Figure P.7.44.**

**7.45.** What is the *maximum* height $x$ of a step so that the force $P$ will roll the 50-lb cylinder over the step with no slipping at $a$? Take $\mu_s = .3$.

**Figure P.7.45.**

**7.46.** The rod $AD$ is pulled at $A$ and it moves to the left. If the coefficient of dynamic friction for the rod at $A$ and $B$ is .4, what must the *minimum* value of $W_2$ be to prevent the block from tipping when $\alpha = 20°$? With this value of $W_2$, determine the minimum coefficient of static friction between the block and the supporting plane needed to just prevent the block from sliding. $W_1$ is 100 N.

**Figure P.7.46.**

**7.47.** If we neglect friction at the rollers, and the coefficient of static friction is .2 for all surfaces, ascertain whether the 5,000-lb weight will go up, go down, or stay stationary.

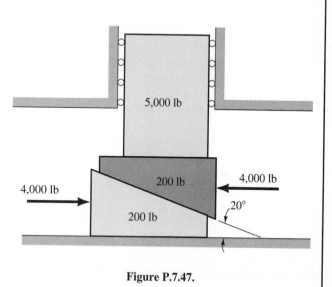

**Figure P.7.47.**

# 7.5 Complex Surface Contact Friction Problems

In the examples undertaken heretofore, the nature of the relative impending or actual motion between the plane surfaces of contact was quite simple—that of motion without rotation. We shall now examine more general types of contacts between bodies. In Example 7.5, we have a plane contact surface but with varying direction of impending or slipping motion for the area elements as a result of rotation. In such problems we shall have to apply Coulomb's laws locally to infinitesimal areas of contact and to integrate the results, for reasons explained in Section 7.2. To do this, we must ascertain the distribution of the normal force at the contact surface, an undertaking that is usually difficult and well beyond the capabilities of rigid-body statics, as explained in Chapter 5. However, we can at times *approximately* compute frictional effects by *estimating* the manner of distribution of the normal force at the surface of contact. We now illustrate this.

# Example 7.5

Compute the frictional resistance to rotation of a rotating solid cylinder with an attached pad $A$ pressing against a flat dry surface with a force $P$ (see Fig. 7.10). The pad $A$ and the stationary flat dry surface constitute a dry *thrust bearing*.

The direction of the frictional forces distributed over the contact surface is no longer simple. We therefore take an infinitesimal area for examination. This area is shown in Fig. 7.10, where the element has been formed from polar-coordinate differentials so as to be related simply to the boundaries. The area $dA$ is equal to $r\,d\theta\,dr$. We shall *assume* that the normal force $P$ is uniformly distributed over the entire area of contact. The normal force on the area element is then

$$dN = \frac{P}{\pi D^2/4} r\,d\theta\,dr \qquad\qquad (a)$$

The friction force associated with this force during motion is

$$df = \mu_d \frac{P}{\pi D^2/4} r\,d\theta\,dr \qquad\qquad (b)$$

**Figure 7.10.** Dry thrust bearing.

The direction of $df$ must oppose the relative motion between the surfaces. The relative motion is rotation of concentric circles about the centerline, so the direction of a force $df_1$ (Fig. 7.11) must lie tangent to a circle of radius $r$. At 180° from the position of the area element for $df_1$, we may carry out a similar calculation for a force $df_2$, which for the same $r$ must be equal and opposite to $df_1$, thus forming a couple. Since the entire area may be decomposed in this way, we can conclude that there are only couples in the plane of contact. If we take moments of all infinitesimal forces about the center, we get the magnitude of the total frictional couple moment. The direction of the couple moment is along the shaft axis. First, consider area elements on the ring of radius $r$:

$$dM = \int_0^{2\pi} r\mu_d \frac{P}{\pi D^2/4} r\,d\theta\,dr \qquad\qquad (c)$$

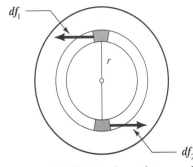

**Figure 7.11.** Friction forces form couples.

**Example 7.5 (Continued)**

Taking $\mu_d$ as constant and holding $r$ constant, we have on integration with respect to $\theta$:

$$dM = \mu_d \frac{P}{\pi D^2/4} 2\pi r^2 \, dr$$

We thus account for all area elements on the ring of radius $r$. To account for all the rings of the contact surface, we next integrate with respect to $r$ from zero to $D/2$. Clearly, this gives us the total resisting torque $M$. Thus,

$$M = \mu_d \frac{8P}{D^2} \int_0^{D/2} r^2 \, dr = \boxed{\frac{PD\mu_d}{3}} \qquad (d)$$

What we have performed in the last three steps is *multiple integration*, which we introduced in Chapter 4 when dealing with rectangular coordinates.

# 7.6  Belt Friction

A flexible belt is shown in Fig. 7.12 wrapped around a portion of a drum, with the amount of wrap indicated by angle $\beta$. The angle $\beta$ is called the *angle of wrap*. Assume that the drum is stationary and tensions $T_1$ and $T_2$ are such that motion is impending between the belt and the drum. We shall take the impending motion of the belt to be clockwise relative to the drum, and therefore the tension $T_1$ exceeds tension $T_2$.

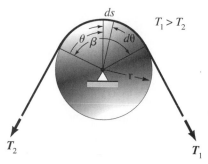

**Figure 7.12.** Belt wrapped around drum.

Consider an infinitesimal segment of the belt as a free body. This segment subtends an angle $d\theta$ at the drum center as shown in Fig. 7.13. Summing force components in the radial and transverse directions and equating them to zero as per *equilibrium*, we get the following scalar equations:

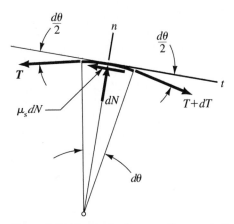

**Figure 7.13.** Free-body diagram of segment of belt; impending slippage.

$$\underline{\sum F_t = 0:}$$

$$-T \cos \frac{d\theta}{2} + (T + dT) \cos \frac{d\theta}{2} - \mu_s dN = 0$$

Therefore,

$$dT \cos \frac{d\theta}{2} = \mu_s \, dN$$

$$\underline{\sum F_n = 0:}$$

$$-T \sin \frac{d\theta}{2} - (T + dT) \sin \frac{d\theta}{2} + dN = 0$$

Therefore,

$$-2T \sin \frac{d\theta}{2} - dT \sin \frac{d\theta}{2} + dN = 0$$

The sine of a very small angle approximately equals the angle itself in radians. Furthermore, to the same degree of accuracy, the cosine of a small angle approaches unity. (That these relations are true may be seen by expanding the sine and cosine in a power series and then retaining only the first terms.) The preceding equilibrium equations then become

$$dT = \mu_s \, dN \tag{7.2a}$$

$$-T d\theta - dT \frac{d\theta}{2} + dN = 0 \tag{7.2b}$$

In the last equation, we have an expression involving the product of two infinitesimals. This quantity may be considered negligible compared to the other terms of the equation involving only one differential. Thus, we have for this equation:

$$T \, d\theta = dN \tag{7.3}$$

From Eqs. 7.2a and 7.3, we may form an equation involving $T$ and $\theta$. Thus, by eliminating $dN$ from the equations, we have

$$dT = \mu_s T \, d\theta$$

Hence,

$$\frac{dT}{T} = \mu_s \, d\theta$$

Integrating both sides around the portion of the belt in contact with the drum,

$$\int_{T_2}^{T_1} \frac{dT}{T} = \int_0^\beta \mu_s \, d\theta$$

we get

$$\ln \frac{T_1}{T_2} = \mu_s \beta$$

or

$$\frac{T_1}{T_2} = e^{\mu_s \beta} \tag{7.4}$$

We therefore have established a relation between the tensions on each part of the belt at a condition of impending motion between the belt and the drum. The same relation can be reached for a *rotating* drum with impending slippage between the belt and the drum *if we neglect centrifugal effects on the belt*. Furthermore, by using the *dynamic* coefficient of friction in the formula above, we have the case of the belt slipping at constant speed over either a rotating or stationary drum (again neglecting centrifugal effects on the belt). Thus, for all such cases, we have

$$\frac{T_1}{T_2} = e^{\mu \beta} \tag{7.5}$$

where the proper coefficient of friction must be used to suit the problem, and the angle $\beta$ must be expressed in *radians*. Note that *the ratio of tensions depends only on the angle of wrap $\beta$ and the coefficient of friction $\mu$*. Thus, if the drum $A$ is forced to the right, as shown in Fig. 7.14, the tensions will increase, but if $\beta$ is not affected by the action, the ratio of $T_1/T_2$ for impending or actual constant speed slippage is *not* affected by this action. However, the *torque* developed by the belt on the drum as a result of friction *is* affected by the force $F$. The torque is easily determined by using the drum and the portion of the belt in contact with the drum as a free body, as is shown in Fig. 7.14.

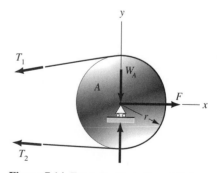

**Figure 7.14.** Force $F$ affects $T_1$ and $T_2$ but not $T_1/T_2$.

Thus,

$$\text{torque} = T_1 r - T_2 r = (T_1 - T_2)r \qquad (7.6)$$

If we pull the drum to the right without disturbing the angle of wrap, we can see from Eq. 7.5 that the tensions $T_1$ and $T_2$ must increase by the same factor to maintain the condition of impending or actual constant speed slippage. And if we call this factor $H$, the new tensions become $HT_1$ and $HT_2$, respectively. Substituting into Eq. 7.6, we see that the frictional torque is also increased by the same factor:

$$\text{torque} = H(T_1 - T_2)r = H(\text{torque})_{\text{original}}$$

If we sum forces on the free body diagram in Fig. 7.14, we can determine the force $F$ needed to maintain the tensions $T_1$ and $T_2$ as follows:

$$(T_1)_x + (T_2)_x = F \qquad (7.7)$$

At impending slippage or at constant speed slippage, Eq. 7.5 is valid, and with Eq. 7.7 and knowing $F$, we can solve for $T_1$ and $T_2$. With Eq. 7.6, the torque that the belt is capable of developing on the drum now becomes a simple computation.

Accordingly, we have three equations at our disposal. Thus, there is the geometry determining for us the angle of wrap and there is the coefficient of friction. These can be combined and used in the *first* equation [Eq. (7.5)] to get the correct *ratio* of tensions for the conditions of impending slippage or for actual slippage. Note, because of the use of *equilibrium* in the derivation of this equation, both of these conditions (impending slippage and actual slippage) require that the belt be stationary or be moving at constant speed in inertial space. These results serve as benchmarks for the design engineer, particularly the impending slippage condition. If the ratio of the belt tensions is less than $e^{\mu_s \beta}$ and the belt has not started to slip, we can say that there will be no slipping until this ratio of the belt tensions is exceeded,[4] at which time for steady speed of the belt, the ratio of the belt tensions will have to become equal to the value $e^{\mu_d \beta}$. In the *second* of the three equations [Eq. (7.6)], the torque on or from the drum is related to the belt tensions via equilibrium. Finally, the *third* equation [Eq. (7.7)], again from equilibrium, gives us the required force $F$ on the drum needed for the aforestated torque. Note that it is desirable to get the ratio of the tensions close to the condition of impending slippage since this means a smaller $F$ and hence a longer life for the belt. This also will mean fewer expensive stoppages for assembly lines.

---

[4]This is like the example given at the outset of moving a piece of furniture where, as the driving force is increased from zero, no movement occurs until the maximum static friction force has been exceeded.

## Example 7.6

A drum (see Fig. 7.15) requires a torque of 200 N-m to get it to start rotating. If the static coefficient of friction $\mu_s$ between the belt and the drum is .35, what is the *minimum* axial force $F$ on the drum required to create enough tension in the belt to start the rotation of the drum?

Torque required = 200 N-m
Radius = 250  mm

**Figure 7.15.** Drum is driven by a belt.

The angle of wrap $\beta$ clearly is $\pi$ radians. To get the minimum force $F$, we use the condition of impending slippage between belt and drum so that we can say

$$\frac{T_1}{T_2} = e^{.35\pi} = 3.00 \qquad (a)$$

For the condition of starting rotation we have

$$\sum M_0 = \text{torque required} = 200 \text{ N-m}$$

$$\therefore (T_1 - T_2)(.25) = 200 \qquad (b)$$

It is now a simple matter to solve the previous two equations simultaneously to get

$$T_1 = 1,200 \text{ N} \qquad T_2 = 400 \text{ N}$$

Finally we can determine the minimum value of $F$ for this problem by summing forces in the $x$ direction as follows:

$$\sum F_x = 0:$$

$$F_{min} = 1,200 + 400 = \boxed{1,600 \text{ N}}$$

## Example 7.7

A conveyor is moving ten 50-lb boxes at a constant speed at a 45° setting (Fig. 7.16). The dynamic coefficient of friction between the belt and the bed of the conveyor is .05. Furthermore, the static coefficient of friction between the driving pulley and the belt is .4. The idler pulley is moved along the direction of the conveyor by a crank mechanism so that the idler pulley is subject to a force $F$ of 500 lb. Compute the maximum tension found in the belt and ascertain if there will be slipping on the driving pulley. Neglect the weight of the belt.

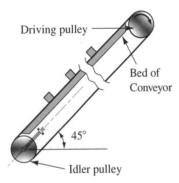

**Figure 7.16.** Conveyor.

In Fig. 7.17, we have shown free-body diagrams of various parts of the conveyor.[5] Consider the portion of the belt on the conveyor frame. From *equilibrium* we set the sums of forces normal and tangent to the belt equal to zero. Thus

$$\sum F_n = 0:$$

$$N - (10)(50)(.707) = 0$$

Therefore,

$$N = 354 \text{ lb} \tag{a}$$

$$\sum F_t = 0:$$

$$T_1 - T_3 - (10)(50)(.707) - (.05)(354) = 0$$

Therefore,

$$T_1 - T_3 = 371 \text{ lb} \tag{b}$$

[5]The weights of the pulleys have been counteracted by supporting forces at the axles and have not been shown.

## Example 7.7 (Continued)

For the idler pulley (no resisting or applied torques), we have

$$\sum M_0 = 0:$$

$$T_3 = T_2 \tag{c}$$

$$\sum F_t = 0:$$

$$T_3 + T_2 = 500 \tag{d}$$

From Eqs. (c) and (d), we conclude that

$$T_2 = T_3 = 250 \text{ lb} \tag{e}$$

From Eq. (b) we now get for the maximum tension $T_1$:

$$T_1 = T_3 + 371 = \boxed{621 \text{ lb}} \tag{f}$$

We must next check the driving pulley to ensure that there is no slippage occurring. For the condition of impending slippage, we have, using as $T_2$ the value of 250 lb and solving for $T_1$

$$T_1 = T_2 e^{.4\pi} = (250)(3.51) = 878 \text{ lb}$$

Clearly, since the $T_1$ needed is only 621 lb (see Eq.(f)), we do not have slippage at the driving pulley, and we conclude that the maximum tension is indeed 621 lb.

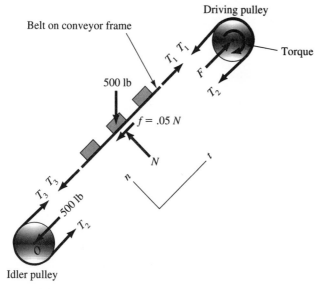

**Figure 7.17.** Various free-body diagrams of parts of conveyor.

## Example 7.8

An electric motor (not shown) in Fig. 7.18 drives at constant speed the pulley $B$, which connects to pulley $A$ by a belt. Pulley $A$ is connected to a compressor (not shown) which requires 700 N-m torque to drive it at constant speed $\omega_A$. If $\mu_s$ for the belt and either pulley is .4, what minimum value of the indicated force $F$ is required to have no slipping anywhere?

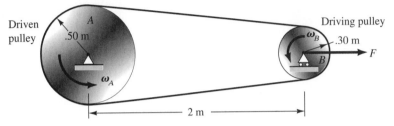

**Figure 7.18.** Belt-driven compressor.

As a first step, we determine the angles of wrap $\beta$ for the respective pulleys. For this purpose, we first compute $\alpha$ (Fig. 7.19). Note that the radii $O_A D$ and $O_B E$, being perpendicular to the same line $DE$, are therefore parallel to each other. Drawing $EC$ parallel to $O_A O_B$, we then form $\alpha$ in the shaded triangle. Hence, we can say:

$$\alpha = \sin^{-1}\frac{CD}{CE} = \sin^{-1}\frac{r_A - r_B}{O_A O_B} = \sin^{-1}\frac{.50 - .30}{2} = 5.74°$$

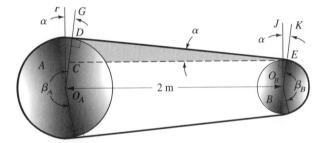

**Figure 7.19.** Find angles of wrap.

Note in Fig. 7.19 that $O_A F$ is perpendicular to $CE$ and that $O_A G$ is perpendicular to $DE$. Therefore, the included angle between $O_A F$ and $O_A G$ must equal the included angle between $CE$ and $DE$. This angle is $\alpha$. Clearly, the angle between $O_B J$ and $O_B K$ is also this angle $\alpha$. We can now express the angles of wrap for both pulleys as follows:

$$\beta_A = 180° + 2(5.74) = 191.5°$$
$$\beta_B = 180° - 2(5.74) = 168.5°$$

## Example 7.8 (Continued)

Now consider pulley $A$ as a free body in Fig. 7.20. Note that the minimum force $F$ corresponds to the condition of impending slippage. Accordingly, for this condition at $A$, we have

$$\frac{(T_1)_A}{(T_2)_A} = e^{\mu_s \beta_A} = e^{(.4)[(191.5/360)2\pi]} = 3.81 \qquad \text{(a)}$$

**Figure 7.20.** Free-body diagram of $A$.

Also, summing moments about the center of the pulley, we have

$$[(T_1)_A - (T_2)_A](.50) - 700 = 0 \qquad \text{(b)}$$

Therefore,

$$(T_1)_A - (T_2)_A = 1,400$$

Solving Eqs. (a) and (b) simultaneously, we get

$$(T_1)_A = 1,898 \text{ N}; \qquad (T_2)_A = 498 \text{ N}$$

From *equilibrium*, we can compute force $F_A$ as follows:

$$(1,898 + 498) \cos 5.74° - F_A = 0 \qquad \text{(c)}$$

Therefore,

$$F_A = 2,384 \text{ N}$$

Now go to pulley $B$ to see what minimum force $F_B$ is needed so that the belt does not slip on it during operations. Consider in *Fig. 7.21* the free-body diagram of pulley $B$. For impending slipping on pulley $B$, we have

$$\frac{(T_1)_B}{(T_2)_B} = e^{\mu_s \beta_B} = e^{(.4)[(168.5/360)2\pi]} = 3.24 \qquad \text{(d)}$$

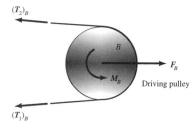

**Figure 7.21.** Free-body diagram of $B$.

## Example 7.8 (Continued)

The torque for pulley $B$ needed to develop 700 N-m on pulley $A$ is next computed. Thus,[6]

$$M_B = \frac{r_B}{r_A} M_A = \frac{.30}{.50} M_A = \frac{.30}{.50} (700)$$

Therefore,

$$M_B = 420 \text{ N-m}$$

Summing moments in Fig. 7.21 about the center of $B$, we then have on using the above result

$$- [(T_1)_B - (T_2)_B](.30) + 420 = 0$$

Therefore,

$$(T_1)_B - (T_2)_B = 1,400 \qquad \text{(e)}$$

Solving Eqs. (d) and (e) simultaneously, we get

$$(T_1)_B = 2,025 \text{ N}; \qquad (T_2)_B = 625 \text{ N}$$

Hence, the minimum $F_B$ needed for pulley $B$ is

$$F_B = (2,025 + 625) \cos 5.74° = \boxed{2,637 \text{ N}}$$

Note that the ratios of the tensions for *impending slippage* for the two cylinders are

$$\text{Driven cylinder } A \qquad \frac{T_1}{T_2} = 3.81 \qquad F_A = 2,384 \text{ N}$$

$$\text{Driving cylinder } B \qquad \frac{T_1}{T_2} = 3.24 \qquad F_B = 2,637 \text{ N}$$

If we use impending slippage for driven cylinder $A$, we will have slipping for driving cylinder $B$, which we cannot tolerate. Thus, if we use the larger force of 2,637 N, we will be well under the impending slippage condition of the driven cylinder $A$. Clearly, the optimum result to *avoid slippage* and to *minimize the belt tension* for greater life of the belt is to have a force somewhere above 2,637 N. Here is a place for good engineering judgement and experience.

---

[6]Note that the ratio of transmitted torques $M_2/M_1$ between directly connected pulleys and gears will equal $r_2/r_1$ or $D_2/D_1$ of the pulleys or gears. Can you verify this yourself?

# PROBLEMS

**7.48.** Compute the frictional resisting torque for the concentric dry thrust bearing. The coefficient of friction is taken as $\mu_d$.

**Figure P.7.48.**

**7.49.** The support end of a dry thrust bearing is shown. Four pads form the contact surface. If a shaft creates a 100-N thrust uniformly distributed over the pads, what is the resisting torque for a dynamic coefficient of friction of .1?

**Figure P.7.49.**

**7.50.** In Example 7.5, the normal force distribution at the contact surface is not uniform but, as a result of wear, is inversely proportional to the radius $r$. What, then, is the resisting torque $M$?

**7.51.** Compute the frictional torque needed to rotate the truncated cone relative to the fixed member. The cone has a 20-mm-diameter base and a 60° cone angle and is cut off 3 mm from the cone tip. The dynamic coefficient of friction is .2.

**Figure P.7.51.**

**7.52.** A 1,000-N block is being lowered down an inclined surface. The block is pinned to the incline at $C$, and at $B$ a cord is played out so as to cause the body to rotate at uniform speed about $C$. Taking $\mu_d$ to be .3 and assuming the contact pressure is uniform along the base of the block, compute $T$ for the configuration shown in the diagram.

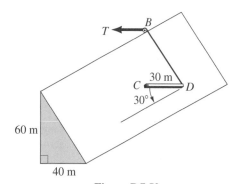

**Figure P.7.52.**

**7.53.** A thrust-bearing supports a force $P$ of 1,000 N. Two rings of the system are in contact with the base of the system. The larger ring has a uniform pressure which is half the value of the uniform pressure on the inner ring. The shaft is rotating with an angular speed $\omega$. If the dynamic coefficient of friction $\mu_d$ is 0.3, what is the resisting torque generated by this thrust bearing? Consider $P$ to include the weights of the shaft and the pads attached to it.

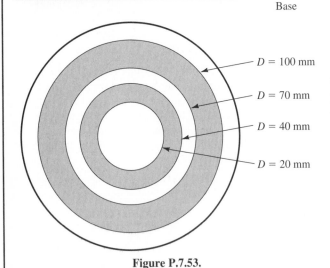

*Figure P.7.53.*

**7.54.** A shaft has three dry thrust-bearing surfaces which push, respectively, on three stationary rigid surfaces, namely rigid plates $A$ and $B$ and the rigid base $C$. The total load $P$ coming onto the shaft including weights of all parts is 6,000 N. What is the minimal torque $T$ needed to start rotation if the pressure on each disc contact surface is uniform for each disc but is proportional to the outside diameter squared of the disc? The static coefficient of friction between the bearing surfaces and the plates $A$ and $B$ as well as the base $C$ is 0.2. The proportionality constant for pressure is the *same value* for all 3 discs. The shaft passes through $A$ and $B$ but is welded to the top surface of $C$.

*Figure P.7.54.*

**7.55.** A pulley requires 200 N-m torque to get it rotating. $\mu_s$ is known to be .25. What is the *minimum* horizontal force $F$ required to create enough tension in the belt so that it can rotate the pulley?

*Figure P.7.55.*

**7.56.** If in Problem 7.55, $\theta = 0$ and the belt is wrapped $2\frac{1}{2}$ times around the pulley, what is the *minimum* horizontal force $F$ needed to rotate the pulley?

**7.57.** The seaman pulls with 100-N force and wants to stop the motorboat from moving away from the dock under power. How few wraps $n$ of the rope must he make around the post if the motorboat develops 3,500 N of thrust and the static coefficient of friction between the rope and the post is .2?

**Figure P.7.57.**

**7.58.** A length of belt rests on a flat surface and runs over a quarter of the drum. A load $W$ rests on the horizontal portion of the belt, which in turn is supported by a table. If the static coefficient of friction for all surfaces is .3, compute the *maximum* weight $W$ that can be moved by rotating the drum.

**Figure P.7.58.**

**7.59.** The rope holding the 50-lb weight $E$ passes over the drum and is attached at A. The weight of C is 60 lb. What is the *minimum* static coefficient of friction between the rope and the drum to maintain equilibrium of the drum?

**Figure P.7.59.**

**7.60.** What is the *maximum* weight that can be supported by the system in the position shown? Pulley B *cannot* turn. Bar AC is fixed to cylinder A, which weighs 500 N. The coefficient of static friction for all contact surfaces is .3.

**Figure P.7.60.**

**7.61.** A mountain climber of weight $W$ hangs freely suspended by one rope that is fastened at one end to his waist, wrapped one-half turn about a rock with $\mu_s = .2$, and held at the other end in his hand. What *minimum* force in terms of $W$ must he pull with to maintain his position? What minimum force must he pull the rope into himself with to gain altitude?

**Figure P.7.61.**

**7.62.** Pulley B is turned by a diesel engine and drives pulley A connected to a generator. If the torque that A must transmit to the generator is 500 N-m, what is the *minimum* static coefficient of friction between the belt and pulleys for the case where the force F is 2,000 N?

**Figure P.7.62.**

313

**7.63.** A hand brake is shown. If $\mu_d = .4$, what is the resisting torque when the shaft is rotating? What are the supporting forces on the rod AB.

**Figure P.7.63.**

**7.64.** A conveyor is shown with two driving pulleys $A$ and $B$. Driver $A$ has an angle of wrap of 330°, whereas $B$ has a wrap of 180°. If the dynamic coefficient of friction between the belt and the bed of the conveyor is .1, and the total weight to be transported is 10,000 N, what is the *smallest* static coefficient of friction between the belt and the driving pulleys? One-fifth of the load can be assumed to be between the two pulleys at all times, and the tension in the slack side (underneath) is 2,000 N. There is a free-wheeling pulley at the left end of the conveyor. You will have to solve an equation by trial and error.

**Figure P.7.64.**

**7.65.** A freely turning idler pulley is used to increase the angle of wrap for the pulleys shown. If the tension in the slack side above is 200 lb, find the *maximum* torque that can be transmitted by the pulleys for a static coefficient of friction of .3.

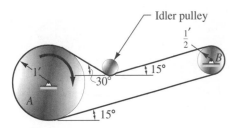

**Figure P.7.65.**

**7.66.** (a) What force $P$ is needed to develop a resisting torque of 65 N-m on the rotating drum? The dynamic coefficient of friction $\mu_d$ is 0.4.

(b) With the same force $P$ from part (a), what must the value of $\mu_d$ be in order to increase the resisting torque by 10 N-m?

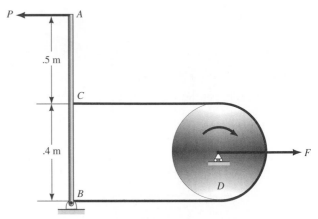

**Figure P.7.66.**

**7.67.** What is the maximum value of axial load $P$ to maintain a rotational speed $\omega = 5$ rad/s of the dry thrust bearing? The sum of the belt forces is 200 N. The contact surface between the belt and drum and between the dry thrust bearing and base have the same $\mu_s$ and $\mu_d$ of .4 and .3, respectively.

**Figure P.7.67.**

314

**7.68.** Rod $AB$ weighing 200 N is supported by a cable wrapped around a semicylinder having a coefficient of friction $\mu_s$ equal to .2. A weight $C$ having a mass of 10 kg can slide on rod $AB$. What is the maximum range $x$ from the centerline that the center of $C$ can be placed without causing slippage?

**Figure P.7.68.**

**7.69.** The cable mechanism shown is similar to that used to move the station indicator on a radio. If the indicator jams, what force is developed at the indicator base to free the jam when the required torque applied to the turning nob is 10 lb-in? Also, what are the forces in the various regions of the cable? The static coefficient of friction is .15.

**Figure P.7.69.**

**7.70.** What are the *minimum* possible supporting force components needed for pulley $B$ as a result of the action of the belt? The static coefficient of friction between the belt and pulley $B$ is .3 and between the belt and pulley $A$ is .4. The torque that the belt delivers to $A$ is 200 N-m.

**Figure P.7.70.**

**7.71.** From first principles, show that the normal force per unit length, $w$, acting on a drum from a belt is given as

$$w = \frac{T_2}{r}\,e^{\mu\theta}$$

Use the indicated diagram as an aid. [*Hint:* Start with Eq. 7.2(a) and use Eq. 7.4 for any point $a$.]

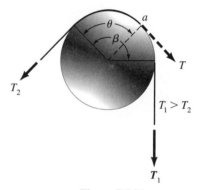

**Figure P.7.71.**

**7.72.** What *minimum* force $F$ is needed so that drum $A$ can transmit a clockwise torque of 500 N-m without slipping? The coefficient of friction, $\mu_s$, between $A$ and the belt is .4. What *minimum* coefficient of static friction is needed between drum $B$ and belt for no slipping?

**Figure P.7.72.**

315

**7.73.** What is the *minimum* weight B that will prevent rotation induced by body C weighing 500 N? The weight of A is 100 N. The static coefficient of friction between the belts and A is .4, and between A and the walls is .1. Neglect friction at pulley G.

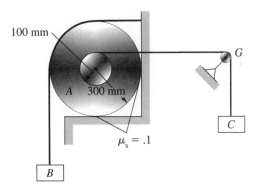

100 mm

$\mu_s = .1$

**Figure P.7.73.**

**7.74.** A V-belt is shown. Show that

$$\frac{T_1}{T_2} = e^{\mu_s \beta / \sin(\alpha/2)}$$

for impending slippage. Use a development analogous to that of the flat belt in Section 7.6.

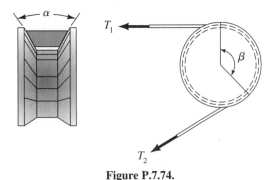

**Figure P.7.74.**

**7.75.** An electric motor drives a pulley B, which drives three V-belts having the cross-section shown. These V-belts then drive a compressor through pulley A. If the torque needed to drive the compressor is 1,000 N-m, what *minimum* force F is needed to do the job? The static coefficient of friction between belts and pulleys is .5. See Problem 7.74 before doing this problem.

**Figure P.7.75.**

**7.76.** A drum of radius r is rotated by a belt with a constant speed ω rad/sec. What is the relation between $T_1$ and $T_2$ for the case of impending slipping between drum and belt if *centrifugal* effects are accounted for? The belt has a mass per unit length of m kg/m. Recall from physics that the centrifugal force of a particle of mass M is $Mr\omega^2$, where r is the distance from the axis of rotation. Assume the belt is thin compared to the radius of the drum. The desired result is

$$\frac{T_1 - r^2\omega^2 m}{T_2 - r^2\omega^2 m} = e^{\mu_s \beta}$$

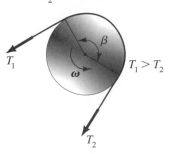

**Figure P.7.76.**

**7.77.** A pulley A is driven by an outside agent at a speed ω of 100 rpm. A belt weighing 30 N/m is driven by the pulley. If $T_2 = 200$ N, what is the *maximum* possible tension $T_1$ computed without considering centrifugal effects? Compute $T_1$ accounting for centrifugal effects, and give the percentage error incurred by not including centrifugal effects. The static coefficient of friction between the belt and the pulley is .3. See Problem 7.76 before doing this problem.

**Figure P.7.77.**

# 7.7   The Square Screw Thread

We shall now consider the action of a nut on a screw that has square threads (Fig. 7.22). Let us take $r$ as the mean radius from the centerline of the screw to the thread. The *pitch*, $p$, is the distance along the screw between adjacent threads, and the *lead*, $L$, is the distance that a nut will advance in the direction of the axis of the screw in one revolution. For screw threads that are single-threaded, $L$ equals $p$. For an $n$-threaded screw, the lead $L$ is $np$.

Forces are transmitted from screw to nut over several revolutions of thread, and hence we have a distribution of normal and friction forces. However, because of the narrow width of the thread, we may consider the distribution to be confined at a distance $r$ from the centerline, thus forming a "loading" strip winding around the centerline of the screw. Figure 7.22 illustrates infinitesimal normal and frictional forces on an infinitesimal part of the strip. The local slope $\tan \alpha$ as one looks in radially is determined by considering the definition of $L$, the lead. Thus,

$$\text{slope} = \tan \alpha = \frac{L}{2\pi r} = \frac{np}{2\pi r}$$

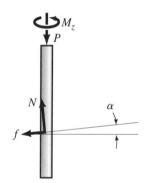

**Figure 7.22.** Square screw thread.

All elements of the proposed distribution have the same inclination (direction cosine) relative to the $z$ direction. In the summation of forces in this direction, therefore, we can consider the distribution to be replaced by a single normal force $N$ and a single friction force $f$ at the inclinations shown in Fig. 7.23 at a position anywhere along the thread. And, since the elements of the distribution have the same moment arm about the centerline in addition to the common inclination, we may use the concentrated forces mentioned above in taking moments about the centerline. There is thus a "limited equivalence" between $N$ and $f$ and the force distribution from the nut onto the screw. The other forces on the screw will be considered as an axial load $P$ and a torque $M_z$ collinear with $P$ (Fig. 7.23). For equilibrium at a condition of *impending motion* to raise the screw, we then have the following scalar equations:[7]

**Figure 7.23.** Free-body diagram.

$$\sum F_z = 0:$$

$$-P + N \cos \alpha - \mu_s N \sin \alpha = 0 \qquad \text{(a)}$$

$$\sum M_z = 0:$$

$$-\mu_s N \cos \alpha \, r - N \sin \alpha \, r + M_z = 0 \qquad \text{(b)}$$

[7]The equations also apply to *steady rotation* of the nut on the screw, in which case one uses the dynamic coefficient of friction $\mu_d$ in the equations.

These equations may be used to eliminate the force $N$ and so get a relation between $P$ and $M_z$ that will be of practical significance. This may readily be done by solving for $N$ in (a) and substituting into (b). The result is

$$M_z = \frac{Pr(\mu_s \cos \alpha + \sin \alpha)}{\cos \alpha - \mu_s \sin \alpha} \qquad (7.8)$$

An important question arises when we employ the screw and nut in the form of a jack as shown in Fig. 7.24. Once having raised a load $P$ by applying the torque $M_z$ to the jackscrew, does the device maintain the load at the raised position when the applied torque is released, or does the screw unwind

**Figure 7.24.** Jackscrew.

under the action of the load and thus lower the load? In other words, is this a *self-locking* device? To examine this, we go back to the equations of equilibrium. Setting $M_z = 0$ and changing the direction of the friction forces, we have the condition for impending "unwinding" of the screw. Eliminating $N$ from the equations, we get

$$\frac{Pr(-\mu_s \cos \alpha + \sin \alpha)}{\cos \alpha + \mu_s \sin \alpha} = 0$$

This requires that

$$-\mu_s \cos \alpha + \sin \alpha = 0$$

Therefore,

$$\mu_s = \tan \alpha \qquad (7.9)$$

We can conclude that, if the coefficient of friction $\mu_s$ equals or exceeds $\tan \alpha$, we will have a self-locking condition. If $\mu_s$ is less than $\tan \alpha$, the screw will unwind and will not support a load $P$ without the proper external torque.

## Example 7.9

A jackscrew with a double thread of mean diameter 2 in. is shown in Fig. 7.24. The pitch is .2 in. If a force $F$ of 40 lb is applied to the device, what load $W$ can be raised? With this load on the device, what will happen if the applied force $F$ is released? Take $\mu_s = .3$ for the surfaces of contact.

The applied torque $M_z$ is clearly:

$$M_z = \tfrac{8}{12}(40) = 26.7 \text{ lb-ft} \tag{a}$$

The angle $\alpha$ for this screw is given as

$$\tan \alpha = \frac{(2)(.2)}{(2\pi)(1)} = .0636 \tag{b}$$

Therefore,

$$\alpha = 3.64°$$

Using Eq. 7.8 we can solve for $P$. Thus,

$$P = \frac{M_z(\cos \alpha - \mu_s \sin \alpha)}{r(\mu_s \cos \alpha + \sin \alpha)}$$

$$= \frac{(26.7)[.998 - (.3)(.0635)]}{\tfrac{1}{12}[(.3)(.998) + .0635]} = \boxed{864 \text{ lb}} \tag{c}$$

The load $W$ is 864 lb. The device is self-locking since $\mu_s$ exceeds tan $\alpha = .0636$.

$$\boxed{\mu_s > \tan \alpha \quad \therefore \text{ self-locking}}$$

To lower the load requires a reverse torque. We may readily compute this torque by using Eq. 7.8 with the friction forces reversed. Thus,

$$(M_z)_{\text{down}} = \frac{864(\tfrac{1}{12})[-(.3)(.998) + .0635]}{.998 + (.3)(.0635)} = -16.71 \text{ lb-ft} \tag{d}$$

## *7.8   Rolling Resistance

Let us now consider the situation where a hard roller moves without slipping along a horizontal surface while supporting a load $W$ at the center. Since we know from experience that a horizontal force $P$ is required to maintain uniform motion, some sort of resistance must be present. We can understand this resistance if we examine the deformation shown in an exaggerated manner in

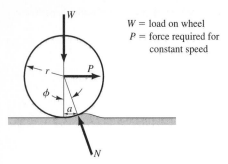

$W$ = load on wheel
$P$ = force required for constant speed

**Figure 7.25.** Rolling resistance model.

Fig. 7.25. If force $P$ is along the centerline as shown, the equivalent force system coming onto the roller from the region of contact must be that of a force $N$ whose line of action also goes through the center of the roller since, you will recall from Chapter 5, three nonparallel forces must be concurrent for equilibrium. In order to develop a resistance to motion, clearly $N$ must be oriented at an angle $\phi$ with the vertical direction, as is shown in Fig. 7.25. The scalar equations of *equilibrium* become

$$W = N \cos \phi; \qquad P = N \sin \phi$$

Therefore,

$$\frac{P}{W} = \tan \phi \qquad (7.10)$$

Since the area of contact is small, we note that $\phi$ is a small angle and that $\tan \phi \approx \sin \phi$. The $\sin \phi$ is seen to be $a/r$ from Fig. 7.25. Therefore, we may say that

$$\frac{P}{W} = \frac{a}{r} \qquad (7.11a)$$

Solving for $P$, we get

$$P = \frac{Wa}{r} \qquad (7.11b)$$

The distance $a$ in these equations is called the *coefficient of rolling resistance*.

Coulomb suggested that for *variable* loads $W$, the ratio $P/W$ is constant for given materials and a given geometry ($r$ = constant). Looking at Eq. 7.11a, we see that $a$ must then be a constant for given geometry and materials. Coulomb added that, for given materials and *variable* radius, the ratio $P/W$ varies inversely as $r$; that is, as the radius of the cylinder is increased, the resistance to uniform motion for a given load $W$ decreases. Thus, considering Eq. 7.11a again, we may conclude that, for given materials, $a$ is also constant for all sizes of rollers and loads. However, other investigators have contested both statements, particularly the latter one, and there is a need for further investigation in this area. Lacking better data, we present the following list of rolling coefficients for your use, but we must caution that you should not expect great accuracy from this general procedure.

## Coefficients of Rolling Resistance

|  | $a$ (in.) |
|---|---|
| Steel on steel | .007 – .015 |
| Steel on wood | .06 – .10 |
| Pneumatic tires on smooth road | .02 – .03 |
| Pneumatic tires on mud road | .04 – .06 |
| Hardened steel on hardened steel | .0002 – .0005 |

## ■ Example 7.10

What is the rolling resistance of a railroad freight car weighing 100 tons? The wheels have a diameter of 30 in. The coefficient of rolling resistance between wheel and track is .01 in. Compare the resistance to that of a truck and trailer having the same total weight and with tires having a diameter of 4 ft. The coefficient of rolling resistance $a$ between the truck tires and road is .025 in.

   We can use Eq. 7.11b directly for the desired results. Thus, for the railroad freight car, we have[8]

$$P_1 = \frac{(100)(2,000)(.01)}{15} = \boxed{133.3 \text{ lb}} \tag{a}$$

For the truck, we get

$$P_2 = \frac{(100)(2,000)(.025)}{24} = \boxed{208 \text{ lb}} \tag{b}$$

We see a decided differences between the two vehicles, with clear advantage toward the railroad freight car.

---

[8]The number of wheels $n$ plays no role here since we divide the load by $n$ to get the load per wheel and then multiply by $n$ to get the total resistance.

# PROBLEMS

**7.78.** A simple C-clamp is used to hold two pieces of metal together. The clamp has a single square thread with a pitch of . 12 in. and a mean diameter of .75 in. The static coefficient of friction is .30. Find the torque required if a 1,000-lb compressive load is required on the blocks. If the thread is a double thread, what is the required torque?

**Figure P.7.78.**

**7.79.** The mast of a sailboat is held by wires called shrouds, as shown in the diagram. Racing sailors are careful to get the proper tension in the shrouds by adjusting the turnbuckle at the bottom of the shrouds. When we do this we say we are "tuning" the boat. If a tension of 150 N exists in the shroud, what torque is needed to start tightening further by turning the turnbuckle? The pitch of the single threaded screw is 1.5 mm and the mean diameter is 8.0 mm. The static coefficient of friction is .2.

**Figure P.7.79.**

**7.80.** Forces $F$ of 50 lb are applied to the jackscrew shown. The thread diameter is 2 in. and the pitch is $\frac{1}{2}$ in. The static coefficient of friction for the thread is .05. The weight $W$ and collar are not permitted to rotate and so the collar must rotate on the shaft of the screw. If the static coefficient of friction between the collar and shaft is .1, determine the weight $W$ that can be lifted by this system.

**Figure P.7.80.**

**7.81.** A brake is shown. Force is developed at the brake shoes by turning $A$, which has a single right-handed square screw thread at $B$ and a single left-handed screw thread at $C$. The diameter of the screw thread is $1\frac{1}{2}$ in. and the pitch is .3 in. If the static coefficient of friction is .1 for the thread and the dynamic coefficient of friction is .4 for the brake shoes, what resisting torque is developed on the wheel by a 100 in.-lb torque at $A$?

**Figure P.7.81.**

**7.82.** A triangular-threaded screw is shown. In a manner paralleling the development of the square-thread formulation in Section 7.7, show that

$$M_z = \frac{rP(\mu_s \cos \alpha + \cos \theta \tan \alpha)}{\cos \theta - \mu_s \sin \alpha}$$

where

$$\cos \theta = \frac{1}{\sqrt{\tan^2 \alpha + \tan^2 \gamma + 1}}$$

and

$$\gamma = \beta - \alpha$$

**Figure P.7.82.**

**7.83.** Consider a single-threaded screw where the pitch $p = 4.5$ mm and the mean radius is 20 mm. For a coefficient of friction $\mu_s = .3$, what torque is needed on the nut for it to turn under a load of 1,000 N? Compute this for a square thread and then do it for a triangular thread where the angle $\beta$ is 30°. See Problem 7.82 before doing this problem.

**7.84.** If the coefficient of rolling resistance of a cylinder on a flat surface is .05 in., at what inclination of the surface will the cylinder of radius $r = 1$ ft roll with uniform velocity?

**7.85.** A 65-kN vehicle designed for polar expeditions is on a very slippery ice surface for which the static coefficient of friction between tires and ice is .005. Also, the coefficient of rolling resistance is known to be .8 mm. Will the vehicle be able to move? The vehicle has four-wheel drive.

**Figure P.7.85.**

**7.86.** In Problem 7.85, suppose there is only rear-wheel drive available. What is the minimum static coefficient of friction needed between tires and ground for the vehicle to move?

**7.87.** A roller thrust bearing is shown supporting a force $P$ of 2.5 kN. What torque $T$ is need to turn the shaft $A$ at constant speed if the only resistance is that from the ball bearings? The coefficient of rolling resistance for the balls and the bearing surfaces is .01270 mm. The mean radius from the centerline of the shaft to the balls is 30 mm.

**Figure P.7.87.**

# 7.9 Closure

In this chapter, we have examined the results of two independent experiments: that of impending or actual sliding of one body over another and that of a cylinder or sphere rolling at constant speed over a flat surface. Without any theoretical basis, the results of such experiments must be used in situations that closely parallel the experiments themselves.

In the case of a rolling cylinder, both rolling resistance and sliding resistance are present. However, for a cylinder accelerating with any appreciable magnitude, only sliding friction need be accounted for. With no acceleration on a horizontal surface, only rolling resistance need be considered. Most situations fall into these categories. For very small accelerations, both effects are present and must be taken into account. We can then expect only a crude result for such computations.

Before going further, we must carefully define certain properties of plane surfaces in order to facilitate later computations in mechanics where such properties are most useful. These plane surface properties and other related topics will be studied in Chapter 8.

**7.88.** If the static coefficient of friction for all surfaces is .35, find the force $F$ needed to start the 200-N weight moving to the right.

**Figure P.7.88.**

**7.89.** A loaded crate is shown. The crate weighs 500 lb with a center of gravity at its geometric center. The contact surface between crate and floor has a static coefficient of friction of .2. If $\theta = 90°$, show that the crate will slide before one can increase $T$ enough for tipping to occur. If a stop is to be inserted in the floor at $A$ to prevent slipping so that the crate could be tipped, what *minimum* horizontal force will be exerted on the stop?

**Figure P.7.89.**

**7.90.** In Problem 7.89, compute a value of $\theta$ and $T$ where slipping and tipping will occur simultaneously. If the actual angle $\theta$ is smaller than this value of $\theta$, is there any further need of the stop at $A$ to prevent slipping?

**7.91.** A friction drive is shown with $A$ the driver disc and $B$ the driven disc. If force $F$ pressing $B$ onto $A$ is 150 N, what is the *maximum* torque $M_2$ that can be developed? For this torque, what is the torque $M_1$ needed for the drive disc $A$? The static coefficient of friction between $A$ and $B$ is .7. What vertical force must rod $G$ withstand for the action described above?

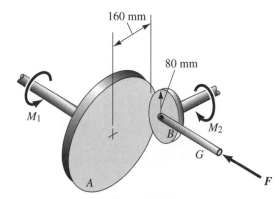

**Figure P.7.91.**

**7.92.** Determine the weight of block $A$ for impending motion to the right. The static coefficient of friction between the cable and the surfaces which it contacts is 0.2. The static coefficient of friction between block $A$ and the surface upon which it rests is 0.4. The two posts are circular with a diameter of 0.25 m.

**Figure P.7.92.**

**7.93.** Do Problem 7.92 for impending slippage of body $A$ to the left.

**7.94.** A tug is pushing a barge into a berth. After the barge turns clockwise and touches the sides of the pilings, what thrust must the tug develop to move it at uniform speed of 2 knots farther into the berth? The dynamic coefficient of friction between the barge and the sides of the berth is .4. The drag from the water is 3,000 N along the centerline of the barge.

**7.96.** The drum is driven by a motor with a maximum torque capability of 500 lb-ft. The static coefficient of friction between the drum and the braking strap (belt) is .4. How much force $P$ must an operator exert to stop the drum if it rotates (1) clockwise and (2) counterclockwise? What are the belt forces in each case?

**Figure P.7.94.**

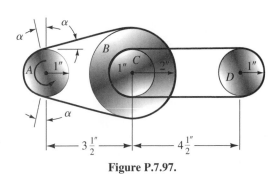

**Figure P.7.96.**

**7.97.** The four drive pulleys shown are used to transmit a torque from pulley $A$ to pulley $D$ on an electric typewriter. If the static coefficient of friction between the belts and the pulleys is .3, what is the torque available at pulley $D$ if 10 lb-in. of torque is input to the shaft of pulley $A$? What are the belt forces?

**7.95.** The static and dynamic coefficients of friction for the upper surface of contact $A$ of the cylinder are $\mu_s = .4$, $\mu_d = .3$, and for the lower surface of contact $B$ are $\mu_s = .1$ and $\mu_d = .08$. What is the *minimum* force $P$ needed to just get the cylinder moving?

**Figure P.7.95.**

**Figure P.7.97.**

**7.98.** A scissors jack is shown lifting the end of a car so that $R = 6.67$ kN. What torque $T$ is needed for this operation? Note that $A$ is merely a bearing and at $B$ we have a nut. The screw is single-threaded with a pitch of 3 mm and a mean diameter of 20 mm. The static coefficient of friction between the screw and nut at $B$ is .3. Neglect the weight of the members and evaluate $T$ for $\theta = 45°$ and for $\theta = 60°$.

**7.100.** A hot rectangular metal ingot is to be flattened by passing through cylindrical rollers. If the ingot is to be drawn into the rollers by friction once it touches the rollers, what is the *minimum* thickness $t$ of the ingot that can be achieved by this process on one pass? The static coefficient of friction for the contact between ingot and cylinder is .3. The cylinders rotate as shown with angular speed $\omega$.

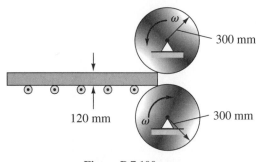

**Figure P.7.100.**

**7.101.** A cone clutch is shown. Assuming that uniform pressures exist between the contact surfaces, compute the *maximum* torque that can be transmitted. The static coefficient of friction is .30 and the activating force $F$ is 100 lb. [*Hint:* Assume that the moving cone transmits its 100-lb axial force to the stationary cone by pressure primarily. That is, we will neglect the friction-force component on the cone surface normal to the transverse direction.]

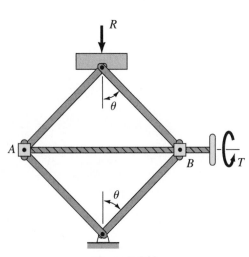

**Figure P.7.98.**

**7.99.** A block $C$ weighing 10 kN is being moved on rollers $A$ and $B$ each weighing 1 kN. What force $P$ is needed to maintain steady motion? Take the coefficient of rolling resistance between the rollers and the ground to be .6 mm and between block $C$ and the rollers to be .4 mm.

**Figure P.7.101.**

**Figure P.7.99.**

**7.102.** In Fig. P.7.18, delete the external forces and couple at point $G$, and replace with a force of 6,000 N at point $G$ at an angle $\alpha$ with the horizontal going from left to right. For the condition of impending slippage in the downward direction, what should the angle $\alpha$ be?

Figure P.7.102.

**7.103.** In Problem 7.39, what is the minimum angle between the supporting links $CD$ and $ED$ to support any weight $W$ if the static coefficient of friction is $\mu_s = .4$?

**7.104.** A shaft $AB$ rotates at constant angular speed in a pair of *dry* journal bearings. The total weight of the shaft and the cylinder it is supporting is $W$. A torque $T$ is needed to maintain the steady angular motion. There is a small clearance between the shaft and the journal, resulting in a *point contact* between these bodies at some position $E$ as shown. Form a two-dimensional force system acting on the shaft by moving $W$, the total friction force, the total normal force on the shaft surface, and the torque to a plane normal to the centerline of the shaft and at a location at the center of the system. View this system along the centerline.

    (a) From considerations of equilibrium, what and where is the resultant force vector from the friction force and the normal force onto the shaft?

    (b) If the angle $\phi$ in the diagram is very small and if the coefficient of dynamic friction is $\mu_d$, explain how we can give the following approximate equation

$$T = W\mu_d r$$

where $r$ is the radius of the shaft.

Figure P.7.104.

**7.105.** Shaft $CD$ rotates at constant angular speed in a set of dry journal bearings (as a result of poor maintenance). Approximately what torque is needed to maintain the angular motion for the following data?

$M_A = 50$ kg    $M_B = 80$ kg    Shaft $CD = 40$ kg    $\mu_d = .2$

See Problem 7.104 before doing this problem.

Figure P.7.105.

**7.106.** Two identical light rods are pinned together at $B$. End $C$ of rod $BC$ is pinned while end $A$ of rod $AB$ rests on a rough floor having a coefficient of friction with the rod of $\mu_d = .5$. The spring requires a force of 5 N/mm of stretch. A load is applied slowly at $B$ and then maintained constant at $F = 300$ N. What is the angle $\theta$ when the system ceases to move? The spring is unstretched when $\theta = 45°$. [*Hint:* You will have to solve an equation by trial and error.]

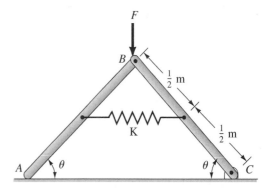

Figure P.7.106.

327

**7.107.** A device for "throwing" baseballs is shown. This device is to be found in amusement parks for batting practice. The two wheels are inflated automobile tires that rotate as shown with a constant speed $\omega$. A baseball is fed to a point where it is touching the wheels. What is the *minimum* separation $d$ of the wheels if the ball is to be drawn into the slot and then ejected on the other side as a pitched ball? The coefficient of static friction at the contact surfaces is .4.

**7.109.** What is the *maximum* angle $\alpha$ for which there will be equilibrium if $A$ weighs 1,000 N and if $\mu_s$ between the supports and the curved rod is .3? The rods are each 1.3 m long. You will have to solve a transcendental equation by trial and error.

**Figure P.7.107.**

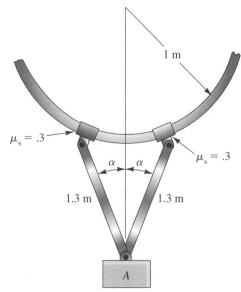

**Figure P.7.109.**

**7.108.** What is the *maximum* angle $\alpha$ for which there will be equilibrium if $A$ has a mass of 50 kg. The static coefficient of friction between the supports and the horizontal rod is equal to .3. What is the force in each of the supporting members?

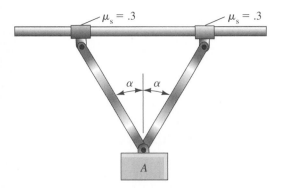

**Figure P.7.108.**

**7.110.** A 500-kN crate is being lowered slowly down an elevator shaft, which as shown is slightly wider than the crate. The cord wraps around a freely turning pulley and then has 2 rotations around a capstan. Find the tension $T$ needed for the operation. The center of gravity of the crate is at its geometric center. [*Hint:* The crate will rotate so as to rub against the elevator shaft at points $A$ and $B$.]

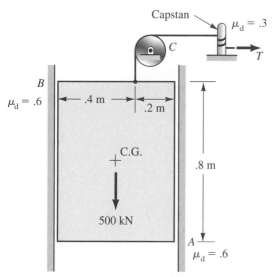

**Figure P.7.110.**

**7.111.** A triangular pile of width 1 ft is being driven into the ground slowly by a force $P$ of 50,000 lb. There is pressure on the lateral surfaces of the pile. This pressure varies linearly from 0 at $A$ to $p_0$ at $B$, as has been shown in the diagram. If the coefficient of dynamic friction between the pile and the soil is 0.6, what is the maximum pressure $p_0$?

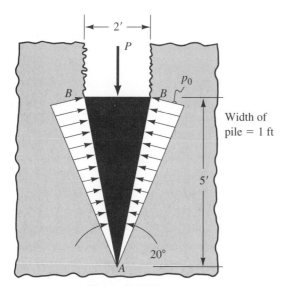

**Figure P.7.111.**

**7.112.** In Problem 7.27, what maximum value should $W_2$ be in order to start moving the system to the left? All other data are unchanged.

**7.113.** A block rests on a surface for which there is a coefficient of friction $\mu_s = .2$. Over what range of angle $\beta$ will there be no movement of the block for the 150-N force? (You will have to solve an equation by trial and error.)

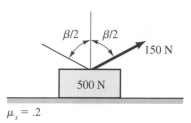

**Figure P.7.113.**

**7.114.** What is the *largest* load that can be suspended without moving blocks $A$ and $B$? The static coefficient of friction for all plane surfaces of contact is .3. Block $A$ weighs 500 N and block $B$ weighs 700 N. Neglect friction in the pulley system.

**Figure P.7.114.**

**7.115.** What is the *minimum* force $F$ to hold the cylinders, each weighing 100 lb? Take $\mu_s = .2$ for all surfaces of contact.

**Figure P.7.115.**

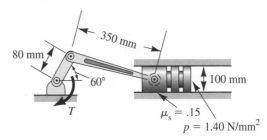

**Figure P.7.116.**

**7.117.** Find the cord tension if block 1 attains maximum friction.

**7.116.** A compressor is shown. If the pressure in the cylinder is 1.40 N/mm² above atmosphere (gage), what *minimum* torque $T$ is needed to initiate motion in the system? Neglect the weight of the crank and connecting rod as far as their contribution toward moving the system. Consider friction only between the piston and cylinder walls where the coefficient of friction $\mu_s = .15$.

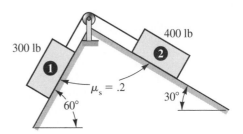

**Figure P.7.117.**

330

# CHAPTER 8

# Properties
# of Surfaces

## 8.1 Introduction

If we are buying a tract of land, we certainly want to consider the size and, with equal interest, the shape and orientation of the earth's surface, and possibly its agricultural, geological, or aesthetic potentials. The size of a surface (i.e., the area) is a familiar concept and has been used in the previous section. Certain aspects of the shape and orientation of a surface will be examined in this chapter. There are a number of formulations that convey meaning about the shape and disposition of a surface relative to some reference. To be sure, these formulations are not used by real estate people, but in engineering work, where a variety of quantitative descriptions are necessary, these formulations will prove most useful. In general, we shall restrict our attention to coplanar surfaces.

## 8.2 First Moment of an Area and the Centroid

A coplanar surface of area $A$ and a reference $xy$ in the plane of the surface are shown in Fig. 8.1. We define the *first moment of area $A$* about the $x$ axis as

$$M_x = \int_A y \, dA \tag{8.1}$$

and the first moment about the $y$ axis as

$$M_y = \int_A x \, dA \tag{8.2}$$

These two quantities convey a certain knowledge of the shape, size, and orientation of the area, which we can use in many analyses of mechanics.

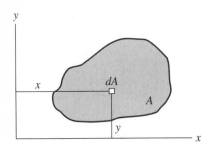

**Figure 8.1.** Plane area.

You will no doubt notice the similarity of the preceding integrals to those which would occur for computing moments about the *x* and *y* axes from a parallel force distribution oriented normal to the area *A* in Fig. 8.1. The moment of such a force distribution has been shown for the purposes of rigid-body calculations to be equivalent to that of a single resultant force located at a particular point $\bar{x}$, $\bar{y}$. Similarly, we can concentrate the entire area *A* at a position $x_c$, $y_c$, called the *centroid*,[1] where, for computations of first moments, this new arrangement is equivalent to the original distribution (Fig. 8.2). The coordinates $x_c$ and $y_c$ are usually called the *centroidal coordinates*. To compute these coordinates, we simply equate moments of the distributed area with that of the concentrated area about both axes:

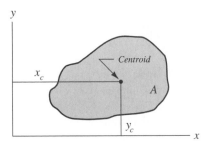

**Figure 8.2.** Centroidal coordinates.

$$Ay_c = \int_A y\, dA; \qquad \text{therefore, } y_c = \frac{\int_A y\, dA}{A} = \frac{M_x}{A} \qquad (8.3a)$$

$$Ax_c = \int_A x\, dA; \qquad \text{therefore, } x_c = \frac{\int_A x\, dA}{A} = \frac{M_y}{A} \qquad (8.3b)$$

The location of the centroid of an area can readily be shown to be *independent* of the *reference axes* employed. That is, the centroid is a *property* only of the *area* itself. We have asked the reader to prove this in Problem 8.1.

If the axes *xy* have their origin at the centroid, then these axes are called *centroidal axes* and clearly the first moments about these axes must be zero.

Finally, we point out that *all* axes going through the centroid of an area are called *centroidal axes* for that area. Clearly, the *first moments of an area about any of its centroidal axes must be zero*. This must be true since the perpendicular distance from the centroid to the centroidal axis must be zero.

---

[1]The concept of the centroid can be used for any geometric quantity. In the next section, we shall consider centroids of volumes and arcs.

# Example 8.1

A plane surface is shown in Fig. 8.3 bounded by the $x$ axis, the curve $y^2 = 25x$, and a line parallel to the $y$ axis. What are the first moments of the area about the $x$ and $y$ axes and what are the centroidal coordinates?

We shall first compute $M_x$ and $M_y$ for this area. Using vertical infinitesimal area elements of width $dx$ and height $y$, we have noting that $y = 5\sqrt{x}$

$$M_y = \int_0^{10} x(y\,dx) = \int_0^{10} x(5\sqrt{x})\,dx$$

$$= \frac{5x^{5/2}}{\frac{5}{2}}\Bigg|_0^{10} = \boxed{632 \text{ ft}^3}$$

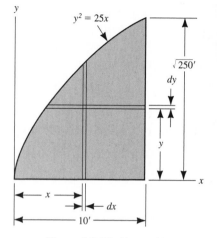

**Figure 8.3.** Find centroid.

To compute $M_x$, we use horizontal area elements of width $dy$ and length $(10-x)$ as shown in the diagram. Thus

$$M_x = \int_0^{\sqrt{250}} y[(10 - x)dy]$$

$$= \int_0^{\sqrt{250}} \left(10y - \frac{y^3}{25}\right) dy$$

$$= \left(5y^2 - \frac{y^4}{100}\right)\Bigg|_0^{\sqrt{250}} = \boxed{625 \text{ ft}^3}$$

We could also have used vertical strips for computing $M_x$ as follows using centroids of the vertical strips:

$$M_x = \int_0^{10} \frac{y}{2}(y\,dx) = \int_0^{10} \frac{25x}{2}\,dx$$

$$= (12.5)\left(\frac{x^2}{2}\right)\Bigg|_0^{10} = 625 \text{ ft}^3$$

To compute the position of the centroid $(x_c, y_c)$, we will need the area $A$ of the surface. Thus, using vertical strips:

$$A = \int_0^{10} y\,dx = \int_0^{10} 5\sqrt{x}\,dx = \frac{5x^{3/2}}{\frac{3}{2}}\Bigg|_0^{10}$$

$$= 105.4 \text{ ft}^2$$

## ■ Example 8.1 (Continued)

The centroidal coordinates are, accordingly,

$$x_c = \frac{M_y}{A} = \frac{632}{105.4} = \boxed{6.00 \text{ ft}}$$

$$y_c = \frac{M_x}{A} = \frac{625}{105.4} = \boxed{5.93 \text{ ft}}$$

To get the moment of the area about an axis $y'$, which is 15 ft to the left of the $y$ axis, simply proceed as follows:

$$M_{y'} = (A)(x_c + 15) = 105.4(6.00 + 15) = 2{,}213 \text{ ft}^3$$

Consider now a plane area with an *axis of symmetry* such as is shown in Fig. 8.4, where the $y$ axis is collinear with the axis of symmetry. In computing $x_c$ for this area, we have

$$x_c = \frac{1}{A} \int_A x \, dA$$

In evaluating the integral above, we can consider area elements in symmetric pairs such as shown in Fig. 8.4, where we have shown a pair of area elements which are mirror images of each other about the axis of symmetry. Clearly, the first moment of such a pair about the axis of symmetry is zero. And, since the entire area can be considered as composed of such pairs, we can conclude that $x_c = 0$. Thus, the centroid of an area with one axis of symmetry must therefore lie somewhere along this axis of symmetry. The axis of symmetry then is a centroidal axis, which is another indication that the first moment of area must be zero about such an axis. With two orthogonal axes of symmetry, the centroid must lie at the intersection of these axes. Thus, for such areas as circles and rectangles, the centroid is easily determined by inspection.

In many problems, the area of interest can be considered formed by the addition or subtraction of simple familiar areas whose centroids are known by inspection as well as by other familiar areas, such as triangles and sectors of circles whose centroids and areas are given in handbooks. We call areas made up of such simple areas *composite* areas. (A listing of familiar areas is given for your convenience on the inside back cover of this text.) For such problems, we can say that

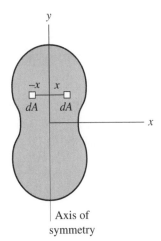

**Figure 8.4.** Area with one axis of symmetry.

$$x_c = \frac{\sum_i A_i \bar{x}_i}{A}$$

$$y_c = \frac{\sum_i A_i \bar{y}_i}{A}$$

where $\bar{x}_i$ and $\bar{y}_i$, (with proper signs) are the centroidal coordinates to simple area $A_i$, and where $A$ is the total area.

# Example 8.2

Find the centroid of the shaded section shown in Fig. 8.5.

**Figure 8.5.** Composite area.

We may consider four separate areas. These are the triangle (1), the circle (2), and the rectangle (3) all cut from an original rectangular $200 \times 140$ mm$^2$ area which we denote as area (4). In composite-area problems, we urge you to set up a format of the kind we shall now illustrate. Using the positions of the centroid of a right triangle as given in the inside covers of this text, we have:

| $A_i$ | $\bar{x}_i$ | $A_i\bar{x}_i$ | $\bar{y}_i$ | $A_i\bar{y}_i$ |
|---|---|---|---|---|
| $A_1 = -\frac{1}{2}(30)(80) = -1,200$ | 10 | $-12,000$ | 113.3 | $-136,000$ |
| $A_2 = -\pi 50^2 \quad\;\; = -7,850$ | 100 | $-785,000$ | 70 | $-549,780$ |
| $A_3 = -(40)(60) \quad = -2,400$ | 180 | $-432,000$ | 110 | $-264,000$ |
| $A_4 = (200)(140) \quad = \underline{28,000}$ | 100 | $2,800,000$ | 70 | $1,960,000$ |
| $A = 16,550$ mm$^2$ | | $\displaystyle\sum_i A_i\bar{x}_i =$ | | $\displaystyle\sum_i A_i\bar{y}_i =$ |
| | | $1.571 \times 10^6$ mm$^3$ | | $1.011 \times 10^6$ mm$^3$ |

Therefore,

$$x_c = \frac{\sum A_i\bar{x}_i}{A} = \frac{1.571 \times 10^6}{16,550} = \boxed{94.9 \text{ mm}}$$

$$y_c = \frac{\sum A_i\bar{y}_i}{A} = \frac{1.011 \times 10^6}{16,550} = \boxed{61.1 \text{ mm}}$$

In closing, we would like to point out that the centroid concept can be of use in finding the simplest resultant of a distributed loading. Thus, consider the distributed loading $w(x)$ shown in Fig. 8.6. The resultant force $F_R$ of this loading, also shown in the diagram, is given as

$$F_R = \int_0^L w(x)\,dx \tag{8.4}$$

**Figure 8.6.** Loading curve $w(x)$ and its resultant $F_R$.

From the equation above, we can readily see that the *resultant force equals the area under the loading curve*. To get the position of the *simplest* resultant for the loading, we then say that

$$F_R \bar{x} = \int_0^L xw(x)\,dx$$

Therefore,

$$\bar{x} = \frac{\int_0^L xw(x)\,dx}{F_R} \tag{8.5}$$

The preceding result shows that $\bar{x}$ is actually the centroidal coordinate of the loading curve area from reference $xy$. Thus, the *simplest resultant force of a distributed load acts at the centroid of the area under the loading curve*. Accordingly, for a triangular load such as is shown in Fig. 8.7, we can replace the loading for free bodies on which the entire loading acts by a force $F$ equal to $(\frac{1}{2})(w_0)(b - a)$ at a position $\frac{2}{3}(b - a)$ from the left end of the loading. You will recall that we pointed this out in Chapter 4.

**Figure 8.7.** Triangular loading with simplest resultant.

# PROBLEMS

**8.1.** Show that the centroid of area $A$ is the same point for axes $xy$ and $x'y'$. Thus, the position of the centroid of an area is a property only of the area.

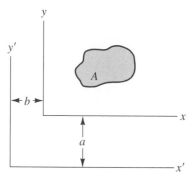
**Figure P.8.1.**

**8.2.** Show that the centroid of the right triangle is $x_c = 2a/3$, $y_c = b/3$.

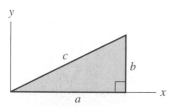
**Figure P.8.2.**

**8.3.** Find the centroid of the area under the half-sine wave. What is the first moment of this area about axis $A$–$A$?

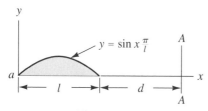
**Figure P.8.3.**

**8.4.** What are the first moments of the area about the $x$ and $y$ axes? The curved boundary is that of a parabola. [*Hint:* The general equation for parabolas of the shape shown is $y^2 = ax + b$.]

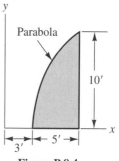
**Figure P.8.4.**

**8.5.** What are the centroidal coordinates for the shaded area? The curved boundary is that of a parabola. [*Hint:* The general equation for parabolas of the shape shown is $y = ax^2 + b$.]

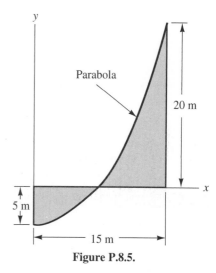
**Figure P.8.5.**

**8.6.** Show that the centroid of the area under a semicircle is as shown in the diagram.

**Figure P.8.6.**

337

**8.7.** What is the first moment of the area under the parabola about an axis through the origin and going through point $r = 6i + 7j$ m. Take $l = 10$ m.

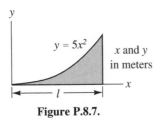

**Figure P.8.7.**

**8.8.** Find the centroidal coordinates of the shaded area.

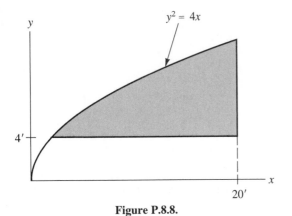

**Figure P.8.8.**

**8.9.** Express the first moment of area of the shaded area as a function of $y$, $b$, and $h$ for $y$ ranging from the top of the rectangle to the bottom.

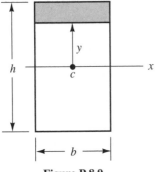

**Figure P.8.9.**

**8.10.** Suppose you had an irregular area and you had to *approximate* the location of the centroid. How could you accomplish this using only a straight edge, a ruler, and a pencil?

**Figure P.8.10.**

**8.11.** Show that the centroid of the triangle is at $x_c = (a + b)/3$, $y_c = h/3$. [*Hint:* Break the triangle into two right triangles for which the centroids are known from Problem 8.2.]

**Figure P.8.11.**

**8.12.** What are the centroidal coordinates for the shaded area? The outer boundary is that of a circle having a radius of 1 m.

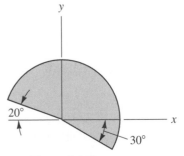

**Figure P.8.12.**

**8.13.** What are the coordinates of the centroid of the shaded area? The parabola is given as $y^2 = 2x$ with $y$ and $x$ in millimeters.

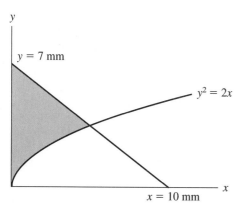

**Figure P.8.13.**

**8.14.** Find the centroid of the shaded area. The equation of the curve is $y = 5x^2$ with $x$ and $y$ in millimeters. What is the first moment of the area about line $AB$?

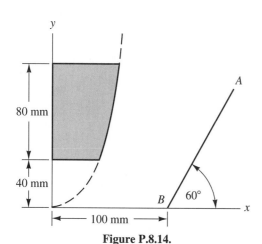

**Figure P.8.14.**

**8.15.** Find the centroid of the shaded area. What is the first moment of this area about line $A$–$A$. The upper boundary is a parabola $y^2 = 50x$ with $x$ and $y$ in millimeters.

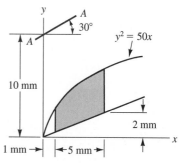

**Figure P.8.15.**

*In the remaining problems of this section, use centroidal positions of simple areas as found in the inside covers.*

**8.16.** Find the centroid of the end shield of a bulldozer blade.

**Figure P.8.16.**

**8.17.** In Example 2.6, determine $z_C$ for the centroid of the triangular faces of the pyramid plus the base area $ABCE$. The height of the pyramid is 300 ft.

**8.18.** A parallelogram and an ellipse have been cut from a rectangular plate, What are the centroidal coordinates of what is left of the plate? What is $M_x$ for this area? Use formulas given on the inside of the back covers.

**Figure P.8.18.**

339

**8.19.** A *median* axis has the same area on one side of the axis as it does on the other side. Find the distance between the horizontal median line and a parallel centroidal axis.

**Figure P.8.19.**

**8.20.** Find the centroid of the truss gusset plate.

**Figure P.8.20.**

**8.21.** Find the centroid of the indicated area.

**Figure P.8.21.**

**8.22.** Find the centroidal coordinates for the shaded area shown. Give the results in meters. [*Hint:* See Fig. P.8.6.]

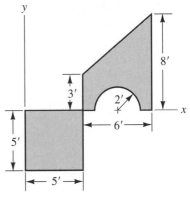

**Figure P.8.22.**

**8.23.** Find the centroid of the end of the bucket of a small front-end loader.

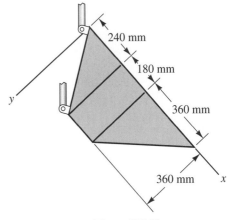

**Figure P.8.23.**

**8.24.** Where is the centroid of the airplane's vertical stabilizer (whole area)?

**Figure P.8.24.**

**8.25.** What is the first moment of the shaded area about the diagonal *A–A*? [*Hint:* Consider symmetry.]

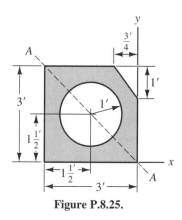

**Figure P.8.25.**

**8.27.** A wide-flange I beam (identified as 14WF202 I beam) is shown with two reinforcing plates on top. At what height above the bottom is the centroid of the beam located?

**Figure P.8.27.**

**8.26.** A built-up beam is shown with four 120-mm by 120-mm by 20-mm angles. Find the vertical distance above the base for the centroid of the cross-section.

**8.28.** Compute the position of the centroid of the shaded area. [*Hint:* See Fig. P.8.6.]

**Figure P.8.26.**

**Figure P.8.28.**

**8.29.** Find the centroid of the sheet metal cover of a centrifugal blower (shown shaded).

**Figure P.8.29.**

**8.30.** What is the position from the left end of the simplest resultant force of the distribution shown?

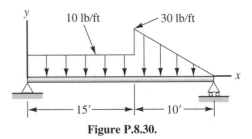

**Figure P.8.30.**

## 8.3 Other Centers

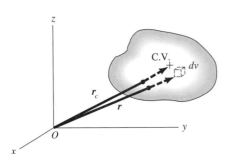

**Figure 8.8.** Center of volume, C.V., of a body.

We employ the concepts of moments and centroids in mechanics for three-dimensional bodies as well as for plane areas. Thus, we introduce now the first moment of a volume, $V$, of a body (see Fig. 8.8) about a point $O$ where we have shown a reference $xyz$. We say that the first *moment of volume V* about $O$ is

$$\text{moment vector of volume} \equiv \iiint_V r \, dv \qquad (8.6)$$

The *center of volume*, $r_c$, is then defined as follows:

$$Vr_c = \iiint_V r \, dv$$

Therefore,

$$r_c = \frac{1}{V} \iiint_V r \, dv \qquad (8.7)$$

We see that the center of volume is the point where we could hypothetically concentrate the entire volume of a body for purposes of computing the first moment of the volume of the body about some point $O$. The components of Eq. 8.7 give the *centroid distances* of volume $x_c$, $y_c$, and $z_c$. Thus, we have

$$x_c = \frac{\iiint x \, dv}{\iiint dv}, \quad y_c = \frac{\iiint y \, dv}{\iiint dv}, \quad z_c = \frac{\iiint z \, dv}{\iiint dv} \qquad (8.8)$$

342

The integral $\iiint x \, dv$, it should be noted, gives the first moment of volume about the $yz$ plane, etc.

If we replace $dv$ by $dm = \rho \, dv$ in Eq. 8.6, where $\rho$ is the mass *density*, we get the *first moment of mass* about $O$. That is,

$$\text{moment vector of mass} \equiv \iiint_V \mathbf{r} \, \rho \, dv \qquad (8.9)$$

The *center of mass* $\mathbf{r}_c$ is then given as

$$\mathbf{r}_c = \frac{1}{M} \iiint_V \mathbf{r} \, \rho \, dv \qquad (8.10)$$

where $M$ is the total mass of the body. The center of mass is the point in space where hypothetically we could concentrate the entire mass for purposes of computing the first moment of mass about a point $O$. Using the components of Eq. 8.10, we can say that

$$x_c = \frac{\iiint x\rho \, dv}{\iiint \rho \, dv}, \quad y_c = \frac{\iiint y\rho \, dv}{\iiint \rho \, dv}, \quad z_c = \frac{\iiint z\rho \, dv}{\iiint \rho \, dv}$$

In our work in dynamics, we shall consider the center of mass of a system of $n$ particles (see Fig. 8.9). We will then say:

$$\left( \sum_{i=1}^{n} m_i \right) \mathbf{r}_c = \sum_{i=1}^{n} m_i \mathbf{r}_i$$

Therefore,

$$\mathbf{r}_c = \frac{\displaystyle\sum_{i=1}^{n} m_i \mathbf{r}_i}{M} \qquad (8.11)$$

where $M$ is the total mass of the system. Clearly, if the particles are of infinitesimal mass and constitute a continuous body, we get back Eq. 8.10.

Finally, if we replace $dv$ by $\gamma \, dv$, where $\gamma (= \rho g)$ is the *specific weight*, we arrive at the concept of *center of gravity* discussed in Chapter 4. We have used the center of gravity of a body in many calculations thus far as a point to concentrate the entire weight of a body.

You should have no trouble in concluding from Eq. 8.10 that if $\rho$ is constant throughout a body, the center of mass coincides with the center of volume. Furthermore, if $\gamma (= \rho g)$ is constant throughout a body, the center of gravity of the body corresponds to the center of volume of the body. If, finally, $\rho$ and $g$ are each constant for a body, all three points coincide for the body.

We now illustrate the computation of the center of volume. Computation for the center of mass follows similar lines, and we have already computed centers of gravity in Chapter 4.

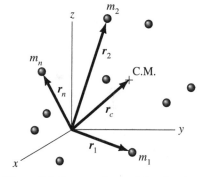

**Figure 8.9.** System of $n$ particles showing center of mass, C.M.

### Example 8.3

Consider a volume of revolution formed by revolving the area shown in Fig. 8.3 about the $x$ axis. This volume has been shown in Fig. 8.10. Clearly, the centroid of this volume must lie somewhere along the $x$ axis. Determine the centroidal distance $x_c$.

Using $r$, $\theta$, and $x$ as coordinates (cylindrical coordinates), we then have, using slices of thickness $dx$ as volume elements:

$$V = \int_0^{10} (\pi r^2)\, dx = \int_0^{10} (\pi)(25x)\, dx$$

where we have replaced $r^2$ with $25x$ according to the equation for the boundary of the generating area. Integrating, we get

$$V = 25\pi \left.\frac{x^2}{2}\right|_0^{10} = 3{,}927 \text{ ft}^3$$

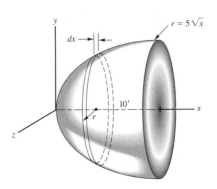

**Figure 8.10.** Body of revolution.

Now we compute $x_c$ by using infinitesimal slices of the body of the kind employed for the computation of $V$. The centroid of each such slice is at the intercept of the slice with the $x$ axis. Thus, we have

$$x_c = \frac{1}{V}\int_0^{10} x(\pi r^2\, dx) = \frac{1}{3{,}927}\int_0^{10} x(\pi)(25x)\, dx$$

$$= \frac{25\pi}{3{,}927} \left.\frac{x^3}{3}\right|_0^{10} = \boxed{6.67 \text{ ft}}$$

Many volumes are composed of a number of simple familiar shapes whose centers of volume are either known by inspection or can be found in handbooks (also see the inside front cover page). Such volumes may be called *composite volumes*. To find the centroid of such a volume, we use the known centroids of the composite parts. Thus, for $x_c$ of the composite body whose total volume is $V$, we have

$$x_c = \frac{\displaystyle\sum_i \bar{x}_i V_i}{V}$$

where $\bar{x}_i$ is the $x$ coordinate to the centroid of the $i$th composite body of volume $V_i$. Similarly,

$$y_c = \frac{\displaystyle\sum_i \bar{y}_i V_i}{V}$$

$$z_c = \frac{\displaystyle\sum_i \bar{z}_i V_i}{V}$$

We now illustrate the use of these formulas.

# Example 8.4

What is the coordinate $x_c$ for the center of volume of the body of revolution shown in Fig. 8.11? Note that a cone has been cut away from the left end while, at the right end, we have a hemispherical region.

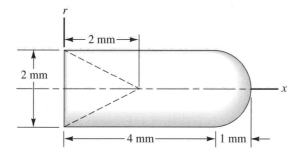

**Figure 8.11.** Composite volume.

We have a composite body consisting of three simple domains—a cone (body 1), a cylinder (body 2), and a hemisphere (body 3). Using formulas from the inside covers, we have:

| $V_i$ (mm³) | | $\bar{x}_i$ (mm) | $V_i\bar{x}_i$ (mm⁴) |
|---|---|---|---|
| 1. $-(\frac{1}{3})(\pi)(1^2)(2)$ | $= -2.09$ | $\frac{2}{4}$ | $-1.047$ |
| 2. $(\pi)(1^2)(4)$ | $= 12.57$ | $2$ | $25.14$ |
| 3. $\frac{2}{3}(\pi)(1^3)$ | $= \underline{\phantom{0}2.09}$ | $4 + \frac{3}{8}(1) = 4.38$ | $\underline{9.15}$ |
| | $V = 12.57$ | | $\sum_i V_i\bar{x}_i = 33.24$ |

Therefore,

$$x_c = \frac{\sum_i V_i\bar{x}_i}{V} = \frac{33.24}{12.57} = \boxed{2.64 \text{ mm}}$$

We have presented a number of three-dimensional problems for determining the center of volume, center of mass, and the center of gravity of composite bodies. We will leave it to the student to work his/her way through these problems, working from first principles, without the need of examples. However, we ask that you follow the following format, which clearly is an extension of what we have been doing up to this point.

| Area, Volume, Mass, or Weight **No.** | Area, Volume, Mass, or Weight **Value** | $(r_c)_i$ | $(r_c)_i \begin{Bmatrix} A_i \\ V_i \\ M_i \\ W_i \end{Bmatrix}$ |
|:---:|:---:|:---:|:---:|
| 1 | • | • | • |
| 2 | • | • | • |
| • | • | • | • |
| • | • | • | • |
| • | • | • | • |
| • | • | • | • |
| • | • | • | • |
| • | • | • | • |

$$\sum_i \begin{Bmatrix} A_i \\ V_i \\ M_i \\ W_i \end{Bmatrix} = \qquad\qquad \sum_i (r_c)_i \begin{Bmatrix} A_i \\ V_i \\ M_i \\ W_i \end{Bmatrix} =$$

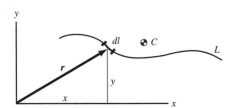

**Figure 8.12.** Centroid for curved line.

In closing, we wish to point out further that curved surfaces and lines have centroids. Since we shall have occasion in the next section to consider the centroid of a line, we simply point out now (see Fig. 8.12) that

$$x_c = \frac{\int x \, dl}{L} \tag{8.12a}$$

$$y_c = \frac{\int y \, dl}{L} \tag{8.12b}$$

where $L$ is the length of the line. Note that the centroid $C$ will not generally lie along the line.

Consider next a curve made up of simple curves each of whose centroids is known. Such is the case shown in Fig. 8.13, made up of straight lines. The line segment $L_1$, has for instance centroid $C_1$ with coordinates $\bar{x}_1, \bar{y}_1$, as has been shown in the diagram. We can then say for the entire curve that

**Figure 8.13.** Centroid for composite line.

$$x_c = \frac{\sum_i \bar{x}_i L_i}{L}$$

$$y_c = \frac{\sum_i \bar{y}_i L_i}{L} \tag{8.13}$$

# *8.4 Theorems of Pappus–Guldinus

The theorems of Pappus–Guldinus were first set forth by Pappus about 300 A.D. and then restated by the Swiss mathematician Paul Guldinus about 1640. These theorems are concerned with the relation of a surface of revolution to its generating curve, and the relation of a volume of revolution to its generating area.

The first of the theorems may be stated as follows:

> *Consider a coplanar generating curve and an axis of revolution in the plane of this curve (see Fig. 8.14). The generating curve can touch but must not cross the axis of revolution. The surface of revolution developed by revolving the generating curve about the axis of revolution has an* area *equal to the product of the* length *of the* generating curve *times the* circumference *of the* circle *formed by the* centroid *of the* generating curve *in the process of generating a surface of revolution.*

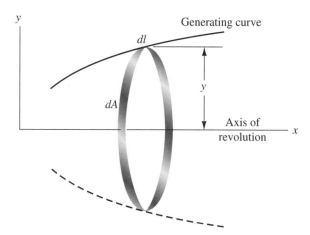

**Figure 8.14.** Coplanar generating curve.

To prove this theorem, consider first an element *dl* of the generating curve shown in Fig. 8.14. For a single revolution of the generating curve about the *x* axis, the line segment *dl* traces an area

$$dA = 2\pi y\ dl$$

For the entire curve this area becomes the surface of revolution given as

$$A = 2\pi \int y\,dl = 2\pi y_c L \tag{8.14}$$

where *L* is the length of the curve and $y_c$ is the centroidal coordinate of the curve. But $2\pi y_c$ is the circumferential length of the circle formed by having the centroid of the curve rotate about the *x* axis. The first theorem is thus proved.

Another way of interpreting Eq. 8.14 is to note that the area of the body of revolution is equal to $2\pi$ times the *first moment* of the generating curve about the axis of revolution. If the generating curve is composed of simple curves, $L_i$, whose centroids are known, such as the case shown in Fig. 8.13, then we can express $A$ as follows:

$$A = 2\pi \left( \sum_i L_i \bar{y}_i \right) \tag{8.15}$$

where $\bar{y}_i$ is the centroidal coordinate to the $i$th line segment $L_i$.

The second theorem may be stated as follows:

*Consider a plane surface and an axis of revolution coplanar with the surface but oriented such that the axis can intersect the surface only as a tangent at the boundary or have no intersection at all. The volume of the body of revolution developed by rotating the plane surface about the axis of revolution equals the product of the area of the surface times the circumference of the circle formed by the centroid of the surface in the process of generating the body of revolution.*

To prove the second theorem, consider a plane surface $A$ as shown in Fig. 8.15. The volume generated by revolving $dA$ of this surface about the $x$ axis is

$$dV = 2\pi y \ dA$$

The volume of the body of revolution formed from $A$ is then

$$V = 2\pi \int_A y \ dA = 2\pi y_c A \tag{8.16}$$

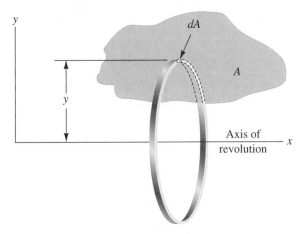

**Figure 8.15.** Plane surface $A$ coplanar with $xy$ plane.

Thus, the volume $V$ equals the area of the generating surface $A$ times the circumferential length of the circle of radius $y_c$. The second theorem is thus also proved.[2]

Another way to interpret Eq. 8.16 is to note that $V$ equals $2\pi$ times the *first moment* of the generating area $A$ about the axis of revolution. If this area $A$ is made up of simple areas $A_i$, we can say that

$$V = 2\pi \left( \sum_i A_i \bar{y}_i \right) \tag{8.17}$$

where $\bar{y}_i$, is the centroidal coordinate to the $i$th area $A_i$.

We now illustrate the use of the theorems of Pappus and Guldinus. As we proceed, it will be helpful to remember the theorems by noting that you multiply a length (or area) of the generator by the distance moved by the centroid of the generator.

---

[2]It is to be pointed out that the centroid of a volume of revolution will not be coincident with the centroid of a longitudinal cross-section taken along the axis of the volume. Example: a cone and its triangular, longitudinal cross-section.

---

## Example 8.5

Determine the surface area and volume of the bulk materials trailer shown in Fig. 8.16.

**Figure 8.16.** Bulk materials trailer.

We shall first determine the surface area by considering the first moment about the centerline A–A (see Fig. 8.17) of the generating curve of

**Figure 8.17.** Generating curve for surface of revolution.

## Example 8.5 (Continued)

the surface of revolution. This curve is a set of 5 straight lines each of whose centroids is easily known by inspection. Accordingly we may use Eq. 8.15. For clarity, we use a column format for the data as follows:

| $L_i$ (ft) | $\bar{y}_i$ (ft) | $L_i\bar{y}_i$ (ft²) |
|---|---|---|
| 1. 3 | 1.5 | 4.5 |
| 2. $\sqrt{8^2 + 1^2} = 8.06$ | 3.5 | 28.21 |
| 3. 20 | 4 | 80 |
| 4. 8.06 | 3.5 | 28.21 |
| 5. 3 | 1.5 | 4.5 |
| | | $\sum_i L_i\bar{y}_i = 145.43$ |

Therefore,

$$A = (2\pi)(145.43) = \boxed{914 \text{ ft}^2}$$

To get the volume, we next show in Fig. 8.18 the generating area for the body of revolution. Notice it has been decomposed into simple composite areas. We shall employ Eq. 8.17 and hence we shall need the first moment of area about the axis $A$–$A$ of the composite areas. Again, we shall employ a column format for the data.

**Figure 8.18.** Generating area for body of revolution.

| $A_i$ (ft²) | $\bar{y}_i$ (ft) | $A_i\bar{y}_i$ (ft³) |
|---|---|---|
| 1. 24 | 1.5 | 36 |
| 2. $\frac{1}{2}(8)(1) = 4$ | $3 + \frac{1}{3} = 3.33$ | 13.33 |
| 3. 80 | 2 | 160 |
| 4. 4 | 3.333 | 13.33 |
| 5. 24 | 1.5 | 36 |
| | | $\sum_i A_i\bar{y}_i = 258.7$ |

Therefore,

$$V = 2\pi\sum_i A_i\bar{y}_i = (2\pi)(258.7) = \boxed{1{,}625 \text{ft}^3}$$

The theorems of Pappus and Guldinus have enabled us to compute the surface area and the volume of the bulk materials trailer quickly and easily.

**8.31.** If $r^2 = ax$ in the body of revolution shown, compute the centroidal distance $x_c$ of the body.

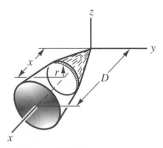

**Figure P.8.31.**

**8.32.** Using vertical elements of volume as shown, compute the centroidal coordinates $x_c$, $y_c$ of the body. Then, using horizontal elements, compute $z_c$.

**Figure P.8.32.**

**8.33.** Compute the center of volume of a right circular cylinder of height $h$ and radius at the base $r$.

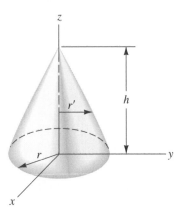

**Figure P.8.33.**

**8.34.** Determine the position of the center of mass of the solid hemisphere having a uniform mass density $\rho$ and with a radius $a$.

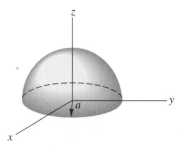

**Figure P.8.34.**

**8.35.** Find the center of mass for the paraboloid of revolution having a uniform density $\rho$.

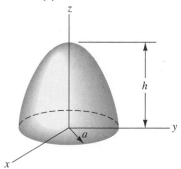

**Figure P.8.35.**

**8.36.** A small bomb has exploded at position $O$. Four pieces of the bomb move off at high speed. At $t = 3$ sec, the following data apply:

| $m$ | (kg) | $r$ (m) |
|---|---|---|
| 1. | .2 | $2i + 3j + 4k$ |
| 2. | .1 | $4i + 4j - 6k$ |
| 3. | .15 | $-3i + 2j - 3k$ |
| 4. | .22 | $2i - 3j + 2k$ |

What is the position of the center of mass?

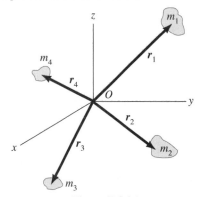

**Figure P.8.36.**

**8.37.** A plate of uniform thickness and density has for its curved edge a rectangular hyperbola ($xy$ = constant). Find the centroid of the upper surface. Find the centers of mass, volume, and gravity for the plate.

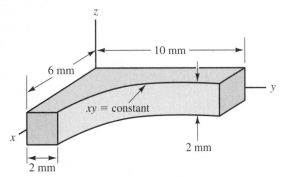

**Figure P.8.37.**

*In Problems 8.38 through 8.46, use the formulas on the inside covers for simple shapes.*

**8.38.** Where must a lifting hook be cast in a tapered concrete beam so that the beam always stays horizontal when lifted?

**Figure P.8.38.**

**8.39.** Two solid semicylinders are glued together. Body $A$ has a uniform mass density of 6.54 kN/m³, while body $B$ has a uniform mass density of 10 kN/m³. Determine:

(a) Center of volume
(b) Center of mass
(c) Center of gravity

**Figure P.8.39.**

**8.40.** What is the centroid of the body shown? It consists of a cylinder $A$ of length 2 m and diameter 6 m, a shaft $B$ of diameter 2 m and length 8 m, and a block $C$ of length 4 m and height and width of 7 m. The $x$ axis is a centerline for the arrangement. Origin $O$ is at the geometric center of cylinder $A$.

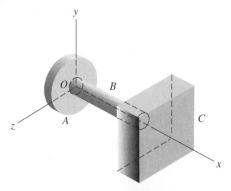

**Figure P.8.40.**

**8.41.** Find the center of volume for the cone–cylinder shown. Note that there is a cylindrical hole of length 16 ft and diameter 4 ft cut into the body.

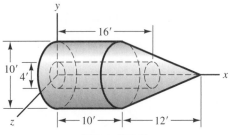

**Figure P.8.41.**

**8.42.** A cylinder has cut out of it a cone; a small cylinder directed through the centerline; and a half-sphere as has been shown in the diagram. Determine the center of gravity.

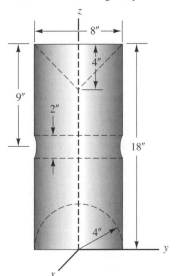

**Figure P.8.42.**

**8.43.** Find the center of gravity of the bent plate. The rectangular cutout occurs at the geometric center of the surface in the $xz$ plane.

**Figure P.8.43.**

**8.44.** A bent aluminum rod weighing 30 N/m is fitted into a plastic cylinder weighing 200 N, as shown. What are the centers of volume, mass, and gravity?

**Figure P.8.44.**

**8.45.** An aluminum cylinder fits snugly into a brass block. The brass weighs 43.2 kN/m³ and the aluminum weighs 30 kN/m³. Find the center of volume, the center of mass, and the center of gravity.

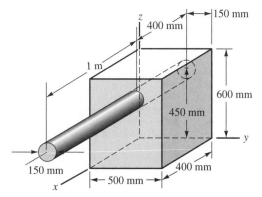

**Figure P.8.45.**

**8.46.** Two thin plates are welded together. One has a circle of radius 200 mm cut out as shown. If each plate weighs 450 N/m², what is the position of the center of mass?

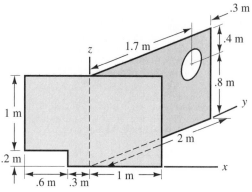

**Figure P.8.46.**

**8.47.** Where is the center of mass of the bent wire if it weighs 10 N/m?

**Figure P.8.47.**

**8.48.** Find the center of mass of the bent wire shown in the $zy$ plane. The wire weighs 15 N/m.

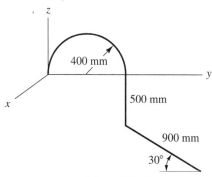

**Figure P.8.48.**

**8.49.** In Problem 8.41, involving a wooden cone–cylinder with a cylindrical hole, find the center of mass for the case where the cylinder has a density of 46.0 lbm/ft³ and the cone has a density of 30.0 lbm/ft³.

**8.50.** The volume of an ellipsoidal body of revolution is known from calculus to be $\frac{1}{6}\pi ab^2$. If the area of an ellipse is $\pi ab/4$, find the centroid of the area for a semiellipse.

**Figure P.8.50.**

**8.51.** Find the centroidal coordinate $y_c$ of the shaded area shown, using the theorems of Pappus and Guldinus.

**Figure P.8.51.**

**8.52.** The cutting tool of a lathe is programmed to cut along the dashed line as shown. What are the volume and the area of the body of revolution formed on the lathe?

**Figure P.8.52.**

**8.53.** Find the surface area and volume of the right conical frustum.

**Figure P.8.53.**

**8.54.** Find the surface area and volume of the Earth entry capsule for an unmanned Mars sampling mission. Approximate the rounded nose with a pointed nose as shown with the dashed lines.

354

**Figure P.8.54.**

**8.55.** Find the volume and surface area of the Apollo spaceship used for lunar exploration.

**Figure P.8.55.**

**8.56.** Find the center of volume $r_c$ for the machine element shown.

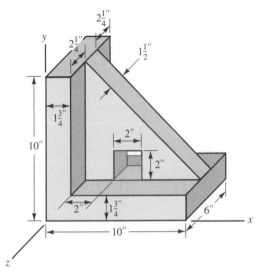

**Figure P.8.56.**

# 8.5   Second Moments and the Product of Area[3] of a Plane Area

We shall now consider other properties of a plane area relative to a given reference. The *second moments* of the area $A$ about the $x$ and $y$ axes (Fig. 8.19), denoted as $I_{xx}$ and $I_{yy}$, respectively, are defined as

$$I_{xx} = \int_A y^2\, dA \qquad\qquad (8.18\text{a})$$

$$I_{yy} = \int_A x^2\, dA \qquad\qquad (8.18\text{b})$$

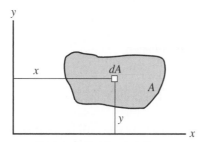

**Figure 8.19.** Plane surface.

The second moment of area cannot be negative, in contrast to the first moment. Furthermore, because the square of the distance from the axis is used, elements of area that are farthest from the axis contribute most to the second moment of area.

In an analogy to the centroid, the entire area may be concentrated at a single point $(k_x, k_y)$ to give the same second moment of area for a given reference. Thus,

$$Ak_x^2 = I_{xx} = \int_A y^2\, dA; \qquad \text{therefore,}\ k_x^2 = \frac{\int_A y^2\, dA}{A}$$

$$Ak_y^2 = I_{yy} = \int_A x^2\, dA; \qquad \text{therefore,}\ k_y^2 = \frac{\int_A x^2\, dA}{A} \qquad (8.19)$$

The distances $k_x$ and $k_y$ are called the *radii of gyration. This point will have a position that depends not only on the shape of the area but also on the position of the reference.* This situation is unlike the centroid, whose location is independent of the reference position.

The *product of area* relates an area directly to a set of axes and is defined as

$$I_{xy} = \int_A xy\, dA \qquad\qquad (8.20)$$

---

[3]We often use the expressions *moment* and *product of inertia* for second moment and product of area, respectively. However, we shall also use the former expressions in Chapter 9 in connection with mass distributions.

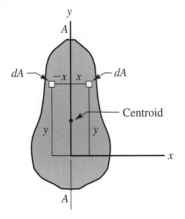

**Figure 8.20.** Area symmetric about $y$ axis.

This quantity may be negative.

If the area under consideration has an axis of symmetry, the product of area for this axis and any axis orthogonal to this axis must be zero. You can readily reach this conclusion by considering the area in Fig. 8.20, which is symmetrical about the axis $A$–$A$. Notice that the centroid is somewhere along this axis. (Why ?) The axis of symmetry has been indicated as the $y$ axis, and an arbitrary $x$ axis coplanar with the area has been shown. Also indicated are two elemental areas that are positioned as mirror images about the $y$ axis. The contribution to the product of area of each element is $xy \, dA$, but with opposite signs, and so the net result is zero. Since the entire area can be considered to be composed of such pairs, it becomes evident that the product of area for such cases is zero. This *should not* be taken to mean that a nonsymmetric area cannot have a zero product of area about a set of axes. We shall discuss this last condition in more detail later.

## 8.6    Transfer Theorems

We shall now set forth a theorem that will be of great use in computing second moments and products of area for areas that can be decomposed into simple parts (composite areas). With this theorem, we can find second moments or products of area about any axis in terms of second moments or products of area about a *parallel* set of axes going through the *centroid* of the area in question.

An $x$ axis is shown in Fig. 8.21 parallel to and at a distance $d$ from an axis $x'$ going through the centroid of the area. The latter axis you will recall is a *centroidal axis*. The second moment of area about the $x$ axis is

$$I_{xx} = \int_A y^2 \, dA = \int_A (y' + d)^2 \, dA$$

where the distance $y$ has been replaced by $(y' + d)$. Carrying out the squaring operation and integrating, leads to the result

$$I_{xx} = \int_A y'^2 \, dA + 2d \int_A y' \, dA + Ad^2$$

The first term on the right-hand side is by definition $I_{x'x'}$. The second term involves the first moment of area about the $x'$ axis. But the $x'$ axis here is a centroidal axis, and so the second term is zero. We can now state the transfer theorem (frequently called the parallel-axis theorem):

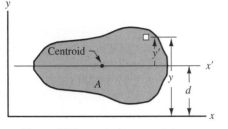

**Figure 8.21.** $x$ and $x'$ are parallel axes.

$$I_{\text{about any axis}} = I_{\substack{\text{about a parallel} \\ \text{axis at centroid}}} + Ad^2 \qquad (8.21)$$

where $d$ is the perpendicular distance between the axis for which $I$ is being computed and the parallel centroidal axis.

In strength of materials, a course generally following statics, second moments of area about noncentroidal axes are commonly used. The areas involved are complicated and not subject to simple integration. Accordingly, in structural handbooks, the areas and second moments about various centroidal axes are listed for many of the practical configurations with the understanding that designers will use the parallel-axis theorem for axes not at the centroid.

Let us now examine the product of area in order to establish a parallel-axis theorem for this quantity. Accordingly, two references are shown in Fig. 8.22, one $(x', y')$ at the centroid and the other $(x, y)$ positioned arbitrarily but *parallel* relative to $x'y'$. Note that $c$ and $d$ are the $x$ and $y$ *coordinates*, respectively, of the centroid of $A$ as measured from reference $xy$. These coordinates accordingly must have the proper signs, dependent on what quadrant the centroid of $A$ is in relative to $xy$. The product of area about the noncentroidal axes $xy$ can then be given as

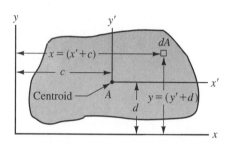

**Figure 8.22.** $c$ and $d$ measured from $xy$.

$$I_{xy} = \int_A xy \, dA = \int_A (x' + c)(y' + d) \, dA$$

Carrying out the multiplication, we get

$$I_{xy} = \int_A x'y' \, dA + c\int_A y' \, dA + d\int_A x' \, dA + Adc$$

Clearly, the first term on the right side by definition is $I_{x'y'}$, whereas the next two terms are zero since $x'$ and $y'$ are centroidal axes. Thus, we arrive at a parallel-axis theorem for products of area of the form:

$$I_{xy \text{ for any set of axes}} = I_{x'y' \text{ for a parallel set of axes at centroid}} + Adc \qquad (8.22)$$

It is important to remember that $c$ and $d$ are measured *from the xy axes to the centroid* and must have the appropriate sign. This will be carefully pointed out again in the examples of Section 8.7.

# 8.7   Computations Involving Second Moments and Products of Area

We shall examine examples for the computation of second moments and products of an area.

## Example 8.6

A rectangle is shown in Fig. 8.23. Compute the second moments and products of area about the centroidal $x'y'$ axes as well as about the $xy$ axes.

**Figure 8.23.** Rectangle: base $b$, height $h$.

$\underline{I_{x'x'}, I_{y'y'}, I_{x'y'}}$. For computing $I_{x'x'}$, we can use a strip of width $dy'$ at a distance $y'$ from the $x'$ axis. The area $dA$ then becomes $b\,dy'$. Hence, we have

$$I_{x'x'} = \int_{-h/2}^{+h/2} y'^2 b\,dy' = b\,\frac{y'^3}{3}\bigg|_{-h/2}^{+h/2} = \frac{b}{3}\left(\frac{h^3}{8} + \frac{h^3}{8}\right) = \boxed{\frac{1}{12}bh^3} \quad \text{(a)}$$

This is a common result and should well be remembered since it occurs so often. Verbally, for such an axis, the second moment of area is equal to $\frac{1}{12}$ the base $b$ times the height $h$ cubed. The second moment of area for the $y'$ axis can immediately be written as

$$I_{y'y'} = \tfrac{1}{12}hb^3 \quad \text{(b)}$$

where the base and height have simply been interchanged.

As a result of the previous statements on symmetry, we immediately note that

$$I_{x'y'} = 0 \quad \text{(c)}$$

$\underline{I_{xx}, I_{yy}, I_{xy}}$. Employing the transfer theorems, we get

$$I_{xx} = \tfrac{1}{12}bh^3 + bhe^2 \quad \text{(d)}$$

$$I_{yy} = \tfrac{1}{12}hb^3 + bhd^2 \quad \text{(e)}$$

In computing the product of area, we must be careful to employ the proper signs for the transfer distances. In checking the derivation of the transfer theorem, we see that these distances are measured from the noncentroidal axes to the centroid $C$. Therefore, in this problem the transfer distances are $(+e)$ and $(-d)$. Hence, the computation of $I_{xy}$ becomes

$$I_{xy} = 0 + (bh)(+e)(-d) = \boxed{-bhed} \quad \text{(f)}$$

and is thus a negative quantity.

## Example 8.7

What are $I_{xx}$, $I_{yy}$, and $I_{xy}$ for the area under the parabolic curve shown in Fig. 8.24?

To find $I_{xx}$, we may use horizontal strips of width $dy$ as shown in Fig. 8.25. We can then say for $I_{xx}$:

$$I_{xx} = \int_0^{10} y^2 [dy(10 - x)]$$

But

$$x = \sqrt{10} y^{1/2}$$

Therefore,

$$I_{xx} = \int_0^{10} y^2 (10 - \sqrt{10} y^{1/2})\, dy$$

$$= \left[ 10 \frac{y^3}{3} - \sqrt{10} y^{7/2} \left( \frac{2}{7} \right) \right]_0^{10}$$

$$= \frac{10(10^3)}{3} - \sqrt{10}(10^{7/2})\left( \frac{2}{7} \right) = \boxed{476.2 \text{ mm}^4}$$

As for $I_{yy}$, we use vertical infinitesimal strips as shown in Fig. 8.26. We can, accordingly, say:

$$I_{yy} = \int_0^{10} x^2 (y\, dx) = \int_0^{10} \frac{x^4}{10}\, dx$$

$$= \frac{x^5}{50}\bigg|_0^{10} = \boxed{2{,}000 \text{ mm}^4}$$

Finally, for $I_{xy}$ we use an infinitesimal area element $dx\, dy$ shown in Fig. 8.27. We must now perform multiple integration.[4] Thus, we have

$$I_{xy} = \int_0^{10} \int_{y=0}^{y=x^2/10} xy\, dy\, dx$$

Notice by holding $x$ constant and letting $y$ first run from $y = 0$ to the curve $y = x^2/10$ we cover the vertical strip of thickness $dx$ at position $x$ such as is shown in Fig. 8.26. Then by letting $x$ run from zero to 10, we cover the entire area. Accordingly, we first integrate with respect to $y$ holding $x$ constant. Thus,

$$I_{xy} = \int_0^{10} x \left( \frac{y^2}{2} \right)\bigg|_0^{x^2/10} dx = \int_0^{10} \frac{x^5}{200}\, dx$$

Next, integrating with respect to $x$, we have

$$I_{xy} = \frac{x^6}{1{,}200}\bigg|_0^{10} = \boxed{833 \text{ mm}^4}$$

[4]This multiple integration involves boundaries requiring some *variable* limits, in contrast to previous multiple integrations.

**Figure 8.24.** Plane area.

**Figure 8.25.** Horizontal strip.

**Figure 8.26.** Vertical strip.

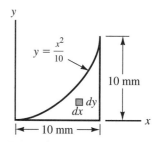

**Figure 8.27.** Element for multiple integration

# Example 8.8

Compute the second moment of area of a circular area about a diameter (Fig. 8.28).

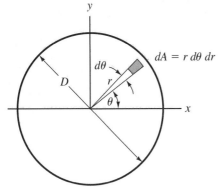

**Figure 8.28.** Circular area with polar coordinates.

Using polar coordinates, we have[5] for $I_{xx}$:

$$I_{xx} = \int_0^{D/2} \int_0^{2\pi} (r \sin \theta)^2 \, r \, d\theta \, dr = \int_0^{D/2} \pi r^3 \, dr$$

Completing the integration, we have

$$I_{xx} = \frac{r^4}{4} \pi \Big|_0^{D/2} = \pi \frac{D^4}{64}$$

The product of area $I_{xy}$ must be zero, owing to symmetry of the area about the $xy$ axes.

[5]The integral $\int_0^{2\pi} \sin^2 \theta \, d\theta$ may be evaluated by methods of substitution or may readily be seen in the following manner. $\int_0^{2\pi} \sin^2 \theta \, d\theta$ equals the area under the curve shown, which is half the area of the dashed rectangle. Hence, this integral equals $\pi$.

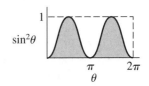

In the previous examples, we computed second moments and products of area using the calculus. Many problems of interest involve an area that may be subdivided into simpler *component* areas. Such an area has been referred to in earlier discussions as a composite area. The second moments and products of area for certain centroidal axes of many simple areas may be found in engineering handbooks (also see the inside covers). Using these formulas combined with the parallel-axis theorems, we can easily compute desired second moments and products of area for composite areas as we have done earlier for first moments of area. The following example illustrates this procedure.

## Example 8.9

Find the centroid of the area of the unequal-leg Z section shown in Fig. 8.29. Next, determine the second moment of area about the centroidal axes parallel to the sides of the Z section. Finally, determine the product of area for the aforementioned centroidal axes.

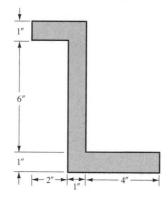

**Figure 8.29.** Unequal-leg Z section.

We shall subdivide the Z section into three rectangular areas, as shown in Fig. 8.30. Also, we shall insert a convenient reference $xy$, as shown in the diagram. To find the centroid, we proceed in the following manner:

| $A_i$ (in.$^2$) | $\bar{x}_i$ (in.) | $\bar{y}_i$ (in.) | $A_i\bar{x}_i$ (in.$^3$) | $A_i\bar{y}_i$ (in.$^3$) |
|---|---|---|---|---|
| 1. $(2)(1) = 2$ | 1 | 7.50 | 2 | 15 |
| 2. $(8)(1) = 8$ | 2.50 | 4 | 20 | 32 |
| 3. $(4)(1) = \underline{4}$ | 5 | .50 | $\underline{20}$ | $\underline{2}$ |
| $\sum_i A_i = 14$ | | | $\sum_i A_i\bar{x}_i = 42$ | $\sum_i A_i\bar{y}_i = 49$ |

**Figure 8.30.** Composite area.

## Example 8.9 (Continued)

Therefore,

$$x_c = \frac{\sum_i A_i \bar{x}_i}{\sum_i A_i} = \frac{42}{14} = \boxed{3 \text{ in.}}$$

$$y_c = \frac{\sum_i A_i \bar{y}_i}{\sum_i A_i} = \frac{49}{14} = \boxed{3.5 \text{ in.}}$$

We have shown the centroidal axes $x_c y_c$ in Fig. 8.31. We now find $I_{x_c x_c}$ and $I_{y_c y_c}$ using the parallel-axis theorem and the formulas $\frac{1}{12} bh^3$ and $\frac{1}{12} hb^3$ for the second moments of area about centroidal axes of symmetry of a rectangle.

$$I_{x_c x_c} = \underbrace{\left[ (\tfrac{1}{12})(2)(1^3) + (2)(4^2) \right]}_{①} + \underbrace{\left[ (\tfrac{1}{12})(1)(8^3) + (8)(\tfrac{1}{2})^2 \right]}_{②}$$

$$+ \underbrace{\left[ (\tfrac{1}{12})(4)(1^3) + (4)(3^2) \right]}_{③} = \boxed{113.2 \text{ in.}^4}$$

$$I_{y_c y_c} = \underbrace{\left[ (\tfrac{1}{12})(1)(2^3) + (2)(2^2) \right]}_{①} + \underbrace{\left[ (\tfrac{1}{12})(8)(1^3) + (8)(\tfrac{1}{2})^2 \right]}_{②}$$

$$+ \underbrace{\left[ (\tfrac{1}{12})(1)(4^3) + (4)(2^2) \right]}_{③} = \boxed{32.67 \text{ in.}^4}$$

**Figure 8.31.** Centroidal axes $x_c y_c$.

Finally, we consider the product of area $I_{x_c y_c}$. Here we must be cautious in using the parallel-axis theorem. Remember that $x_c y_c$ are centroidal axes for the *entire* area of the Z section. In using the parallel-axis theorem for a *subarea*, we must note that $x_c y_c$ are not centroidal axes for the subarea. The centroidal axes to be used in this problem for subareas are the axes of symmetry of each subarea. In short, $x_c y_c$ are simply axes about which we are computing the product of area of each subarea. Therefore, in the parallel-axis theorem, the transfer distances $c$ and $d$ are measured *from the $x_c y_c$ axes to the centroid* in each subarea, as noted in the development of the parallel-axis theorem. The proper sign must be assigned each time to the transfer distances with this in mind. We have for $I_{x_c y_c}$:

$$I_{x_c y_c} = \underbrace{\left[ 0 + (2)(-2)(4) \right]}_{①} + \underbrace{\left[ 0 + (8)(-\tfrac{1}{2})(\tfrac{1}{2}) \right]}_{②} + \underbrace{\left[ 0 + (4)(2)(-3) \right]}_{③} = \boxed{-42 \text{ in.}^4}$$

**8.57.** Find $I_{xx}$, $I_{yy}$, and $I_{xy}$ for the triangle shown. Give the results in feet.

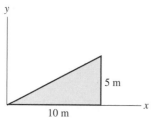
Figure P.8.57.

**8.58.** What are the second moments and products of area of the ellipse for reference $xy$? [*Hint:* Can you work with one quadrant and then multiply by 4 for the second moments?]

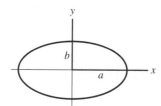
Figure P.8.58.

**8.59.** Find $I_{xx}$ and $I_{yy}$ for the quarter circle of radius 5 m.

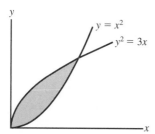
Figure P.8.59.

**8.60.** Find $I_{xx}$, $I_{yy}$, and $I_{xy}$ for the shaded area.

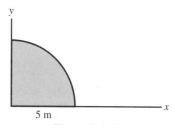
Figure P.8.60.

**8.61.** Find $I_{yy}$ for the shaded area. You must first determine the constant $c$.

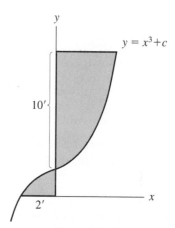
Figure P.8.61.

**8.62.** Find $I_{yy}$ for the area between the curves

$$y = 2 \sin x \text{ ft}$$
$$y = \sin 2x \text{ ft}$$

from $x = 0$ to $x = \pi$ ft.

**8.63.** Find $I_{yy}$ for the areas enclosed between curves $y = \cos x$ and $y = \sin x$ and the lines $x = 0$ and $x = \pi/2$.

**8.64.** Show that $I_{xx} = bh^3/12$, $I_{yy} = b^3h/12$, and $I_{xy} = b^2h^2/24$ for the right triangle.

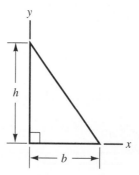
Figure P.8.64.

**8.65.** Find $I_{xx}$, $I_{yy}$, and $I_{xy}$ for the shaded area.

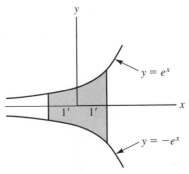

**Figure P.8.65.**

**8.66.** Find $I_{xx}$, $I_{yy}$, and $I_{xy}$ for the area of Problem 8.4. The equation of the curve is $y^2 = 20x - 60$.

**8.67.** Find $I_{xx}$, $I_{yy}$, and $I_{xy}$ for the cross-section shown.

**Figure P.8.67.**

**8.68.** Find $I_{xx}$ and $I_{yy}$ for the area of Problem 8.5. The equation of the parabola is $y = (x^2/9) - 5$. [*Hint:* The area of a vertical element in the region below the $x$ axis is $(0 - y)\,dx$.]

**8.69.** In Problem 8.68 determine $I_{xy}$ using multiple integration.

**8.70.** Find the second moments of area about axes $xy$ for the shaded area shown.

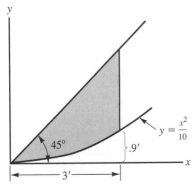

**Figure P.8.70.**

**8.71.** If the second moment of area about axis $A$–$A$ is known to be 600 ft$^4$, what is the second moment of area about a parallel axis $B$–$B$ a distance 3 ft from $A$–$A$, for an area of 10 ft$^2$? The centroid of this area is 4 ft from $B$–$B$.

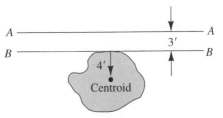

**Figure P.8.71.**

**8.72.** Using the results of Problem 8.64, show that $I_{x_c x_c} = bh^3/36$, $I_{y_c y_c} = hb^3/36$, and $I_{x_c y_c} = -b^2 h^2/72$ for the right triangle shown.

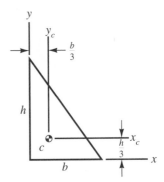

**Figure P.8.72.**

**8.73.** Show that $I_{xx} = bh^3/12$, $I_{yy} = (hb/12)(b^2 + ab + a^2)$, and $I_{xy} = (h^2 b/24)(2a + b)$ for the triangle. [*Hint:* Break the triangle into two right triangles for which the various moments are known. (See Problem 8.72.)]

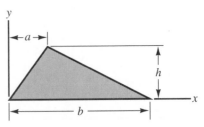

**Figure P.8.73.**

**8.74.** In Problem 8.73, show that $I_{x_c x_c} = bh^3/36$, $I_{y_c y_c} = (bh/36)(b^2 - ab + a^2)$, and $I_{x_c y_c} = (h^2 b/72)(2a - b)$ for the triangle. [*Hint:* Use the results of Problems 8.11 and 8.73 and the parallel-axis theorem.]

**8.75.** Find $I_{xx}$, $I_{yy}$, and $I_{xy}$ of the extruded section. Disregard all rounded edges. Do this problem using 4 areas. Check using 2 areas.

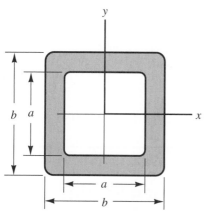

**Figure P.8.75.**

**8.76.** Find the second moment of area of the rectangle (with a hole) about the base of the rectangle. Also, determine the product of area about the base and left side.

**Figure P.8.76.**

**8.77.** Find $I_{xx}$, $I_{yy}$, $I_{x_c x_c}$, and $I_{y_c y_c}$ for the structural "hat" section. Disregard all rounded edges.

**Figure P.8.77.**

**8.78.** Find $I_{xx}$, $I_{yy}$, and $I_{xy}$ of the hexagon.

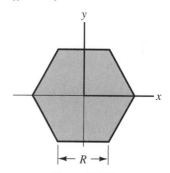

**Figure P.8.78.**

**8.79.** A beam cross-section is made up of an I-shaped section with an additional thick plate welded on. Find the second moments of area for the centroidal axes $x_c y_c$ of the beam cross-section. What is $I_{x_c y_c}$? Give the results in millimeters.

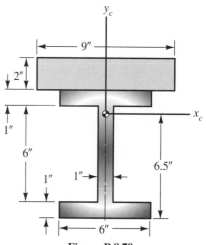

**Figure P.8.79.**

365

**8.80.** Find the second moments of the area shown about centroidal axes parallel to the $x$ and $y$ axes. That is, find $I_{x_c x_c}$ and $I_{y_c y_c}$. Give the results in millimeters.

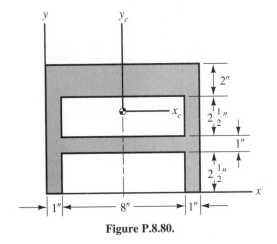

**Figure P.8.80.**

**8.81.** The following configuration is given for a plane area.
(a) Find the first moments of area about the $x$ and $y$ axes.
(b) Find the second moments of area for these axes.
(c) Find the product of area for these axes.
(d) What are the radii of gyration, $k_x$ and $k_y$, for this area?

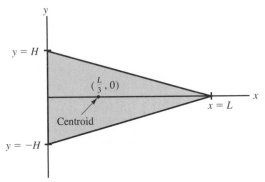

**Figure P.8.81.**

## 8.8 Relation Between Second Moments and Products of Area

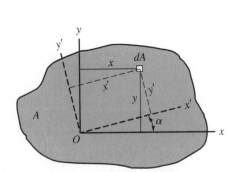

**Figure 8.32.** Rotation of axes.

We shall now show that we can ascertain second moments and product of area relative to a rotated reference $x'y'$ if we know these quantities for reference $xy$ that has the *same origin*. Such a reference $x'y'$ rotated an angle $\alpha$ from $xy$ (counterclockwise as positive) is shown in Fig. 8.32. We shall assume that the second moments and product of area for the unprimed reference are known.

Before proceeding, we must know the relation between the coordinates of the area elements $dA$ for the two references. From Fig. 8.32, you may show that

$$x' = x \cos \alpha + y \sin \alpha \tag{8.23a}$$
$$y' = -x \sin \alpha + y \cos \alpha \tag{8.23b}$$

With relation 8.23b, we can express $I_{x'x'}$ in the following manner:

$$I_{x'x'} = \int_A (y')^2 \, dA = \int_A (-x \sin \alpha + y \cos \alpha)^2 \, dA \tag{8.24}$$

Carrying out the square, we have

$$I_{x'x'} = \sin^2 \alpha \int_A x^2 \, dA - 2 \sin \alpha \cos \alpha \int_A xy \, dA + \cos^2 \alpha \int_A y^2 \, dA$$

Therefore,

$$I_{x'x'} = I_{yy} \sin^2 \alpha + I_{xx} \cos^2 \alpha - 2I_{xy} \sin \alpha \cos \alpha \tag{8.25}$$

A more common form of the desired relation can be formed by using the following trigonometric identities:

$$\cos^2 \alpha = \tfrac{1}{2}(1 + \cos 2\alpha) \tag{a}$$

$$\sin^2 \alpha = \tfrac{1}{2}(1 - \cos 2\alpha) \tag{b}$$

$$2 \sin \alpha \cos \alpha = \sin 2\alpha \tag{c}$$

We then have[6]

$$I_{x'x'} = \frac{I_{xx} + I_{yy}}{2} + \frac{I_{xx} - I_{yy}}{2} \cos 2\alpha - I_{xy} \sin 2\alpha \tag{8.26}$$

To determine $I_{y'y'}$, we need only replace the $\alpha$ in the preceding result by $(\alpha + \pi/2)$. Thus,

$$I_{y'y'} = \frac{I_{xx} + I_{yy}}{2} + \frac{I_{xx} - I_{yy}}{2} \cos(2\alpha + \pi) - I_{xy} \sin(2\alpha + \pi)$$

Note that $\cos(2\alpha + \pi) = -\cos 2\alpha$ and $\sin(2\alpha + \pi) = -\sin 2\alpha$. Hence, the equation above becomes

$$I_{y'y'} = \frac{I_{xx} + I_{yy}}{2} - \frac{I_{xx} - I_{yy}}{2} \cos 2\alpha + I_{xy} \sin 2\alpha \tag{8.27}$$

Next, the product of area $I_{x'y'}$ can be computed in a similar manner:

$$I_{x'y'} = \int_A x'y' \, dA = \int_A (x \cos \alpha + y \sin \alpha)(-x \sin \alpha + y \cos \alpha) \, dA$$

This becomes

$$I_{x'y'} = \sin \alpha \cos \alpha \,(I_{xx} - I_{yy}) + (\cos^2 \alpha - \sin^2 \alpha)I_{xy}$$

Utilizing the previously defined trigonometric identities, we get

$$I_{x'y'} = \frac{I_{xx} - I_{yy}}{2} \sin 2\alpha + I_{xy} \cos 2\alpha \tag{8.28}$$

Thus, we see that, if we know the quantities $I_{xx}$, $I_{yy}$, and $I_{xy}$ for some reference $xy$ at point $O$, the second moments and products of area for every set of axes at point $O$ can be computed. And if, in addition, we employ the transfer theorems, we can compute second moments and products of area for *any* reference in the plane of the area.

---

[6]Equations 8.26, 8.27, and 8.28 are called *transformation* equations. They will appear in the next chapter and in your upcoming solid mechanics course for variables other than second moments and products of area. In the remaining portions of this chapter, you will see that a number of important properties of second moments and products of area are deducible *directly* from these transformation equations. This primarily accounts for the importance of the transformation equations. Chapter 9 will give you additional insight into this topic.

## Example 8.10

Find $I_{x'x'}$, $I_{y'y'}$ and $I_{x'y'}$ for the cross section of the beam shown in Fig. 8.33. The origin of $x'y'$ is at the centroid of the cross-section.

We shall consider three rectangles labeled in Fig. 8.34. The dimensions of these rectangles are listed below.

Rect. (1)   100 mm $\times$ 40 mm (outside outline)
Rect. (2)    20 mm $\times$ 30 mm
Rect. (3)    20 mm $\times$ 30 mm

We shall find $I_{xx}$, $I_{yy}$, and $I_{xy}$ as a first step. Thus

$$I_{xx} = \left(\tfrac{1}{12}\right)(40)(100)^3 - 2\left[\tfrac{1}{12}(30)(20)^3 + (20)(30)(40)^2\right] = 1.373 \times 10^6 \ \text{mm}^4$$

$$I_{yy} = \left(\tfrac{1}{12}\right)(100)(40)^3 - 2\left[\tfrac{1}{12}(20)(30)^3\right] = 4.43 \times 10^5 \ \text{mm}^4$$

$$I_{xy} = 0 \quad \text{(symmetry)}$$

**Figure 8.33.** A composite area.

Hence, we can now go to the transformation equations for the desired information.[7]

$$I_{x'x'} = \frac{1.373 \times 10^6 + 4.43 \times 10^5}{2} + \frac{1.373 \times 10^6 - 4.43 \times 10^5}{2} \cos 60° + 0$$

$$= \boxed{1.141 \times 10^6 \ \text{mm}^4}$$

$$I_{y'y'} = \frac{1.373 \times 10^6 + 4.43 \times 10^5}{2} - \frac{1.373 \times 10^6 - 4.43 \times 10^5}{2} \cos 60° + 0$$

$$= \boxed{6.76 \times 10^5 \ \text{mm}^4}$$

**Figure 8.34.** Composite area with three subareas.

$$I_{x'y'} = \frac{1.373 \times 10^6 - 4.43 \times 10^5}{2} \sin 60° = \boxed{4.03 \times 10^5 \ \text{mm}^4}$$

---

[7]It will be helpful to note that in going from the computation of $I_{x'x'}$ to the computation of $I_{y'y'}$, one needs only to change the sign associated with the second and third expressions for $I_{x'x'}$.

# 8.9   Polar Moment of Area

In the previous section, we saw that the second moments and product of area for an orthogonal reference determined all such quantities for *any* orthogonal reference having the same origin. We shall now show that the sum of the pairs of second moments of area is a constant for all such references at a point. Thus, in Fig. 8.35 we have a reference $xy$ associated with point $a$. Summing $I_{xx}$ and $I_{yy}$ we have

$$I_{xx} + I_{yy} = \int_A y^2 \, dA + \int_A x^2 \, dA$$

$$= \int_A (x^2 + y^2) \, dA = \int_A r^2 \, dA$$

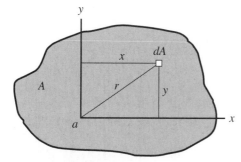

**Figure 8.35.** $J = I_{xx} + I_{yy}$.

Since $r^2$ is independent of the orientation of the coordinate system, the sum $I_{xx} + I_{yy}$ is independent of the orientation of the reference. Therefore, the sum of second moments of area about orthogonal axes is a function only of the position of the origin $a$ for the axes. This sum is termed the *polar moment of area, J*.[8] We can then consider $J$ to be a scalar field. Mathematically, this statement is expressed as

$$J = J(x', y') \tag{8.29}$$

where $x'$ and $y'$ are the coordinates as measured from some convenient reference $x'y'$ for the point of interest.

That the quantity $(I_{xx} + I_{yy})$ does not change on rotation of axes can also be deduced by summing transformation equations 8.26 and 8.27 as we suggest you do. This group of terms is accordingly termed an *invariant*. Parenthetically, we can similarly show that $(I_{xx}I_{yy} - I_{xy}^2)$ is also invariant under a rotation of axes.

---

[8]Quite often $I_p$ is used for the polar moment of area.

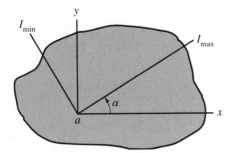

**Figure 8.36.** Principal axes.

## 8.10   Principal Axes

Still other conclusions may be drawn about second moments and products of area associated with a point in an area. In Fig. 8.36 an area is shown with a reference $xy$ having its origin at point $a$. We shall assume that $I_{xx}$, $I_{yy}$, and $I_{xy}$ are known for this reference, and shall ask at what angle $\alpha$ we shall find an axis having the *maximum* second moment of area. Since the sum of the second moments of area is constant for any reference with origin at $a$, the *minimum* second moment of area must then correspond to an axis at *right angles* to the axis having the maximum second moment. Since second moments of area have been expressed in Eqs. 8.26 and 8.27 as functions of the variable $\alpha$ at a point, these extremes may readily be determined by setting the partial derivative of $I_{x'x'}$ with respect to $\alpha$ equal to zero. Thus,

$$\frac{\partial I_{x'x'}}{\partial \alpha} = (I_{xx} - I_{yy})(-\sin 2\alpha) - 2I_{xy}\cos 2\alpha = 0$$

If we denote the value of $\alpha$ that satisfies the equation above as $\tilde{\alpha}$, we have

$$(I_{yy} - I_{xx})\sin 2\tilde{\alpha} - 2I_{xy}\cos 2\tilde{\alpha} = 0$$

Hence,

$$\tan 2\tilde{\alpha} = \frac{2I_{xy}}{I_{yy} - I_{xx}} \tag{8.30}$$

This formulation gives us the angle $\tilde{\alpha}$, which corresponds to an extreme value of $I_{x'x'}$ (i.e., to a maximum or minimum value). Actually, there are two possible values of $2\tilde{\alpha}$ which are $\pi$ radians apart that will satisfy the equation above. Thus,

$$2\tilde{\alpha} = \beta \quad \text{where } \beta = \tan^{-1}\frac{2I_{xy}}{I_{yy} - I_{xx}}$$

or

$$2\tilde{\alpha} = \beta + \pi$$

This means that we have two values of $\tilde{\alpha}$, given as

$$\tilde{\alpha}_1 = \frac{\beta}{2}, \qquad \tilde{\alpha}_2 = \frac{\beta}{2} + \frac{\pi}{2}$$

Thus, there are two axes orthogonal to each other having extreme values for the second moment of area at $a$. On one of these axes is the maximum second moment of area and, as pointed out earlier, the minimum second moment of area must appear on the other axis. These axes are called the *principal axes*.

Let us now substitute the angle $\tilde{\alpha}$ into Eq. 8.28 for $I_{x'y'}$:

$$I_{x'y'} = \frac{I_{xx} - I_{yy}}{2} \sin 2\tilde{\alpha} + I_{xy} \cos 2\tilde{\alpha} \qquad (8.31)$$

If we now form a right triangle with legs $2I_{xy}$ and $(I_{yy} - I_{xx})$ and angle $2\tilde{\alpha}$ such that Eq. (8.30) is satisfied we can readily express the sine and cosine expressions needed in the preceding equation. Thus

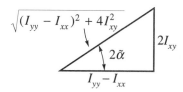

$$\sin 2\tilde{\alpha} = \frac{2I_{xy}}{\sqrt{\left(I_{yy} - I_{xx}\right)^2 + 4I_{xy}^2}}$$

$$\cos 2\tilde{\alpha} = \frac{I_{yy} - I_{xx}}{\sqrt{\left(I_{yy} - I_{xx}\right)^2 + 4I_{xy}^2}}$$

By substituting these results into Eq. 8.31, we get

$$I_{x'y'} = -\left(I_{yy} - I_{xx}\right) \frac{I_{xy}}{\left[\left(I_{yy} - I_{xx}\right)^2 + 4I_{xy}^2\right]^{1/2}} + I_{xy} \frac{I_{yy} - I_{xx}}{\left[\left(I_{yy} - I_{xx}\right)^2 + 4I_{xy}^2\right]^{1/2}}$$

Hence,

$$I_{x'y'} = 0$$

Thus, we see that the *product of area corresponding to the principal axes is zero*. If we set $I_{x'y'}$ equal to zero in Eq. 8.28, you can demonstrate the converse of the preceding statement by solving for $\alpha$ and comparing the result with Eq. 8.30. That is, if the product of area is zero for a set of axes at a point, these axes must be the principal axes at that point. Consequently, if one axis of a set of axes at a point is symmetrical for the area, the axes are principal axes at that point.

The concept of principal axes will appear again in the following chapter in connection with the inertia tensor. Thus, the concept is not an isolated occurrence but is characteristic of a whole family of quantities. We shall, then, have further occasion to examine some of the topics introduced in this chapter from a more general viewpoint.

# Example 8.11

Find the principal second moments of area at the centroid of the Z section of Example 8.9.

We have from this example the following results that will be of use to us:

$$I_{x_c x_c} = 113.2 \text{ in.}^4$$
$$I_{y_c y_c} = 32.67 \text{ in.}^4$$
$$I_{x_c y_c} = -42.0 \text{ in.}^4$$

Hence, we have

$$\tan 2\tilde{\alpha} = \frac{2 I_{x_c y_c}}{I_{y_c y_c} - I_{x_c x_c}} = \frac{(2)(-42.0)}{32.67 - 113.2} = 1.043$$

$$2\tilde{\alpha} = 46.21°; \ 226.2°$$

For $2\tilde{\alpha} = 46.21°$:

$$I_1 = \frac{113.2 + 32.67}{2} + \frac{113.2 - 32.67}{2} \cos(46.21°) - (-42) \sin 46.21°$$

$$= 72.9 + 27.9 + 30.3 = \boxed{131.1 \text{ in.}^4}$$

For $2\tilde{\alpha} = 226.2°$:

$$I_2 = 72.9 - 27.9 - 30.3 = \boxed{14.75 \text{ in.}^4}$$

As a check on our work, we note that the sum of the second moments of area are invariant at a point for a rotation of axes. This means that

$$I_{x_c x_c} + I_{y_c y_c} = I_1 + I_2$$
$$113.2 + 32.7 = 131.1 + 14.75$$

Therefore,

$$145.9 = 145.9$$

We thus have a check on our work.

Before closing, we wish to point out that there is a graphical construction called *Mohr's circle* relating second moments and products of area for all possible axes at a point. However, in this text we shall use the analytical relations thus far presented rather than Mohr's circle. You will see Mohr circle construction in your strength of materials course where its use in conjunction with the important topics of plane stress and plane strain is very helpful.[9]

[9]See I. H. Shames, *Introduction to Solid Mechanics*, Second Edition, Prentice-Hall, Inc., Englewood Cliffs, N.J., 1989.

**8.82.** It is known that area $A$ is 10 ft$^2$ and has the following moments and products of area for the centroidal axes shown:

$$I_{xx} = 40 \text{ ft}^4, \qquad I_{yy} = 20 \text{ ft}^4, \qquad I_{xy} = -4 \text{ ft}^4$$

Find the moments and products of area for the $x'y'$ reference at point a.

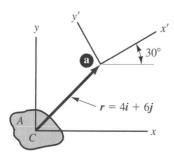

**Figure P.8.82.**

**8.83.** The cross-section of a beam is shown. Compute $I_{x'x'}$, $I_{y'y'}$, and $I_{x'y'}$ in the simplest way without using formulas for second moments and products of area for a triangle.

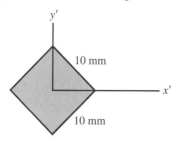

**Figure P.8.83.**

**8.84.** Find $I_{xx}$, $I_{yy}$, and $I_{xy}$ for the rectangle. Also, compute the polar moment of area at points $a$ and $b$.

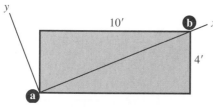

**Figure P.8.84.**

**8.85.** Express the polar moment of area of the square as a function of $x$, $y$, the coordinates of points about which the polar moment is taken.

**Figure P.8.85.**

**8.86.** Use the calculus to show that the polar moment of area of a circular area of radius $r$ is $\pi r^4/2$ at the center.

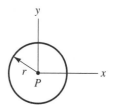

**Figure P.8.86.**

**8.87.** Find the direction of the principal axes for the angle section at point $A$.

**Figure P.8.87.**

**8.88.** What are the principal second moments of area at the origin for the area of Example 8.7?

**8.89.** Find the principal second moments of area at the centroid for the area shown.

**Figure P.8.89.**

**8.90.** Determine the principal second moments of area at point $A$.

**Figure P.8.90.**

**8.91.** A rectangular area has two holes cut out. What is the maximum second moment of area at $A$? What is it at $B$ halfway between the holes?

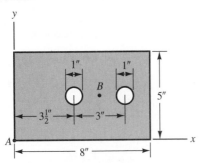

**Figure P.8.91.**

**8.92.** Show that the axes for which the product of area is a maximum are rotated from $xy$ by an angle $\alpha$ so that

$$\tan 2\alpha = \frac{I_{xx} - I_{yy}}{2I_{xy}}$$

**Figure P.8.92.**

**8.93.** What is the value of the angle $\alpha$ for the principal axes at A.

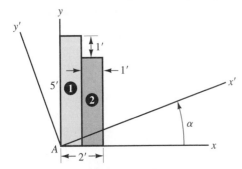

**Figure P.8.93.**

**8.94.** Find the principal second moments of area at $A$.

**Figure P.8.94.**

**8.95.** (a) Find $I_{xx}$, $I_{yy}$, and $I_{xy}$ for the $xy$ axes shown.

(b) What are the algebraically largest and smallest second moments of area at position $A$? What are the orientations of the corresponding axes? What is the product of area for these axes?

**Figure P.8.95.**

**8.96.** (a) Find the centroidal coordinates relative to axes $xy$.

(b) Find the principal axes and the corresponding maximum and minimum second moments of area at $A$.

**Figure P.8.96.**

# 8.11 Closure

In this chapter, we discussed primarily the first and second moments of plane areas as well as the product of plane areas. These formulations give certain kinds of evaluations of the distribution of area relative to a plane reference $xy$. You will most certainly make much use of these quantities in your later courses in strength of materials.

In this chapter, we have touched on subject matter that you will encounter in the next chapter and also, most assuredly, in later courses. Specifically, in Chapter 9 you will be introduced to the so-called second-order inertia tensor having nine terms which change (or transform) in a certain particular way when we rotate coordinate axes at a point. These particular transformation equations define the inertia tensor. Any other set of nine symmetric terms that transform via the same form of equations, are symmetric second-order tensors. You will also learn that the second moments and products of *area* form a two-dimensional simplification of the inertia tensor. The transformation equations 8.26 through 8.28 are thus special cases of the three-dimensional defining equations of the inertia tensor. Take note that certain vital results emerged from these simplified two-dimensional transformation equations. They included

1. Invariant property at a point for the sum $(I_{xx} + I_{yy})$ on rotation of axes.
2. Principal axes and principal moments of area at a point.
3. A graphical construction called Mohr's circle that depicts the transformation equations. This topic has been omitted in this chapter but will be described in your solids course where you will study the two-dimensional simplifications of the stress and strain tensors. At that time one can make much use of the Mohr's circle construction and it is a simple matter then to describe Mohr's circle for second moments and products of area.

We will emphasize in the next chapter that there are many symmetric sets of nine terms that transform exactly like the terms of the inertia tensor and we classify all of them as second-order tensors. Identifying these sets as second-order tensors immediately yields vital properties common to all of them such as those presented above for the two-dimensional moments and products of area.

There will be more to be said on tensors in a "looking ahead" section of the next chapter which we invite you to examine.

**8.97.** Find the position of the centroid of the shaded area under the curve $y = \sin^2 x$ m. Find $M_x$ and $M_y$ of this area.

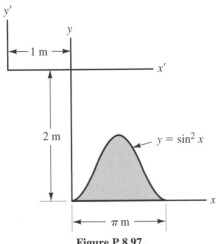

**Figure P.8.97.**

**8.98.** Find the center of volume of the body of revolution with a cylindrical cavity.

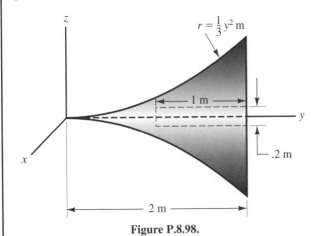

**Figure P.8.98.**

**8.99.** Locate the center of volume, center of mass, and center of gravity of the wooden rectangular block and the plastic semicylinder. The wood weighs .0003 N/mm³ and the plastic weighs .0005 N/mm³.

**Figure P.8.99.**

**8.100.** Find $I_{xx}$, $I_{yy}$, $I_{xy}$, $I_{x_c x_c}$, $I_{y_c y_c}$, and $I_{x_c y_c}$ of the unequal-leg rolled channel section.

**Figure P.8.100.**

**8.101.** Find the centroid of the half cone. Use the known formula for the volume of a half cone, $V = (1/6) \, \pi r^2 h$.

**Figure P.8.101.**

**8.102.** A half body of revolution is shown with the *xz* plane as a plane of symmetry. Determine the centroidal coordinates. The radius at any section *x* varies as the square of *x*.

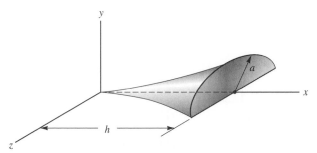

**Figure P.8.102.**

**8.103.** (a) Find $I_{xx}$, $I_{yy}$, and $I_{xy}$ for the *xy* axes at position *A*.

(b) Find the principal second moments of area at point *A*.

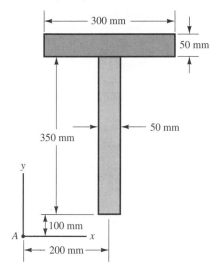

**Figure P.8.103.**

**8.104.** What are the directions of the principal axes at point *A*?

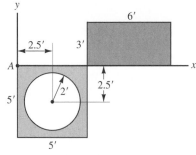

**Figure P.8.104.**

**8.105.** Using the theorems of Pappus and Guldinus, find the centroid of the area of a quarter-circle.

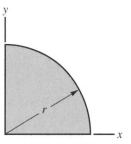

**Figure P.8.105.**

**8.106.** A tank has a semispherical dome at the left end. Using the theorems of Pappus and Guldinus, compute the surface and volume of the tank. Give the results in meters.

**Figure P.8.106.**

**8.107.** Find $I_{x'x'}$, $I_{y'y'}$, and $I_{x'y'}$ for the set of axes at point *A* for the rectangular area.

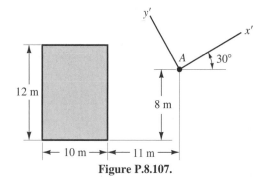

**Figure P.8.107.**

**8.108.** Find the centroid of the area, and then find the second moments of the area about centroidal axes parallel to the sides of the area.

377

**Figure P.8.108.**

**8.109.** Find the principal second moments of area at a point where $I_{xy} = 321$ in.⁴, $I_{xx} = 118.4$ in.⁴, and $I_{yy} = 1,028$ in.⁴.

**8.110.** Find the polar moment of area at point $O$ for the shaded area.

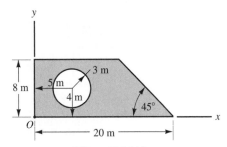

**Figure P.8.110.**

**8.111.** (a) What are the centroidal coordinates of the shaded area?

(b) What are $M_{x'}$, and $M_{y'}$ for axes $x'y'$ at $A$? [*Hint:* Use formulas for the sector of a circle given on the inside back cover.]

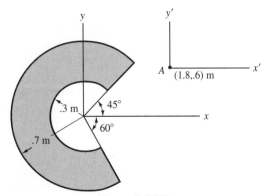

**Figure P.8.111.**

**8.112.** A wide-flanged I beam is shown.[10] For the upper flange, what is the first moment of the shaded area as a function of $s$ measured from $s = 0$ at the right end to where $s$ approaches the web? Next get $M_x$ for the entire area above the position shown as $y$ in the web. Let $y$ go from $C$ to the top of the web.

**Figure P.8.112.**

**8.113.** Find the center of volume of the machine element with a circular hole as shown.

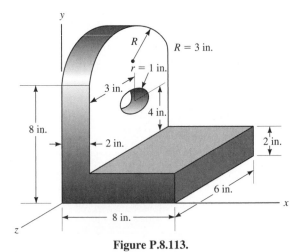

**Figure P.8.113.**

[10]Actually, it is a 14WF202 I beam, wherein the weight is 202 lb/ft.

# CHAPTER 9

# Moments and Products of Inertia[1]

## 9.1 Introduction

In this chapter, we shall consider certain measures of mass distribution relative to a reference. These quantities are vital for the study of the dynamics of rigid bodies. Because these quantities are so closely related to second moments and products of area, we shall consider them at this early stage rather than wait for dynamics. We shall also discuss the fact that these measures of mass distribution—the second moments of inertia of mass and the products of inertia of mass—are components of what we call a second-order tensor. Recognizing this fact early will make more simple and understandable your future studies of stress and strain, since these quantities also happen to be components of second-order tensors.

## 9.2 Formal Definition of Inertia Quantities

We shall now formally define a set of quantities that give information about the distribution of mass of a body relative to a Cartesian reference. For this purpose, a body of mass $M$ and a reference $xyz$ are presented in Fig. 9.1. This reference and the body may have any motion whatever relative to each other. The ensuing discussion then holds for the instantaneous orientation shown at time $t$. We shall consider that the body is composed of a continuum of particles, each of which has a mass given by $\rho\, dv$. We now present the following definitions:

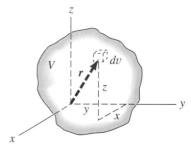

**Figure 9.1.** Body and reference at time $t$.

[1]This chapter may be covered at a later stage when studying dynamics. In that case, it should be covered directly after Chapter 15.

$$I_{xx} = \iiint\limits_{V} (y^2 + z^2)\rho \, dv \qquad (9.1a)$$

$$I_{yy} = \iiint\limits_{V} (x^2 + z^2)\rho \, dv \qquad (9.1b)$$

$$I_{zz} = \iiint\limits_{V} (x^2 + y^2)\rho \, dv \qquad (9.1c)$$

$$I_{xy} = \iiint\limits_{V} xy \, \rho \, dv \qquad (9.1d)$$

$$I_{xz} = \iiint\limits_{V} xz \, \rho \, dv \qquad (9.1e)$$

$$I_{yz} = \iiint\limits_{V} yz \, \rho \, dv \qquad (9.1f)$$

The terms $I_{xx}$, $I_{yy}$, and $I_{zz}$ in the set above are called the *mass moments of inertia* of the body about the $x, y,$ and $z$ axes, respectively.[2] Note that in each such case we are integrating the mass elements $\rho \, dv$, times the *perpendicular distance squared* from the mass elements to the coordinate axis about which we are computing the moment of inertia. Thus, if we look along the $x$ axis toward the origin in Fig. 9.1, we would have the view shown in Fig. 9.2. The quantity $y^2 + z^2$ used in Eq. 9.1a for $I_{xx}$ is clearly $d^2$, the perpendicular distance squared from $dv$ to the $x$ axis (now seen as a dot). Each of the terms with mixed indices is called the *mass product of inertia* about the pair of axes given by the indices. Clearly, from the definition of the product of inertia, we could reverse indices and thereby form three additional products of inertia for a reference. The additional three quantities formed in this way, however, are equal to the corresponding quantities of the original set. That is,

$$I_{xy} = I_{yx}, \qquad I_{xz} = I_{zx}, \qquad I_{yz} = I_{zy}$$

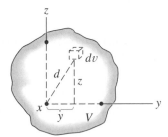

**Figure 9.2.** View of body along $x$ axis.

We now have nine inertia terms at a point for a given reference at this point. The values of the set of six independent quantities will, for a given body,

[2]We use the same notation as was used for second moments and products of area, which are also sometimes called moments and products of inertia. This is standard practice in mechanics. There need be no confusion in using these quantities if we keep the context of discussions clearly in mind.

depend on the *position* and *inclination* of the reference relative to the body. You should also understand that the reference may be established anywhere in space and *need not* be situated in the rigid body of interest. Thus there will be nine inertia terms for reference *xyz* at point $O$ outside the body (Fig. 9.3) computed using Eqs. 9.1, where the domain of integration is the volume $V$ of the body. As will be explained later, the nine moments and products of inertia are components of the inertia tensor.

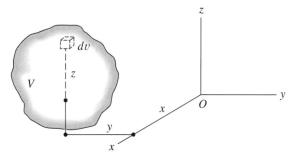

**Figure 9.3.** Origin of *xyz* outside body.

It will be convenient, when referring to the nine moments and products of inertia for reference *xyz* at a point, to list them in a matrix array, as follows:

$$I_{ij} = \begin{pmatrix} I_{xx} & I_{xy} & I_{xz} \\ I_{yx} & I_{yy} & I_{yz} \\ I_{zx} & I_{zy} & I_{zz} \end{pmatrix}$$

Notice that the first subscript gives the row and the second subscript gives the column in the array. Furthermore, the left-to-right downward diagonal in the array is composed of mass moment of inertia terms while the products of inertia, oriented at mirror-image positions about this diagonal, are equal. For this reason we say that the array is *symmetric*.

We shall now show that the sum of the mass moments of inertia for a set of orthogonal axes is independent of the orientation of the axes and depends only on the position of the origin. Examine the sum of such a set of terms:

$$I_{xx} + I_{yy} + I_{zz} = \iiint_V (y^2 + z^2)\rho\, dv + \iiint_V (x^2 + z^2)\rho\, dv + \iiint_V (x^2 + y^2)\rho\, dv$$

Combining the integrals and rearranging, we get

$$I_{xx} + I_{yy} + I_{zz} = \iiint_V 2(x^2 + y^2 + z^2)\rho\, dv = \iiint_V 2|r|^2 \rho\, dv \quad (9.2)$$

But the magnitude of the position vector from the origin to a particle is *independent* of the inclination of the reference at the origin. Thus, *the sum of the moments of inertia at a point in space for a given body clearly is an invariant with respect to rotation of axes.*

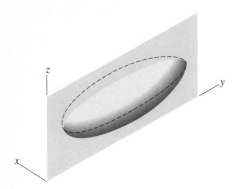

**Figure 9.4.** $zy$ is plane of symmetry.

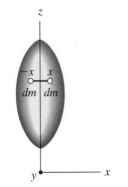

**Figure 9.5.** View along $y$ axis.

Clearly, on inspection of the equations 9.1, it is clear that the moments of inertia must always be positive, while the products of inertia may be positive or negative. Of interest is the case where one of the coordinate planes is a *plane of symmetry* for the mass distribution of the body. Such a plane is the $zy$ plane shown in Fig. 9.4 cutting a body into two parts, which, by definition of symmetry, are mirror images of each other. For the computation of $I_{xz}$, each half will give a contribution of the same magnitude but of opposite sign. We can most readily see that this is so by looking along the $y$ axis toward the origin. The plane of symmetry then appears as a line coinciding with the $z$ axis (see Fig. 9.5). We can consider the body to be composed of pairs of mass elements $dm$ which are mirror images of each other with respect to position and shape about the plane of symmetry. The product of inertia $I_{xz}$ for such a pair is then

$$xz\,dm - xz\,dm = 0$$

Thus, we can conclude that

$$I_{xz} = \underbrace{\int xz\,dm}_{\substack{\text{right} \\ \text{domain}}} - \underbrace{\int xz\,dm}_{\substack{\text{left} \\ \text{domain}}} = 0$$

This conclusion is also true for $I_{xy}$. We can say that $I_{xy} = I_{xz} = 0$. But on consulting Fig. 9.4, you should be able to readily decide that the term $I_{zy}$ will have a positive value. Note that those products of inertia having $x$ as an index are zero and that the $x$ coordinate axis is normal to the plane of symmetry. Thus, we can conclude that *if two axes form a plane of symmetry for the mass distribution of a body, the products of inertia having as an index the coordinate that is normal to the plane of symmetry will be zero.*

Consider next a body of *revolution*. Take the $z$ axis to coincide with the axis of symmetry. It is easy to conclude for the origin $O$ of $xyz$ anywhere along the axis of symmetry that

$$I_{xz} = I_{yz} = I_{xy} = 0$$
$$I_{xx} = I_{yy} = \text{constant}$$

for all possible $xy$ axes formed by rotating about the $z$ axis at $O$. Can you justify these conclusions?

Finally, we define *radii of gyration* in a manner analogous to that used for second moments of area in Chapter 8. Thus:

$$I_{xx} = k_x^2 M$$
$$I_{yy} = k_y^2 M$$
$$I_{zz} = k_z^2 M$$

where $k_x$, $k_y$, and $k_z$ are the radii of gyration and $M$ is the total mass.

# Example 9.1

Find the nine components of the inertia tensor of a rectangular body of uniform density $\rho$ about point $O$ for a reference $xyz$ coincident with the edges of the block as shown in Fig. 9.6.

We first compute $I_{xx}$. Using volume elements $dv = dx\,dy\,dz$, we get on using simple multiple integration:

$$I_{xx} = \int_0^a \int_0^b \int_0^c (y^2 + z^2)\rho\,dx\,dy\,dz$$

$$= \int_0^a \int_0^b (y^2 + z^2)c\rho\,dy\,dz = \int_0^a \left(\frac{b^3}{3} + z^2 b\right)c\rho\,dz \qquad (a)$$

$$= \left(\frac{ab^3 c}{3} + \frac{a^3 bc}{3}\right)\rho = \frac{\rho V}{3}(b^2 + a^2)$$

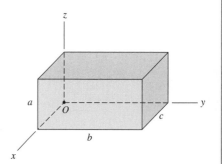

**Figure 9.6.** Find $I_{ij}$ at $O$.

where $V$ is the volume of the body. Note that the $x$ axis about which we are computing the moment of inertia $I_{xx}$ is *normal* to the plane having sides of length $a$ and $b$, i.e., along the $z$ and $y$ axes. Similarly:

$$I_{yy} = \frac{\rho V}{3}(c^2 + a^2) \qquad (b)$$

$$I_{zz} = \frac{\rho V}{3}(b^2 + c^2) \qquad (c)$$

We next compute $I_{xy}$.

$$I_{xy} = \int_0^a \int_0^b \int_0^c xy\,\rho\,dx\,dy\,dz = \int_0^a \int_0^b \frac{c^2}{2} y\rho\,dy\,dz$$

$$= \int_0^a \frac{c^2 b^2}{4}\rho\,dz = \frac{ac^2 b^2}{4}\rho = \frac{\rho V}{4}cb \qquad (d)$$

Note for $I_{xy}$, we use the lengths of the sides along the $x$ and $y$ axes.

$$I_{xz} = \frac{\rho V}{4}ac \qquad (e)$$

$$I_{yz} = \frac{\rho V}{4}ab \qquad (f)$$

We accordingly have, for the inertia tensor:

$$I_{ij} = \begin{pmatrix} \frac{\rho V}{3}(b^2 + a^2) & \frac{\rho V}{4}cb & \frac{\rho V}{4}ac \\ \frac{\rho V}{4}cb & \frac{\rho V}{3}(c^2 + a^2) & \frac{\rho V}{4}ab \\ \frac{\rho V}{4}ac & \frac{\rho V}{4}ab & \frac{\rho V}{3}(b^2 + c^2) \end{pmatrix} \qquad (g)$$

# Example 9.2

Compute the components of the inertia tensor at the center of a solid sphere of uniform density $\rho$ as shown in Fig. 9.7.

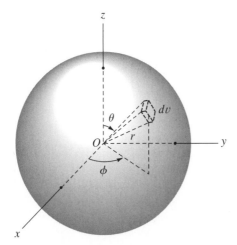

**Figure 9.7.** Find $I_{ij}$ at $O$.

We shall first compute $I_{yy}$. Using spherical coordinates, we have[3]

$$
\begin{aligned}
I_{yy} &= \iiint_V (x^2 + z^2)\rho \, dv \\
&= \int_0^R \int_0^{2\pi} \int_0^\pi \left[ (r \sin \theta \cos \phi)^2 + (r \cos \theta)^2 \right] \rho \left( r^2 \sin \theta \, d\theta \, d\phi \, dr \right) \\
&= \int_0^R \int_0^{2\pi} \int_0^\pi \left( r^4 \sin^3 \theta \cos^2 \phi \right) \rho \, d\theta \, d\phi \, dr \\
&\qquad\qquad + \int_0^R \int_0^{2\pi} \int_0^\pi \left( r^4 \cos^2 \theta \sin \theta \right) \rho \, d\theta \, d\phi \, dr \\
&= \rho \int_0^R \int_0^{2\pi} \left( r^4 \cos^2 \phi \right) \left( \int_0^\pi \sin^3 \theta \, d\theta \right) d\phi \, dr \\
&\qquad\qquad + \rho \int_0^R \int_0^{2\pi} r^4 \left( \int_0^\pi \cos^2 \theta \sin \theta \, d\theta \right) d\phi \, dr
\end{aligned}
$$

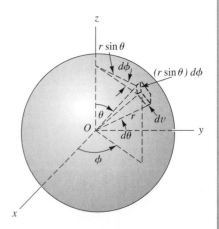

**Figure 9.8.** $dv = (r \sin \theta \, d\phi)(dr)(r \, d\theta) = r^2 \sin \theta \, d\theta \, d\phi \, dr$.

[3]For those unfamiliar with spherical coordinates, we have shown in Fig. 9.8 a more detailed study of the volume element used. The volume $dv$ is simply the product of the three edges of the element shown in the diagram.

# Example 9.2 (Continued)

With the aid of integration formulas from Appendix I, we have

$$I_{yy} = \rho \int_0^R \int_0^{2\pi} r^4 \cos^2 \phi \left[ -\tfrac{1}{3} \cos \theta (\sin^2 \theta + 2) \right] \Big|_0^{\pi} d\phi \, dr$$

$$+ \rho \int_0^R \int_0^{2\pi} r^4 \left( -\frac{\cos^3 \theta}{3} \right) \Big|_0^{\pi} d\phi \, dr$$

$$= \rho \int_0^R \int_0^{2\pi} r^4 \cos^2 \phi \, \tfrac{4}{3} \, d\phi \, dr + \rho \int_0^R \int_0^{2\pi} (r^4) \tfrac{2}{3} \, d\phi \, dr$$

Integrating next with respect to $\phi$, we get

$$I_{yy} = \rho \int_0^R (r^4) \left( \tfrac{4}{3} \right) (\pi) \, dr + \rho \int_0^R (r^4) \left( \tfrac{2}{3} \right) (2\pi) \, dr$$

Finally, we get

$$I_{yy} = \rho \frac{R^5}{5} \frac{4}{3} \pi + \rho \frac{R^5}{5} \frac{4}{3} \pi$$

$$\therefore I_{yy} = \frac{8}{15} \rho \pi R^5$$

But

$$M = \rho \tfrac{4}{3} \pi R^3$$

Hence,

$$I_{yy} = \tfrac{2}{5} MR^2$$

Because of the point symmetry about point $O$, we can also say that

$$I_{xx} = I_{zz} = \tfrac{2}{5} MR^2$$

Because the coordinate planes are all planes of symmetry for the mass distribution, the products of inertia are zero. Thus, the inertia tensor can be given as

$$I_{ij} = \begin{pmatrix} \tfrac{2}{5} MR^2 & 0 & 0 \\ 0 & \tfrac{2}{5} MR^2 & 0 \\ 0 & 0 & \tfrac{2}{5} MR^2 \end{pmatrix}$$

## 9.3   Relation Between Mass-Inertia Terms and Area-Inertia Terms

We now relate the second moment and product of area studied in Chapter 8 with the inertia tensor. To do this, consider a plate of constant thickness $t$ and uniform density $\rho$ (Fig. 9.9) A reference is selected so that the $xy$ plane is in the midplane of this plate. The components of the inertia tensor are rewritten for convenience as

$$
\begin{aligned}
I_{xx} &= \rho \iiint_V (y^2 + z^2)\ dv, & I_{xy} &= \rho \iiint_V xy\ dv \\
I_{yy} &= \rho \iiint_V (x^2 + z^2)\ dv, & I_{xz} &= \rho \iiint_V xz\ dv \\
I_{zz} &= \rho \iiint_V (x^2 + y^2)\ dv, & I_{yz} &= \rho \iiint_V yz\ dv
\end{aligned}
\tag{9.3}
$$

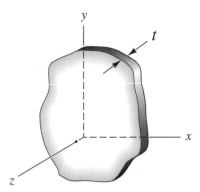

**Figure 9.9.** Plate of thickness $t$

Now consider that the thickness $t$ is *small* compared to the lateral dimensions of the plate. This means that $z$ is restricted to a range of values having a small magnitude. As a result, we can make two simplifications in the equations above. First, we shall set $z$ equal to zero whenever it appears on the right side of the equations above. Second, we shall express $dv$ as

$$dv = t\ dA$$

where $dA$ is an area element on the *surface* of the plate, as shown in Fig. 9.10. Equations 9.3 then become

$$I_{xx} = \rho t \iint_A y^2 \ dA, \qquad\qquad I_{xy} = \rho t \iint_A xy \ dA$$

$$I_{yy} = \rho t \iint_A x^2 \ dA, \qquad\qquad I_{xz} = 0$$

$$I_{zz} = \rho t \iint_A (x^2 + y^2) \ dA, \qquad I_{yz} = 0$$

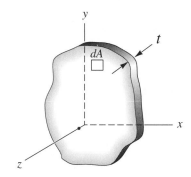

**Figure 9.10.** Use volume elements $t \ dA$.

Notice, now, that the integrals on the right sides of the equations above are moments and products of *area* as presented in Chapter 8. Denoting mass-moment and product of inertia terms with a subscript $M$ and second moment and product of area terms with a subscript $A$, we can then say for the nonzero expressions:

$$(I_{xx})_M = \rho t (I_{xx})_A$$
$$(I_{yy})_M = \rho t (I_{yy})_A$$
$$(I_{zz})_M = \rho t (J)_A$$
$$(I_{xy})_M = \rho t (I_{xy})_A$$

Thus, for a thin plate with a constant product $\rho t$ throughout, we can compute the inertia tensor components for reference $xyz$ (see Fig. 9.9) by using the second moments and product of area of the surface of the plate relative to axes $xy$.

It is important to point out that $\rho t$ is the *mass per unit area* of the plate. Imagine next that $t$ goes to zero and simultaneously $\rho$ goes to infinity at rates such that the product $\rho t$ becomes unity in the limit. One might think of the resulting body to be a *plane area*. By this approach, we have thus formed a plane area from a plate and in this way we can think of a plane area as a special mass. This explains why we use the same notation for mass moments and products of inertia as we use for second moments and products of area. However the units clearly will be different. We now examine a plate problem.

## Example 9.3

Determine the inertia tensor components for the thin plate (Fig. 9.11) relative to the indicated axes $xyz$. The weight of the plate is .002 N/mm.² For the top edge, $y = 2\sqrt{x}$ with $x$ and $y$ in millimeters.

It is clear that for $\rho t$ we have, remembering that this product represents mass per unit area:

$$\rho t = \frac{.002}{9.81} = .000204 \text{ kg/mm}^2 \qquad \text{(a)}$$

**Figure 9.11.** Plate of thickness $t$.

We now examine the second moments and product of area for the surface of the plate about axes $xy$. Thus,[4]

$$(I_{xx})_A = \int_0^{100} \int_{y=0}^{y=2\sqrt{x}} y^2 \, dy \, dx$$

$$= \int_0^{100} \frac{y^3}{3} \Big|_0^{2\sqrt{x}} dx = \int_0^{100} \frac{8}{3} x^{3/2} dx$$

$$= \frac{8}{3} \frac{x^{5/2}}{\frac{5}{2}} \Big|_0^{100} = \left(\frac{8}{3}\right)\left(\frac{2}{5}\right)\left(100^{5/2}\right)$$

$$= 1.067 \times 10^5 \text{ mm}^4$$

$$(I_{yy})_A = \int_0^{100} \int_{y=0}^{y=2\sqrt{x}} x^2 \, dy \, dx$$

$$= \int_0^{100} x^2 y \Big|_0^{2\sqrt{x}} dx = \int_0^{100} x^2 \left(2\sqrt{x}\right) dx$$

$$= 2 \frac{x^{7/2}}{\frac{7}{2}} \Big|_0^{100} = 2\left(\frac{2}{7}\right)\left(100^{7/2}\right)$$

$$= 5.71 \times 10^6 \text{ mm}^4$$

$$(I_{xy})_A = \int_0^{100} \int_{y=0}^{y=2\sqrt{x}} xy \, dy \, dx$$

$$= \int_0^{100} x \frac{y^2}{2} \Big|_0^{2\sqrt{x}} dx = \int_0^{100} 2x^2 \, dx$$

$$= 2\left(\frac{100^3}{3}\right) = 6.67 \times 10^5 \text{ mm}^4$$

---

[4]Note we have multiple integration where one of the boundaries is variable. The procedure to follow should be evident from the example.

## ■ Example 9.3 (Continued)

Using Eq. (a), we can then say for the nonzero inertia tensor components:

$$(I_{xx})_M = (.000204)(1.067 \times 10^5) = 21.76 \text{ kg-mm}^2$$
$$(I_{yy})_M = (.000204)(5.71 \times 10^6) = 1,165 \text{ kg-mm}^2$$
$$(I_{xy})_M = (.000204)(6.67 \times 10^5) = 136.1 \text{ kg-mm}^2$$

Note that the nonzero inertia tensor components for a reference $xyz$ on a plate (see Fig. 9.9) are *proportional* through $\rho t$ to the corresponding area-inertia terms for the plate surface. This means that all the formulations of Chapter 8 apply to the aforementioned nonzero inertia tensor components. Thus, on rotating the axes about the $z$ axis we may use the transformation equations of Chapter 8. Consequently, the concept of *principal axes* in the midplane of the plate at a point applies. For such axes, the product of inertia is zero. One such axis then gives the maximum moment of inertia for all axes in the midplane at the point, the other the minimum moment of inertia. We have presented such problems at the end of this section.

What about principal axes for the inertia tensor at a point in a general three-dimensional body? Those students who have time to study Section 9.7 will learn that there are *three principal axes* at a point in the general case. These axes are *mutually orthogonal* and the *products of inertia are all zero* for such a set of axes at a point.[5] Furthermore, one of the axes will have a maximum moment of inertia, another axis will have a minimum moment of inertia, while the third axis will have an intermediate value. The sum of these three inertia terms must have a value that is common for all sets of axes at the point.

If, perchance, a set of axes $xyz$ at a point is such that $xy$ and $xz$ form *two planes of symmetry* for the mass distribution of the body, then, as we learned earlier, since the $z$ axis and the $y$ axis are normal to the planes of symmetry, $I_{xy} = I_{xz} = I_{yz} = 0$. Thus, all products of inertia are zero. This would also be true for *any* two sets of axes of $xyz$ forming two planes of symmetry. Clearly, axes forming two planes of symmetry must be *principal axes*. This information will suffice in most instances when we have to identify principal axes. On the other hand, consider the case where there is only *one plane of symmetry* for the mass distribution of a body at some point $A$. Let the $xy$ plane at $A$ form this plane of symmetry. Then, clearly, the products of inertia between the $z$ axis that is normal to the plane of symmetry $xy$ and *any axis* in the $xy$ plane at $A$ must be zero, as pointed out earlier. Obviously, the $z$ axis must be a principal axis. The other two principal axes must be in the plane of symmetry, but generally cannot be located by inspection.

---

[5]The third principal axis for a plate at a point in the midplane is the $z$ axis normal to the plate. Note that $(I_{zz})_M$ must always equal  const. $+[I_{xx})_M + (I_{yy})_{M}.]$ Why?

**9.1.** A uniform homogeneous slender rod of mass $M$ is shown. Compute $I_{xx}$ and $I_{x'x'}$.

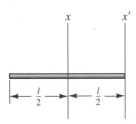

**Figure P.9.1.**

**9.2.** Find $I_{xx}$ and $I_{x'x'}$ for the thin rod of Problem 9.1 for the case where the mass per unit length at the left end is 5 lbm/ft and increases linearly so that at the right end it is 8 lbm/ft. The rod is 20 ft in length.

**9.3.** Compute $I_{xy}$ for the thin homogeneous hoop of mass $M$.

**Figure P.9.3.**

**9.4.** Compute $I_{xx}$, $I_{yy}$, $I_{zz}$, and $I_{xy}$ for the homogeneous rectangular parallelepiped.

**Figure P.9.4.**

**9.5.** A wire having the shape of a parabola is shown. The curve is in the $yz$ plane. If the mass of the wire is .3 N/m, what are $I_{yy}$ and $I_{xz}$? [*Hint:* Replace $ds$ along the wire by $\sqrt{(dy/dz)^2 + 1}\,dz$.]

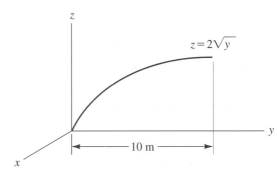

**Figure P.9.5.**

**9.6.** Compute the moment of inertia, $I_{BB}$, for the half-cylinder shown. The body is homogeneous and has a mass $M$.

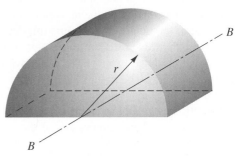

**Figure P.9.6.**

**9.7.** Find $I_{zz}$ and $I_{xx}$ for the homogeneous right circular cylinder of mass $M$.

**Figure P.9.7.**

**9.8.** For the cylinder in Problem 9.7, the density increases linearly in the $z$ direction from a value of .100 grams/mm$^3$ at the left end to a value of .180 grams/mm$^3$ at the right end. Take $r = 30$ mm and $l = 150$ mm. Find $I_{xx}$ and $I_{zz}$.

**9.9.** Show that $I_{zz}$ for the homogeneous right circular cone is $\frac{3}{10} MR^2$.

**Figure P.9.9.**

**9.10.** In Problem 9.9, the density increases as the square of $z$ in the $z$ direction from a value of .200 grams/mm³ at the left end to a value of .400 grams/mm³ at the right end. If $r = 20$ mm and the cone is 100 mm in length, find $I_{zz}$.

**9.11.** A body of revolution is shown. The radial distance $r$ of the boundary from the $x$ axis is given as $r = .2x^2$ m. What is $I_{xx}$ for a uniform density of 1,600 kg/m³?

**Figure P.9.11.**

**9.12.** A thick hemispherical shell is shown with an inside radius of 40 mm and an outside radius of 60 mm. If the density $\rho$ is 7,000 kg/m,³ what is $I_{yy}$?

**Figure P.9.12.**

**9.13.** Find the mass moment of inertia $I_{xx}$ for a very thin plate forming a quarter-sector of a circle. The plate weighs .4 N. What is the second moment of area about the $x$ axis? What is the product of inertia? Axes are in the midplane of the plate.

**Figure P.9.13.**

**9.14.** Find the second moment of area about the $x$ axis for the front surface of a very thin plate. If the weight of the plate is .02 N/mm², find the mass moments of inertia about the $x$ and $y$ axes. What is the mass product of inertia $I_{xy}$?

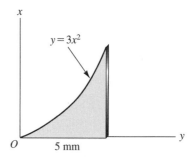

**Figure P.9.14.**

**\*9.15.** A uniform tetrahedron is shown having sides of length $a$, $b$, and $c$, respectively, and a mass $M$. Show that $I_{yz} = \frac{1}{20} Mac$. (*Suggestion:* Let $z$ run from zero to surface $ABC$. Let $x$ run from zero to line $AB$. Finally, let $y$ run from zero to $B$. Note that the equation of a plane surface is $z = \alpha x + \beta y + \gamma$, where $\alpha$, $\beta$, and $\gamma$ are constants. The mass of the tetrahedron is $\rho abc/6$. It will be simplest in expanding $(1 - x/b - y/c)^2$ to proceed in the form $[(1 - y/c) - (x/b)]^2$, keeping $(1 - y/c)$ intact. In the last integration replace $y$ by $[-c(1 - y/c) + c]$, etc.)

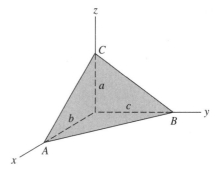

**Figure P.9.15.**

**9.16.** In Problem 9.13, find the three principal mass moments of inertia at $O$. Use the following results from problem 9.13

$$(I_{xx})_M = 101.9 \text{ kg-mm}^2$$
$$(I_{xy})_M = 64.9 \text{ kg-mm}^2$$

**Figure P.9.18.**

**9.17.** In Problem 9.14, compute the values of the three principal mass moments of inertia at $O$. From Problem 9.14 we have the results

$$(I_{xx})_M = 205 \text{ kg-mm}^2$$
$$(I_{yy})_M = 3.82 \text{ kg-mm}^2$$
$$(I_{xy})_M = 23.9 \text{ kg-mm}^2$$

**9.18.** Can you identify by inspection any of the principal axes of inertia at $A$? At $B$? Explain. The density of the material is uniform.

**9.19.** By inspection, identify as many principal axes as you can for mass moments of inertia at positions $A$, $B$, and $C$. Explain your choices. The mass density of the material is uniform throughout.

**Figure P.9.19.**

## 9.4 Translation of Coordinate Axes

In this section, we will compute mass moment and product of inertia quantities for a reference $xyz$ that is displaced under a translation (no rotation) from a reference $x'y'z'$ at the center of mass (Fig. 9.12) for which the inertia terms are presumed known. Let us first compute the mass moment of inertia $I_{zz}$.

Observing Fig. 9.12, we see that

$$r = r_c + r'$$

Hence,

$$x = x_c + x'$$
$$y = y_c + y'$$
$$z = z_c + z'$$

We can now formulate $I_{zz}$ in the following way:

$$I_{zz} = \iiint_V (x^2 + y^2)\rho \, dv = \iiint_V \left[ (x_c + x')^2 + (y_c + y')^2 \right] \rho \, dv \quad (9.4)$$

Carrying out the squares and rearranging, we have

$$I_{zz} = \iiint_V (x_c^2 + y_c^2)\rho \, dv + 2\iiint_V x_c x' \rho \, dv$$
$$+ 2\iiint_V y_c y' \rho \, dv + \iiint_V (x'^2 + y'^2)\rho \, dv \quad (9.5)$$

**Figure 9.12.** $xyz$ translated from $x'y'z'$ at C.M.

392

Note that the quantities bearing the subscript $c$ are constant for the integration and can be extracted from under the integral sign. Thus,

$$I_{zz} = M(x_c^2 + y_c^2) + 2x_c \iiint_V x' \, dm$$

$$+ 2y_c \iiint_V y' \, dm + \iiint_V (x'^2 + y'^2)\rho \, dv \quad (9.6)$$

where $\rho \, dv$ has been replaced in some terms by $dm$, and the integration $\iiint_V \rho \, dv$

in the first integral has been evaluated as $M$, the total mass of the body. The origin of the primed reference being at the center of mass requires of the first moments of mass that $\iiint x' \, dm = \iiint y' \, dm = \iiint z' \, dm = 0$. The middle two terms accordingly drop out of the expression above, and we recognize the last expression to be $I_{z'z'}$. Thus, the desired relation is

$$I_{zz} = I_{z'z'} + M(x_c^2 + y_c^2) \quad (9.7)$$

By observing the body in Fig. 9.12 along the $z$ and $z'$ axes (i.e., from directly above), we get a view as is shown in Fig 9.13. From this diagram, we can see that $y_c^2 + x_c^2 = d,^2$ where $d$ is the perpendicular distance between the $z'$ axis through the center of mass and the $z$ axis about which we are taking moments of inertia. We may then give the result above as

$$I_{zz} = I_{z'z'} + Md^2 \quad (9.8)$$

Let us generalize from the previous statement.

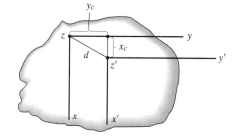

**Figure 9.13.** View along $z$ direction (from above).

*The moment of inertia of a body about any axis equals the moment of inertia of the body about a parallel axis that goes through the center of mass, plus the total mass times the perpendicular distance between the axes squared.*

We leave it to you to show that for products of inertia a similar relation can be reached. For $I_{xy}$, for example, we have

$$I_{xy} = I_{x'y'} + Mx_c y_c \quad (9.9)$$

Here, we must take care to put in the proper signs of $x_c$ and $y_c$ as measured *from* the *xyz* reference. Equations 9.8 and 9.9 comprise the well-known *parallel-axis theorems* analogous to those formed in Chapter 8 for areas. You can use them to advantage for bodies composed of simple familiar shapes, as we now illustrate.

## Example 9.4

Find $I_{xx}$ and $I_{xy}$ for the body shown in Fig. 9.14. Take $\rho$ as constant for the body. Use the formulations for moments and products of inertia at the center of mass as given on the inside front cover page.

We shall consider first a solid rectangular prism having the outer dimensions given in Fig. 9.14, and we shall then subtract the contribution of the cylinder and the rectangular block that have been cut away. Thus, we have, for the overall rectangular block which we consider as body 1,

$$(I_{xx})_1 = (I_{xx})_c + Md^2 = \tfrac{1}{12} M(a^2 + b^2) + Md^2$$
$$= \tfrac{1}{12}\big[(\rho)(20)(8)(15)\big](8^2 + 15^2) + \big[(\rho)(20)(8)(15)\big](4^2 + 7.5^2) \quad \text{(a)}$$
$$= 231{,}200\rho$$

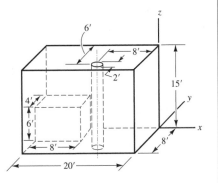

**Figure 9.14.** Find $I_{xx}$ and $I_{xy}$.

From this, we shall take away the contribution of the cylinder, which we denote as body 2. Using the formulas from the inside front cover page,

$$(I_{xx})_2 = \tfrac{1}{12} M(3r^2 + h^2) + Md^2$$
$$= \tfrac{1}{12}\big[\rho\pi(1)^2(15)\big]\big[3(1^2) + 15^2\big] + \big[\rho\pi(1)^2(15)\big]\big[6^2 + 7.5^2\big] \quad \text{(b)}$$
$$= 5{,}243\rho$$

Also, we shall take away the contribution of the rectangular cutout (body 3):

$$(I_{xx})_3 = \tfrac{1}{12} M(a^2 + b^2) + Md^2$$
$$= \tfrac{1}{12}\big[\rho(8)(6)(4)\big](4^2 + 6^2) + \big[\rho(8)(6)(4)\big](2^2 + 3^2) \quad \text{(c)}$$
$$= 3{,}328\rho$$

The quantity $I_{xx}$ for the body with the rectangular and cylindrical cavities is then

$$I_{xx} = (231{,}200 - 5{,}243 - 3{,}328)\rho$$

$$I_{xx} = 223{,}000\rho \quad \text{(d)}$$

We follow the same procedure to obtain $I_{xy}$. Thus, for the block as a whole, we have

$$(I_{xy})_1 = (I_{xy})_c + Mx_c y_c$$

At the center of mass of the block, both the $(x')_1$ and $(y')_1$ axes are normal to planes of symmetry. Accordingly, $(I_{xy})_c = 0$. Hence,

$$(I_{xy})_1 = 0 + [\rho(20)(8)(15)](-4)(-10)$$
$$= 96{,}000\rho \quad \text{(e)}$$

For the cylinder, we note that both the $(x')_2$ and $(y')_2$ axes at the center of mass are normal to planes of symmetry. Hence, we can say that

$$(I_{xy})_2 = 0 + [\rho(\pi)(1^2)(15)](-8)(-6)$$
$$= 2{,}262\rho \quad \text{(f)}$$

## Example 9.4 (Continued)

Finally, for the small cutout rectangular parallelepiped, we note that the $(x')_3$ and $(y')_3$ axes at the center of mass are perpendicular to planes of symmetry. Hence, we have

$$(I_{xy})_3 = 0 + [\rho(8)(6)(4)](-2)(-16)$$
$$= 6{,}144\rho \tag{g}$$

The quantity $I_{xy}$ for the body with the rectangular and cylindrical cavities is then

$$I_{xy} = (96{,}000 - 2{,}262 - 6{,}144)\rho = \boxed{87{,}600\rho} \tag{h}$$

If $\rho$ is given in units of lbm/ft$^3$ the inertia terms have units of lbm-ft$^2$.

## *9.5  Transformation Properties of the Inertia Terms

Let us assume that the six independent inertia terms are known at the origin of a given reference. What is the mass moment of inertia for an axis going through the origin of the reference and having the direction cosines $l$, $m$, and $n$ relative to the axes of this reference? The axis about which we are interested in obtaining the mass moment of inertia is designated as $kk$ in Fig. 9.15.

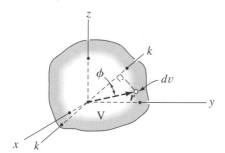

**Figure 9.15.** Find $I_{kk}$.

From previous conclusions, we can say that

$$I_{kk} = \iiint_V \left[|r|(\sin \phi)\right]^2 \rho \, dv \tag{9.10}$$

where $\phi$ is the angle between $kk$ and $r$. We shall now put $\sin^2 \phi$ into a more useful form by considering the right triangle formed by the position vector $r$ and the axis $kk$. This triangle is shown enlarged in Fig. 9.16. The side $a$ of the triangle has a magnitude that can be given by the dot product of $r$ and the unit vector $\epsilon_k$ along $kk$. Thus,

$$a = r \bullet \epsilon_k = (x\boldsymbol{i} + y\boldsymbol{j} + z\boldsymbol{k}) \bullet (l\boldsymbol{i} + m\boldsymbol{j} + n\boldsymbol{k}) \tag{9.11}$$

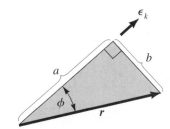

**Figure 9.16.** Right triangle formed by $r$ and $kk$.

Hence

$$a = lx + my + nz$$

Using the Pythagorean theorem, we can now give side $b$ as

$$b^2 = |r|^2 - a^2 = (x^2 + y^2 + z^2)$$
$$- (l^2 x^2 + m^2 y^2 + n^2 z^2 + 2lmxy + 2lnxz + 2mnyz)$$

The term $\sin^2 \phi$ may next be given as

$$\sin^2 \phi = \frac{b^2}{r^2} = \frac{(x^2 + y^2 + z^2) - (l^2 x^2 + m^2 y^2 + n^2 z^2 + 2lmxy + 2lnxz + 2mnyz)}{x^2 + y^2 + z^2}$$
$$\tag{9.12}$$

Substituting back into Eq. 9.10, we get, on canceling terms,

$$I_{kk} = \iiint\limits_V [(x^2 + y^2 + z^2)$$
$$- (l^2 x^2 + m^2 y^2 + n^2 z^2 + 2lmxy + 2lnxz + 2mnyz)]\rho \, dv$$

Since $l^2 + m^2 + n^2 = 1$, we can multiply the first bracketed expression in the integral by this sum:

$$I_{kk} = \iiint\limits_V [(x^2 + y^2 + z^2)(l^2 + m^2 + n^2)$$
$$- (l^2 x^2 + m^2 y^2 + n^2 z^2 + 2lmxy + 2lnxz + 2mnyz)]\rho \, dv$$

Carrying out the multiplication and collecting terms, we get the relation

$$I_{kk} = l^2 \iiint_V (y^2 + z^2)\rho\, dv + m^2 \iiint_V (x^2 + z^2)\rho\, dv + n^2 \iiint_V (x^2 + y^2)\rho\, dv$$

$$- 2lm \iiint_V (xy)\rho\, dv - 2ln \iiint_V (xz)\rho\, dv - 2mn \iiint_V (yz)\rho\, dv$$

Referring back to the definitions presented by Eqs. 9.1, we reach the desired transformation equation:

$$I_{kk} = l^2 I_{xx} + m^2 I_{yy} + n^2 I_{zz} - 2lm I_{xy} - 2ln I_{xz} - 2mn I_{yz} \qquad (9.13)$$

We next put this in a more useful form of the kind you will see in later courses in mechanics. Note first that $l$ is the direction cosine between the $k$ axis and the $x$ axis. It is common practice to identify this cosine as $a_{kx}$ instead of $l$. Note that the subscripts identify the axes involved. Similarly, $m = a_{ky}$ and $n = a_{kz}$. We can now express Eq. 9.13 in a form similar to a matrix array as follows on noting that $I_{xy} = I_{yx}$, etc.

$$
\begin{aligned}
I_{kk} = \quad &I_{xx} a_{kx}^2 &&- I_{xy} a_{kx} a_{ky} &&- I_{xz} a_{kx} a_{kz} \\
&- I_{yx} a_{ky} a_{kx} &&+ I_{yy} a_{ky}^2 &&- I_{yz} a_{ky} a_{kz} \\
&- I_{zx} a_{kz} a_{kx} &&- I_{zy} a_{kz} a_{ky} &&+ I_{zz} a_{kz}^2
\end{aligned}
\qquad (9.14)
$$

This format is easily written by first writing the matrix array of $I$'s on the right side and then inserting the $a$'s remembering to insert minus signs for off-diagonal terms.

Let us next compute the product of inertia for a pair of mutually perpendicular axes, $Ok$ and $Oq$, as shown in Fig. 9.17. The direction cosines of $Ok$ we shall take as $l$, $m$, and $n$, whereas the direction cosines of $Oq$ we shall take as $l'$, $m'$, and $n'$. Since the axes are at right angles to each other, we know that

$$\boldsymbol{\epsilon}_k \boldsymbol{\cdot} \boldsymbol{\epsilon}_q = 0$$

Therefore,

$$ll' + mm' + nn' = 0 \qquad (9.15)$$

Noting that the coordinates of the mass element $\rho\, dv$ along the axes $Ok$ and $Oq$ are $\boldsymbol{r} \boldsymbol{\cdot} \boldsymbol{\epsilon}_k$ and $\boldsymbol{r} \boldsymbol{\cdot} \boldsymbol{\epsilon}_q$, respectively, we have, for $I_{kq}$:

$$I_{kq} = \iiint_V (\boldsymbol{r} \boldsymbol{\cdot} \boldsymbol{\epsilon}_k)(\boldsymbol{r} \boldsymbol{\cdot} \boldsymbol{\epsilon}_q)\rho\, dv$$

Using $xyz$ components of $\boldsymbol{r}$ and the unit vectors, we have

$$I_{kq} = \iiint_V \big[(x\boldsymbol{i} + y\boldsymbol{j} + z\boldsymbol{k}) \boldsymbol{\cdot} (l\boldsymbol{i} + m\boldsymbol{j} + n\boldsymbol{k})\big]$$

$$\times \big[(x\boldsymbol{i} + y\boldsymbol{j} + z\boldsymbol{k}) \boldsymbol{\cdot} (l'\boldsymbol{i} + m'\boldsymbol{j} + n'\boldsymbol{k})\big]\rho\, dv \qquad (9.16)$$

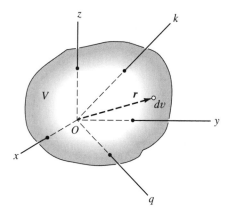

**Figure 9.17.** Find $I_{kq}$.

Carrying out the dot products in the integrand above, we get the following result:

$$I_{kq} = \iiint_V (xl + ym + zn)(xl' + ym' + zn')\rho \, dv$$

Hence,

$$I_{kq} = \iiint_V (x^2 ll' + y^2 mm' + z^2 nn' + xylm' + xzln'$$
$$+ yxml' + yzmn' + zxnl' + zynm')\rho \, dv \tag{9.17}$$

Noting from Eq. 9.15 that $(ll' + mm' + nn')$ is zero, we may for convenience add the term, $(-x^2 - y^2 - z^2)(ll' + mm' + nn')$, to the integrand in the equation above. After canceling some terms, we have

$$I_{kq} = \iiint_V (-x^2 mm' - x^2 nn' - y^2 ll' - y^2 nn' - z^2 ll' - z^2 mm'$$
$$+ xylm' + xzln' + yxml' + yzmn' + zxnl' + zynm')\rho \, dv$$

Collecting terms and bringing the direction cosines outside the integrations, we get

$$I_{kq} = -ll' \iiint_V (y^2 + z^2)\rho \, dv - mm' \iiint_V (x^2 + z^2)\rho \, dv$$
$$- nn' \iiint_V (y^2 + x^2)\rho \, dv + (lm' + ml') \iiint_V xy\rho \, dv \tag{9.18}$$
$$+ (ln' + nl') \iiint_V xz\rho \, dv + (mn' + nm') \iiint_V yz\rho \, dv$$

Noting the definitions in Eq. 9.1, we can state the desired transformation:

$$I_{kq} = -ll'I_{xx} - mm'I_{yy} - nn'I_{zz} + (lm' + ml')I_{xy}$$
$$+ (ln' + nl')I_{xz} + (mn' + nm')I_{yz} \tag{9.19}$$

We can now rewrite the previous equation in a more useful and simple form using $a$'s as direction cosines. Thus, noting that $l' = a_{qx}$, etc., we proceed as in Eq. 9.14 to obtain

$$- I_{kq} = I_{xx}a_{kx}a_{qx} - I_{xy}a_{kx}a_{qy} - I_{xz}a_{kx}a_{qz}$$
$$- I_{yx}a_{ky}a_{qx} + I_{yy}a_{ky}a_{qy} - I_{yz}a_{ky}a_{qz} \tag{9.20}$$
$$- I_{zx}a_{kz}a_{qx} - I_{zy}a_{kz}a_{qy} + I_{zz}a_{kz}a_{qz}$$

Again you will note that the right side can easily be set forth by first putting down the matrix array of $I_{ij}$ and then inserting the $a$'s with easily determined subscripts while remembering to insert minus signs for off-diagonal terms.

# Example 9.5

Find $I_{z'z'}$ and $I_{x'z'}$ for the solid cylinder shown in Fig. 9.18. The reference $x'y'z'$ is found by rotating about the $y$ axis an amount 30°, as shown in the diagram. The mass of the cylinder is 100 kg.

It is simplest to first get the inertia tensor components for reference $xyz$. Thus, using formulas from the inside front cover page we have

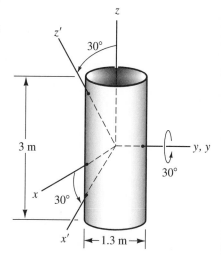

Figure 9.18. Find $I_{z'z'}$ and $I_{x'z'}$.

$$I_{zz} = \tfrac{1}{2} Mr^2 = \tfrac{1}{2}(100)\left(\tfrac{1.3}{2}\right)^2 = 21.13 \text{ kg-m}^2$$

$$I_{xx} = I_{yy} = \tfrac{1}{12} M(3r^2 + h^2)$$

$$= \tfrac{1}{12}(100)\left[(3)\left(\tfrac{1.3}{2}\right)^2 + 3^2\right]$$

$$= 85.56 \text{ kg-m}^2$$

Noting that the $xyz$ coordinate planes are planes of symmetry, we can conclude that

$$I_{xz} = I_{yx} = I_{yz} = 0$$

Next, evaluate the direction cosines of the $z'$ and the $x'$ axes relative to $xyz$. Thus,

For $z'$ axis:

$$a_{z'x} = \cos 60° = .500$$
$$a_{z'y} = \cos 90° = 0$$
$$a_{z'z} = \cos 30° = .866$$

For $x'$ axis:

$$a_{x'x} = \cos 30° = .866$$
$$a_{x'y} = \cos 90° = 0$$
$$a_{x'z} = \cos 120° = -.500$$

First, we employ Eq. 9.14 to get $I_{z'z'}$.

$$I_{z'z'} = (85.56)(.500)^2 + (21.13)(.866)^2$$

$$= \boxed{37.24 \text{ kg-m}^2}$$

Finally, we employ Eq. 9.20 to get $I_{x'z'}$.

$$-I_{x'z'} = (85.56)(.500)(.866) + (21.13)(.866)(-.500)$$

Therefore,

$$\boxed{I_{x'z'} = -27.90 \text{ kg-m}^2}$$

# *9.6    Looking Ahead: Tensors

By making the axis $Ok$ in Fig. 9.17 an $x'$ axis at $O$ and using the direction cosines for this axis ($a_{x'x}$, $a_{x'y}$, $a_{x'z}$), we can formulate $I_{x'x'}$ from Eq. 9.14. By a similar procedure, we can consider axis $Ok$ to be a $y'$ axis at $O$ and we can formulate $I_{y'y'}$ using for this axis direction cosines $a_{y'x}$, $a_{y'y}$, $a_{y'z}$ from Eq. 9.14. Similarly, for $I_{z'z'}$. We can thus get the mass moments of inertia for reference $x'$, $y'$, $z'$ at $O$ rotated arbitrarily relative to $xyz$. Also by considering the $Ok$ and $Oq$ axes to be $x'$ and $y'$ axes, respectively, with $a_{x'x}$, $a_{x'y}$, and $a_{x'z}$ as direction cosines for the $x'$ axis and $a_{y'x}$, $a_{y'y}$, and $a_{y'z}$ as direction cosines for the $y'$ axis, we can evaluate $I_{x'y'}$ at $O$ using Eq. 9.20. This approach can similarly be followed to find $I_{x'z'}$ and $I_{y'z'}$. Thus, employing Eqs. 9.14 and 9.20 as parent equations, we can develop equations for computing the nine inertia quantities for a reference $x'y'z'$ rotated arbitrarily relative to $xyz$ at $O$ in terms of the nine known inertia quantities for reference $xyz$ at $O$. Thus, once the nine inertia quantities are known for one reference at some point, they can be determined for *any* reference at that point. We say that the inertia terms *transform* from one set of components for $xyz$ at some point $O$ to another set of components for $x'y'z'$ at point $O$ by means of certain transformations formed from Eqs. 9.14 and 9.20.

We now define a symmetric,[6] *second-order tensor as a set of nine components*

$$\begin{pmatrix} A_{xx} & A_{xy} & A_{xz} \\ A_{yx} & A_{yy} & A_{yz} \\ A_{zx} & A_{zy} & A_{zz} \end{pmatrix}$$

*which transforms with a rotation of axes according to the following parent equations.* For the diagonal terms,

$$\begin{aligned} A_{kk} = \quad & A_{xx}a_{kx}^2 \quad + A_{xy}a_{kx}a_{ky} + A_{xz}a_{kx}a_{kz} \\ + \, & A_{yx}a_{ky}a_{kx} + A_{yy}a_{ky}^2 \quad + A_{yz}a_{ky}a_{kz} \\ + \, & A_{zx}a_{kz}a_{kx} + A_{zy}a_{kz}a_{ky} + A_{zz}a_{kz}^2 \end{aligned} \tag{9.21}$$

For the off-diagonal terms,

$$\begin{aligned} A_{kq} = \quad & A_{xx}a_{kx}a_{qx} + A_{xy}a_{kx}a_{qy} + A_{xz}a_{kx}a_{kz} \\ + \, & A_{yx}a_{ky}a_{qx} + A_{yy}a_{ky}a_{qy} + A_{yz}a_{ky}a_{kz} \\ + \, & A_{zx}a_{kz}a_{qx} + A_{zy}a_{kz}a_{qy} + A_{zz}a_{kz}a_{qz} \end{aligned} \tag{9.22}$$

On comparing Eqs. 9.21 and 9.22, respectively, with Eqs. 9.14 and 9.20, we can conclude that the array of terms

$$I_{ij} = \begin{pmatrix} I_{xx} & -I_{xy} & -I_{xz} \\ -I_{yx} & I_{yy} & -I_{yz} \\ -I_{zx} & -I_{zy} & I_{zz} \end{pmatrix} \tag{9.23}$$

is a second-order tensor.

---

[6] The word "symmetric" refers to the condition $A_{12} = A_{21}$, etc., that is required if the transformation equation is to have the form given. We can have nonsymmetric second-order tensors, but since they are less common in engineering work, we shall not concern ourselves here with such possibilities.

You will learn that because of the common transformation law identifying certain quantities as tensors, there will be extremely important common characteristics for these quantities which set them apart from other quantities. Thus, in order to learn these common characteristics in an efficient way and to understand them better, we become involved with tensors as an entity in the engineering sciences, physics, and applied mathematics. You will soon be confronted with the stress and strain tensors in your courses in strength of materials.

To explore this point further, we have shown an infinitesimal rectangular parallelepiped extracted from a solid under load. On three orthogonal faces we have shown nine force intensities (i.e., forces per unit area). Those with repeated indices are called *normal stresses* while those with different pairs of indices are called *shear stresses*. You will learn, that knowing nine such stresses, you can readily find three stresses, one normal and two orthogonal shear stresses, on *any* interface at any orientation inside the rectangular parallelepiped. To find such stresses on an interface knowing the stresses shown in Fig. 9.19, we have the *same transformation equations* given by Eqs. 9.21 and 9.22. Thus stress is a *second-order tensor*.

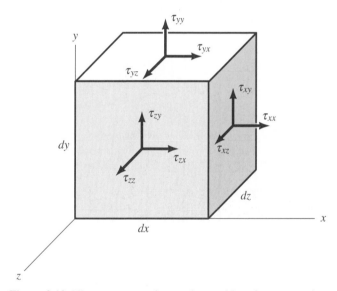

**Figure 9.19.** Nine stresses on three orthogonal interfaces at a point.

A *two-dimensional* simplification of $\tau_{ij}$ involving the quantities $\tau_{xx}$, $\tau_{yy}$, and $\tau_{xy}$ $(= \tau_{yx})$ as the only nonzero stresses is called *plane stress*. This occurs in a thin plate loaded in the plane of symmetry as shown in Fig. 9.20. Plane stress is the direct analog of *second moments and products of area*, which is a two-dimensional simplification of the inertia tensor. Clearly, plane stress and second moments and products of area have the same transformation equations, which are Eqs. 8.26 through 8.28 with $\tau_{xx}$, $\tau_{yy}$, $\tau_{xy}$, $\tau_{x'y'}$ replacing $I_{xx}$, $I_{yy}$, $-I_{xy}$, $-I_{x'y'}$ respectively.

In solid mechanics, you will also learn that there are nine terms $\epsilon_{ij}$ that describe deformation at a point. Thus consider the undeformed an infinitesimal

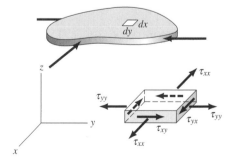

**Figure 9.20.** The case of plane stress.

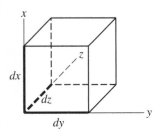

**Figure 9.21.** An infinitesimal rectangular parallelepiped with three edges highlighted.

rectangular parallelepiped in Fig. 9.21. When there is a deformation there are *normal strains* $\epsilon_{xx}$, $\epsilon_{yy}$, $\epsilon_{zz}$ along the direction of the darkened edges which give the changes of length per unit original length of these edges. Furthermore, when there is a deformation, there are six *shear strains* $\epsilon_{xy} = \epsilon_{yx}$, $\epsilon_{xz} = \epsilon_{zx}$, $\epsilon_{yz} = \epsilon_{zy}$ that give the change in angle in radians from that of the right angles of the three darkened edges. Knowing these quantities, we can find any other strains in the rectangular parallelepiped. These other strains can be found by using transformation Eqs. 9.21 and 9.22 and so *strain* is also a *second-order tensor*.

The two-dimensional simplification of $\epsilon_{ij}$ involving the quantities, $\epsilon_{xx}$, $\epsilon_{yy}$, and $\epsilon_{xy}$ $(= \epsilon_{yx})$ as the only nonzero strains is called *plane strain* and represents the strains in a prismatic body constrained at the ends with loading normal to the centerline in which the loading does not vary with $z$ (see Fig. 9.22). Also, the prismatic body must not be subject to bending. Plane strain is an analogous mathematically to plane stress and second moments and products of area. All three are two-dimensional simplifications of second-order symmetric tensors and have the *same transformation equations* as well as other mathematical properties. Finally, in electromagnetic theory and nuclear physics, you will be introduced to the quadruple tensor.[7]

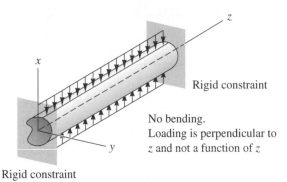

Rigid constraint

No bending.
Loading is perpendicular to $z$ and not a function of $z$

Rigid constraint

**Figure 9.22.** Example of plane strain.

---

[7]Vectors may be defined in terms of the way components of the vector for a new reference are related to the components of the old reference at a point. Thus, for any direction $n$, we have for component $A_n$:

$$A_n = A_x a_{nx} + A_y a_{ny} + A_z a_{nz} \tag{a}$$

Using Eq. (a), we can find components of vector $A$ with respect to $x'y'z'$ rotated arbitrarily relative to $xyz$. Thus, all vectors must transform in accordance with Eq. (a) on rotation of the reference. Obviously, the vector, as seen from this point of view, is a special, simple case of the second-order tensor. We say, accordingly, that vectors are *first-order tensors*.

As for scalars, there is clearly no change in value when there is a rotation of axes at a point. Thus,

$$T(x', y', z') = T(x, y, z) \tag{b}$$

for $x'y'z'$ rotated relative to $xyz$. Scalars are a special form of tensor when considered from a transformation point of view. In fact, they are called *zero-order tensors*.

# PROBLEMS

In the following problems, use the formulas for moments and products of inertia at the mass center to be found in the inside front cover page.

**9.20.** What are the moments and products of inertia for the $xyz$ and $x'y'z'$ axes for the cylinder?

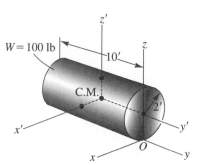

**Figure P.9.20.**

**9.21.** For the uniform block, compute the inertia tensor at the center of mass, at point $a$, and at point $b$ for axes parallel to the $xyz$ reference. Take the mass of the body as $M$ kg.

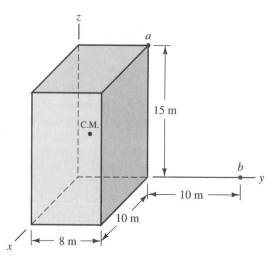

**Figure P.9.21.**

**9.22.** Determine $I_{xx} + I_{yy} + I_{zz}$ as a function of $x$, $y$, and $z$ for all points in space for the uniform rectangular parallelepiped. Note that $xyz$ has its origin at the center of mass and is parallel to the sides.

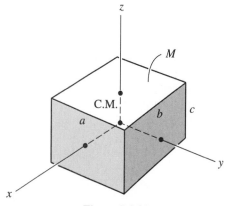

**Figure P.9.22.**

**9.23.** A thin plate weighing 100 N has the following mass moments of inertia at mass center $O$:

$$I_{xx} = 15 \text{ kg-m}^2$$
$$I_{yy} = 13 \text{ kg-m}^2$$
$$I_{xy} = -10 \text{ kg-m}^2$$

What are the moments of inertia $I_{x'x'}$, $I_{y'y'}$, and $I_{z'z'}$ at point $P$ having the position vector:

$$r = .5i + .2j + .6k \text{ m}$$

Also determine $I_{x'z'}$ at $P$.

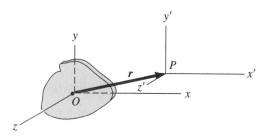

**Figure P.9.23.**

**9.24.** A crate with its contents weighs 20 kN and has its center of mass at

$$r_c = 1.3i + 3j + .8k \text{ m}$$

It is known that at corner $A$,

$$I_{x'x'} = 5,500 \text{ kg-m}^2$$
$$I_{x'y'} = -1,500 \text{ kg-m}^2$$

for primed axes parallel to $xyz$. At point $B$, find $I_{x''x''}$ and $I_{x''y''}$ for double-primed axes parallel to $xyz$.

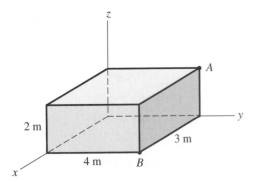

**Figure P.9.24.**

**9.25** A cylindrical crate and its contents weigh 500 N. The center of mass is at

$$r_c = .6i + .7j + 2k \text{ m}$$

It is known that at $A$,

$$(I_{yy})_A = 85 \text{ kg-m}^2$$
$$(I_{yz})_A = -22 \text{ kg-m}^2$$

Find $I_{yy}$ and $I_{zy}$ at $B$.

**Figure P.9.25.**

**9.26.** A block having a uniform density of 5 grams/cm$^3$ has a hole of diameter 40 mm cut out. What are the principal moments of inertia at point $A$ at the centroid of the right face of the block?

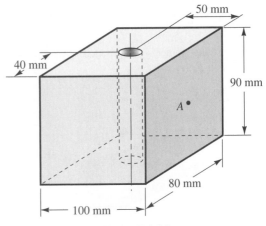

**Figure P.9.26.**

**9.27.** Find maximum and minimum moments of inertia at point $A$. The block weighs 20 N and the cone weighs 14 N.

**Figure P.9.27.**

**9.28.** Solid spheres $C$ and $D$ each weighing 25 N and having radius of 50 mm are attached to a thin solid rod weighing 30 N. Also, solid spheres $E$ and $G$ each weighing 20 N and having radii of 30 mm are attached to a thin rod weighing 20 N. The rods are attached to be orthogonal to each other. What are the principal moments of inertia at point $A$?

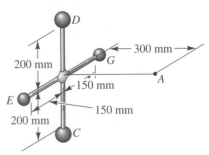

**Figure P.9.28.**

**9.29.** A cylinder is shown having a conical cavity oriented along the axis *A–A* and a cylindrical cavity oriented normal to *A–A*. If the density of the material is 7,200 kg/m³, what is $I_{AA}$?

**Figure P.9.29.**

**9.30.** A flywheel is made of steel having a specific weight of 490 lb/ft.³ What is the moment of inertia about its geometric axis? What is the radius of gyration?

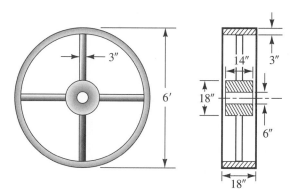

**Figure P.9.30.**

**9.31.** Compute $I_{yy}$ and $I_{xy}$ for the right circular cylinder, which has a mass of 50 kg, and the square rod, which has a mass of 10 kg, when the two are joined together so that the rod is radial to the cylinder. The *x* axis lies along the bottom of the square rod.

**Figure P.9.31.**

**9.32.** Compute the moments and products of inertia for the *xy* axes. The specific weight is 490 lb/ft³ throughout.

**Figure P.9.32.**

**9.33.** A disc *A* is mounted on a shaft such that its normal is oriented 10° from the centerline of the shaft. The disc has a diameter of 2 ft, is 1 in. in thickness, and weighs 100 lb. Compute the moment of inertia of the disc about the centerline of the shaft.

**Figure P.9.33.**

**9.34.** A gear *B* having a mass of 25 kg rotates about axis *C–C*. If the rod *A* has a mass distribution of 7.5 kg/m, compute the moment of inertia of *A* and *B* about the axis *C–C*.

**Figure P.9.34.**

405

**9.35.** A block weighing 100 N is shown. Compute the moment of inertia about the diagonal $D$–$D$.

**Figure P.9.35.**

**9.36.** A solid sphere $A$ of diameter 1 ft and weight 100 lb is connected to the shaft $B$–$B$ by a solid rod weighing 2 lb/ft and having a diameter of 1 in. Compute $I_{z'z'}$ for the rod and ball.

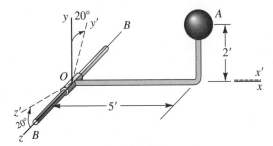

**Figure P.9.36.**

**9.37.** In Problem 9.13, we found the following results for the thin plate:

$$I_{xx} = I_{yy} = .1019 \text{ grams-m}^2$$
$$I_{xy} = .0649 \text{ grams-m}^2$$

Find all components for the inertia tensor for reference $x'y'z'$. Axes $x'y'$ lie in the midplane of the plate.

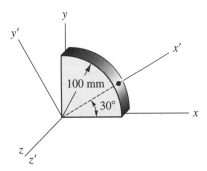

**Figure P.9.37.**

**9.38.** A bent rod weighs .1 N/mm. What is $I_{nn}$ for

$$\epsilon_n = .30i + .45j + .841k?$$

**Figure P.9.38.**

**9.39.** Evaluate the matrix of direction cosines for the primed axes relative to the unprimed axes.

$$a_{ij} = \begin{pmatrix} a_{x'x} & a_{x'y} & a_{x'z} \\ a_{y'x} & a_{y'y} & a_{y'z} \\ a_{z'x} & a_{z'y} & a_{z'z} \end{pmatrix}$$

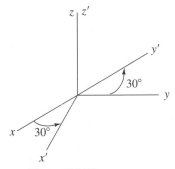

**Figure P.9.39.**

**9.40.** The block is uniform in density and weighs 10 N. Find $I_{y'z'}$.

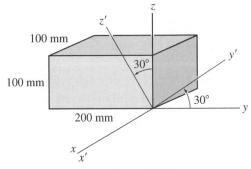

**Figure P.9.40.**

**\*9.41.** A thin rod of length 300 mm and weight 12 N is oriented relative to $x'y'z'$ such that

$$\epsilon_n = .4i' + .3j' + .866k'$$

What is $I_{x'y'}$?

**Figure P.9.41.**

**\*9.42.** Show that the transformation equation for the inertia tensor components at a point when there is a rotation of axes (i.e., Eqs. 9.14 and 9.20) can be given as follows:

$$I_{kq} = \sum_j \sum_i a_{ki} a_{qj} I_{ij}$$

where $k$ can be $x'$, $y'$, or $z'$ and $q$ can be $x'$, $y'$, or $z'$, and where $i$ and $j$ go from $x$ to $y$ to $z$. The equation above is a compact definition of *second-order tensors*. Remember that in the inertia tensor you must have a minus sign in front of each product of inertia term (i.e., $-I_{xy}$, $-I_{yz}$, etc.). [*Hint:* Let $i = x$; then sum over $j$; then let $i = y$ and sum again over $j$; etc.]

**\*9.43.** In Problem 9.42, express the transformation equation to get $I_{y'z'}$ in terms of the inertia tensor components for reference $xyz$ having the same origin as $x'y'z'$.

# \*9.7 The Inertia Ellipsoid and Principal Moments of Inertia

Equation 9.14 gives the moment of inertia of a body about an axis $k$ in terms of the direction cosines of that axis measured from an orthogonal reference with an origin $O$ on the axis, and in terms of six independent inertia quantities for this reference. We wish to explore the nature of the variation of $I_{kk}$ at a point $O$ in space as the direction of $k$ is changed. (The $k$ axis and the body are shown in Fig. 9.23, which we shall call the physical diagram.) To do this, we will employ a geometric representation of moment of inertia at a point that is developed in the following manner. Along the axis $k$, we lay off as a distance the quantity $OA$ given by the relation

$$OA = \frac{d}{\sqrt{I_{kk}/M}} \qquad (9.24)$$

where $d$ is an arbitrary constant that has a dimension of length that will render $OA$ dimensionless, as the reader can verify. The term $\sqrt{I_{kk}/M}$ is the *radius of gyration* and was presented earlier. To avoid confusion, this operation is shown in another diagram, called the inertia diagram (Fig. 9.24), where the new $\xi$, $\eta$, and $\zeta$ axes are *parallel* to the $x$, $y$, and $z$ axes of the physical diagram. Considering all possible directions of $k$, we observe that some surface will be formed about the point $O'$, and this surface is related to the shape of the body through Eq. 9.14. We can express the equation of this surface quite

**Figure 9.23.** Physical diagram.

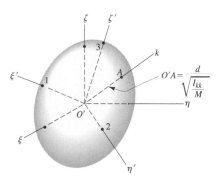

**Figure 9.24.** Inertia diagram.

readily. Suppose that we call $\xi$, $\eta$, and $\zeta$ the coordinates of point $A$. Since $O'A$ is parallel to the line $k$ and thus has the direction cosines $a_{kx}$, $a_{ky}$, and $a_{kz}$ that are associated with this line, we can say that

$$
\begin{aligned}
a_{kx} &= \frac{\xi}{O'A} = \frac{\xi}{d\sqrt{M/I_{kk}}} \\[2mm]
a_{ky} &= \frac{\eta}{O'A} = \frac{\eta}{d\sqrt{M/I_{kk}}} \\[2mm]
a_{kz} &= \frac{\zeta}{O'A} = \frac{\zeta}{d\sqrt{M/I_{kk}}}
\end{aligned}
\tag{9.25}
$$

Now replace the direction cosines in Eq. 9.13, using the relations above:

$$
\begin{aligned}
I_{kk} = {}& \frac{\xi^2}{Md^2/I_{kk}} I_{xx} + \frac{\eta^2}{Md^2/I_{kk}} I_{yy} + \frac{\zeta^2}{Md^2/I_{kk}} I_{zz} \\[2mm]
& + 2\frac{\xi\eta}{Md^2/I_{kk}}(-I_{xy}) + 2\frac{\xi\zeta}{Md^2/I_{kk}}(-I_{xz}) + 2\frac{\eta\zeta}{Md^2/I_{kk}}(-I_{yz})
\end{aligned}
\tag{9.26}
$$

We can see that $I_{kk}$ cancels out of the preceding equation, leaving an equation involving the coordinates $\xi$, $\eta$, and $\zeta$ of the surface and the inertia terms of the body itself. Rearranging the terms, we then have

$$
\begin{aligned}
& \frac{\xi^2}{Md^2/I_{xx}} + \frac{\eta^2}{Md^2/I_{yy}} + \frac{\zeta^2}{Md^2/I_{zz}} \\[2mm]
& + \frac{2\xi\eta}{Md^2}(-I_{xy}) + \frac{2\xi\zeta}{Md^2}(-I_{xz}) + \frac{2\eta\zeta}{Md^2}(-I_{yz}) = 1
\end{aligned}
\tag{9.27}
$$

Considering analytic geometry, we know that the surface is that of an ellipsoid (see Fig. 9.24), and is thus called the *ellipsoid of inertia*. The distance squared from $O'$ to any point $A$ on the ellipsoid is inversely proportional to the moment of inertia (see Eq. 9.24) about an axis in the body at $O$ having the same direction as $O'A$. We can conclude that the inertia tensor for any point of a body can be represented geometrically by such a second-order surface, and this surface may be thought of as analogous to the arrow used to represent a vector graphically. The size, shape, and inclination of the ellipsoid will vary for each point in space for a given body. (Since all second-order tensors may be represented by second-order surfaces, you will, if you study elasticity, also encounter the ellipsoids of stress and strain.)[8]

An ellipsoid has three orthogonal axes of symmetry, which have a common point at the center, $O'$ (see Fig. 9.24). In the figure, these axes are shown as $O'1$, $O'2$, and $O'3$. We pointed out that the shape and inclination of the ellipsoid of inertia depend on the mass distribution of the body about the *origin* of the $xyz$ reference, and they have nothing to do with the choice of *the orientation of the xyz* (and hence the $\xi\eta\zeta$) reference at the point. We can

---

[8]See I. H. Shames, *Mechanics of Deformable Solids*, Prentice-Hall, Inc., Englewood Cliffs, N.J., 1964, Chap. 2. Also, Krieger Publishing Co., N.Y., 1979.

therefore imagine that the $xyz$ reference (and hence the $\xi\eta\zeta$ reference) can be chosen to have directions that coincide with the aforementioned symmetric axes, $O'1$, $O'2$, and $O'3$. If we call such references $x'y'z'$ and $\xi'\eta'\zeta'$, respectively, we know from analytic geometry that Eq. 9.27 becomes

$$\frac{(\xi')^2}{Md^2/I_{x'x'}} + \frac{(\eta')^2}{Md^2/I_{y'y'}} + \frac{(\zeta')^2}{Md^2/I_{z'z'}} = 1 \qquad (9.28)$$

where $\xi',\eta'$, and $\zeta'$ are coordinates of the ellipsoidal surface relative to the new reference, and $I_{x'x'}$, $I_{y'y'}$, and $I_{z'z'}$ are mass moments of inertia of the body about the new axes. We can now draw several important conclusions from this geometrical construction and the accompanying equations. One of the symmetrical axes of the ellipsoid above is the longest distance from the origin to the surface of the ellipsoid, and another axis is the smallest distance from the origin to the ellipsoidal surface. Examining the definition in Eq. 9.24, we must conclude that the minimum moment of inertia for the point $O$ must correspond to the axis having the maximum length, and the maximum moment of inertia must correspond to the axis having the minimum length. The third axis has an intermediate value that makes the sum of the moment of inertia terms equal to the sum of the moment of inertia terms for all orthogonal axes at point $O$, in accordance with Eq. 9.2. In addition, Eq. 9.28 leads us to conclude that $I_{x'y'} = I_{y'z'} = I_{x'z'} = 0$. That is, the products of inertia of the mass about these axes must be zero. Clearly, these axes are the *principal axes* of inertia at the point $O$.

Since the preceding operations could be carried out at any point in space for the body, we can conclude that:

*At each point there is a set of principal axes having the extreme values of moments of inertia for that point and having zero products of inertia.[9] The orientation of these axes will vary continuously from point to point throughout space for the given body.*

All symmetric second-order tensor quantities have the properties discussed above for the inertia tensor. By transforming from the original reference to the principal reference, we change the inertia tensor representation from

$$\begin{pmatrix} I_{xx} & (-I_{xy}) & (-I_{xz}) \\ (-I_{yx}) & I_{yy} & (-I_{yz}) \\ (-I_{zx}) & (-I_{zy}) & I_{zz} \end{pmatrix} \text{ to } \begin{pmatrix} I_{x'x'} & 0 & 0 \\ 0 & I_{y'y'} & 0 \\ 0 & 0 & I_{z'z'} \end{pmatrix} \qquad (9.29)$$

In mathematical parlance, we have "diagonalized" the tensor by the preceding operations.

---

[9]A general procedure for computing principal moments of inertia is set forth in Appendix II.

## 9.8    Closure

In this chapter, we first introduced the nine components comprising the inertia tensor. Next, we considered the case of the very thin flat plate in which the *xy* axes form the midplane of the plate. We found that the mass moments and products of inertia terms $(I_{xx})_M$, $(I_{yy})_M$, and $(I_{xy})_M$ for the plate are proportional, respectively, to $(I_{xx})_A$, $(I_{yy})_A$, and $(I_{xy})_A$, the second moments and product of area of the plate surface. As a result, we could set forth the concept of principal axes for the inertia tensor as an extension of the work in Chapter 8. Thus, we pointed out that for these axes the products of inertia will be zero. Furthermore, one principal axis corresponds to the maximum moment of inertia at the point while another of the principal axes corresponds to the minimum moment of inertia at the point. We pointed out that for bodies with two orthogonal planes of symmetry, the principal axes at any point on the line of intersection of the planes of symmetry must be along this line of intersection and normal to this line in the planes of symmetry.

Those readers who studied the starred sections from Section 9.5 onward will have found proofs of the extensions set forth earlier about principal axes from Chapter 8. Even more important is the disclosure that the inertia tensor components change their values when the axes are rotated at a point in exactly the same way as many other physical quantities having nine components. Such quantities are called second-order tensors. Because of the common transformation equation for such quantities, they have many important identical properties, such as principal axes. In your course in strength of materials you should learn that stress and strain are second-order tensors and hence have principal axes.[10] Additionally, you will find that a two-dimensional stress distribution called *plane stress* is related to the stress tensor exactly as the moments and products of area are related to the inertia tensor. The same situation exists with strain. Consequently, there are similar mathematical formulations for plane stress and the corresponding case for strain (plane strain). Thus, by taking the extra time to consider the mathematical considerations of Sections 9.5 through 9.7, you will find unity between Chapter 9 and some very important aspects of strength of materials to be studied later in your program.

In Chapter 10, we shall introduce another approach to studying equilibrium beyond what we have used thus far. This approach is valuable for certain important classes of statics problems and at the same time forms the groundwork for a number of advanced techniques that many students will study later in their programs.

[10]See I.H. Shames, *Introduction to Solid Mechanics*, 2nd ed., Prentice-Hall, Inc., Englewood Cliffs, New Jersey, 1989.

**9.44.** Find $I_{zz}$ for the body of revolution having uniform density of .2 kg/mm³. The radial distance out from the z axis to the surface is given as

$$r^2 = -4z \text{ mm}^2$$

where z is in millimeters. [*Hint:* Make use of the formula for the moment of inertia about the axis of a disc, $\frac{1}{2}Mr^2$.]

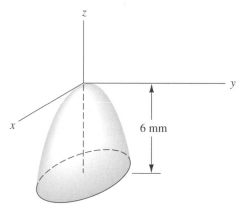

**Figure P.9.44.**

**9.45.** In Problem 9.44, determine $I_{zz}$ without using the disc formula but using multiple integration instead.

**9.46.** What are the inertia tensor components for the thin plate about axes *xyz*? The plate weighs 2 N.

**Figure P.9.46.**

**9.47.** In Problem 9.46, what are the principal axes and the principal moments of inertia for the inertia tensor at *O*?

**9.48.** What are the principal mass moments of inertia at point *O*? Block *A* weighs 15 N. Rod *B* weighs 6 N and solid sphere *C* weighs 10 N. The density in each body is uniform. The diameter of the sphere is 50 mm.

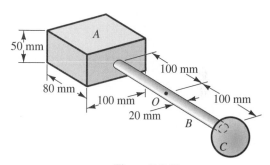

**Figure P.9.48.**

**9.49.** The block has a density of 15 kg/m.³ Find the moment of inertia about axis *AB*.

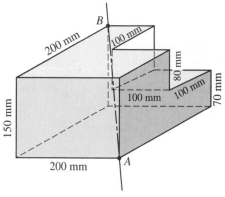

**Figure P.9.49.**

411

**9.50.** A crate and its contents weighs 10 kN. The center of mass of the crate and its contents is at

$$r_c = .40i + .30j + .60k \text{ m}$$

If at $A$ we know that

$$I_{yy} = 800 \text{ kg-m}^2$$
$$I_{yz} = 500 \text{ kg-m}^2$$

find $I_{yy}$ and $I_{yz}$ at $B$.

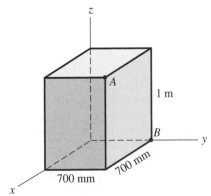

**Figure P.9.50.**

**9.51.** A semicylinder weighs 50 N. What are the principal moments of inertia at $O$? What is the product of inertia $I_{y'z'}$? What conclusion can you draw about the direction of principal axes at $O$?

**Figure P.9.51.**

**9.52.** Find $I_{yy}$ and $I_{yz}$. The diameter of $A$ is 0.3 m. $B$ is the center of the right face of the block. Take $\rho = \rho_0$ kg/m.$^3$

**Figure P.9.52.**

**9.53.** A body is composed of two adjoining blocks. Both blocks have a uniform mass density $\rho$ equal to 10 kg/m.$^3$

    (a) Find the mass moments of inertia, $I_{xx}$ and $I_{zz}$
    (b) Find the product of inertia $I_{xy}$.
    (c) Is the product of inertia $I_{yz} = 0$ (yes or no)? Why?

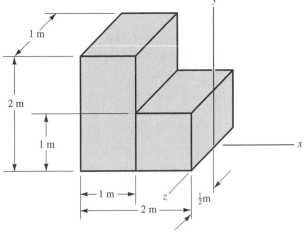

**Figure P.9.53.**

# *Methods of Virtual Work and Stationary Potential Energy

## 10.1 Introduction

In the study of statics thus far, we have followed the procedure of isolating a body to expose certain unknown forces and then formulating either scalar or vector equations of equilibrium that include *all* the forces acting on the body. At this time, alternative methods of expressing conditions of equilibrium, called the *method of virtual work* and, derived from it, the *method of stationary potential energy*, will be presented. These methods will yield equilibrium equations equivalent to those of preceding sections. Furthermore, these new equations include only certain forces on a body, and accordingly in some problems will provide a more simple means of solving for desired unknowns.

Actually, we are making a very modest beginning into a vast field of endeavor called *variational mechanics* or *energy methods* with important applications to both rigid-body and deformable-body solid mechanics. Indeed, more advanced studies in these fields will surely center around these methods.[1]

A central concept for energy methods is the work of a force. A differential amount of work $d\mathcal{W}_k$ due to a force $F$ acting on a particle equals the component of this force in the direction of movement of the particle times the differential displacement of the particle:

$$d\mathcal{W}_k = F \cdot dr \qquad (10.1)$$

And the work $\mathcal{W}_k$ on a particle by force $F$ when the particle moves along some path (see Fig. 10.1) from point 1 to point 2 is then

$$\mathcal{W}_k = \int_{r_1}^{r_2} F \cdot dr \qquad (10.2)$$

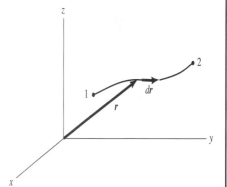

**Figure 10.1.** Path of particle on which $F$ does work.

[1]For a treatment of energy methods for deformable solids, see I. H. Shames and C. Dym, *Energy and Finite Element Methods in Structural Mechanics,* Taylor and Francis Publishers, 1985.

Note that the value and direction of $F$ can vary along the path. This fact must be taken into account during the integration. We shall have more to say about the concept of work in later sections.[2]

# Part A:   Method of Virtual Work

## 10.2   Principle of Virtual Work for a Particle

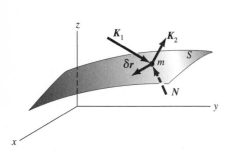

Figure 10.2. Particle on a frictionless surface.

For our introduction to the principle of virtual work, we will first consider a particle acted on by external loads $K_1, K_2,..., K_n$, whose resultant force pushes the particle against a rigid constraining surface $S$ in space (Fig. 10.2). This surface $S$ is assumed to be frictionless and will thus exert a constraining force $N$ on the particle which is normal to $S$. The forces $K_i$ are called *active forces* in connection with the method of virtual work, while $N$ retains the identification of a *constraining force* as used previously. Employing the resultant active force $K_R$, we can give the necessary and sufficient[3] conditions for equilibrium for the particle as

$$K_R + N = 0 \qquad (10.3)$$

We shall now prove that we can express the necessary and sufficient conditions of equilibrium in yet another way. Let us imagine that we give the particle an infinitesimal hypothetical arbitrary displacement that is consistent with the constraints (i.e., along the surface), while keeping the forces $K_R$ and $N$ constant. Such a displacement is termed a *virtual displacement*, and will be denoted by $\delta r$, in contrast to a real infinitesimal displacement, $dr$, which might actually occur during a time interval $dt$. We can then take the dot product of the vector $\delta r$ with the force vectors in the equation above:

$$K_R \cdot \delta r + N \cdot \delta r = 0 \qquad (10.4)$$

Since $N$ is normal to the surface and $\delta r$ is tangential to the surface, the corresponding scalar product must be zero, leaving

$$K_R \cdot \delta r = 0 \qquad (10.5)$$

The expression $K_R \cdot \delta r$ is called the *virtual work* of the system of forces and is denoted as $\delta \mathcal{W}_{\text{Virt}}$. Thus, the virtual work by the active forces on a particle

---

[2]We could have defined work as

$$\mathcal{W}_k = \int_{t_1}^{t_2} F \cdot V \, dt$$

where $V$ is the velocity of the point of application of the force. When the force acts on a particular particle, the result above becomes $\int_{r_1}^{r_2} F \cdot dr$, (with $V dt$ replaced by $dr$) where $r$ is the position vector of the particle. There are times when the force acts on continually *changing* particles as time passes. The more general formulation above can then be used effectively.

[3]The sufficiency condition applies to an initially stationary particle.

with frictionless constraints is *necessarily* zero for a particle in equilibrium for any virtual displacement consistent with the frictionless constraints.

We shall now show that this statement is also *sufficient* to ensure equilibrium for the case of a particle initially at rest (relative to an inertial reference) at the time of application of the active loads. To demonstrate this, *assume that Eq. 10.5 holds but that the particle is not in equilibrium.* If the particle is not in equilibrium, it must move in a direction that corresponds to the direction of the resultant of all forces acting on the particle. Consider that *dr* represents the initial displacement during the time interval *dt*. The work done by the forces must exceed zero for this movement. Since the normal force *N* cannot do work for this displacement,

$$K_R \cdot dr > 0 \qquad (10.6)$$

However, we can choose a *virtual* displacement $\delta r$ to be used in Eq. 10.5 that is *exactly* equal to the proposed *dr* stated above, and so we see that, by admitting nonequilibrium, we arrive at a result (10.6) that is in *contradiction to the starting known condition* (Eq. 10.5). We can then conclude that the conjecture that the particle is not in equilibrium is false. Thus, Eq. 10.3 is not only a necessary condition of equilibrium, but, for an initially stationary particle, is in itself sufficient for equilibrium. Thus, Eq. 10.5 is completely equivalent to the equation of equilibrium, 10.3.

We can now state the principle of virtual work for a particle.

*The necessary and sufficient condition for equilibrium of an initially stationary particle with frictionless constraints requires that the virtual work for all virtual displacements consistent with the constraints be zero.*[4]

The case of a particle that is not constrained is a special case of the situation discussed above. Here $N = 0$, so that Eq. 10.5 is applicable for *all* infinitesimal displacements as a criterion for equilibrium.

## 10.3   Principle of Virtual Work for Rigid Bodies

We now examine a rigid body in equilibrium acted on by active forces $K_i$ and constrained without the aid of friction (Fig. 10.3). The constraining forces $N_i$ arise from direct contact with other immovable bodies (in which case the con-

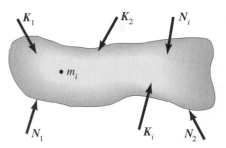

**Figure 10.3.** Rigid body with active forces and ideal constraining forces.

[4]This test breaks down for a particle that is moving. Consider a particle constrained to move in a circular path in a horizontal plane, as shown in the diagram. The particle is moving with constant speed. There are no active forces, and we consider the constraints as frictionless. The work for a virtual displacement consistent with the constraints at any time *t* gives us a zero result. Nevertheless, the particle is not in equilibrium, since clearly there is at time *t* an acceleration toward the center of curvature. Thus, we had to restrict the sufficiency condition to particles that are initially stationary.

straining forces are oriented normal to the contact surface) or from contact with immovable bodies through pin and ball-joint connections. We shall consider the body to be made up of elementary particles for the purposes of discussion.

Now consider a particle of mass $m_i$. Active loads, external constraining forces, and forces from other particles may possibly be acting on the particle. The forces from other particles are internal forces $S_i$ which maintain the rigidity of the body. Using the resultants of these various forces on the particle, we may state from Newton's law that the necessary and sufficient[5] condition for equilibrium of the $i$th particle is

$$(K_R)_i + (N_R)_i + (S_R)_i = \mathbf{0} \tag{10.7}$$

Now, we give the particle a virtual displacement $\delta r_i$ that is consistent with the exterior constraints and with the condition that the body is rigid. Taking the dot product of the vectors in the equation above with $\delta r_i$, we get

$$(K_R)_i \cdot \delta r_i + (N_R)_i \cdot \delta r_i + (S_R)_i \cdot \delta r_i = 0 \tag{10.8}$$

Clearly, $(N_R)_i \cdot \delta r_i$ must be zero, because $\delta r_i$ is normal for $N_i$ for constraint stemming from direct contact with immovable bodies or because $\delta r_i = \mathbf{0}$ for constraint stemming from pin and ball-joint connections with immovable bodies. Let us then sum the resulting equations of the form 10.8 for all the particles that are considered to make up the body. We have, for $n$ particles,

$$\sum_{i=1}^{n} (K_R)_i \cdot \delta r_i + \sum_{i=1}^{n} (S_R)_i \cdot \delta r_i = 0 \tag{10.9}$$

Let us now consider in more detail the internal forces in order to show that the second quantity on the left-hand side of the equation above is zero. The force on $m_i$ from particle $m_j$ will be equal and opposite to the force on particle $m_j$ from particle $m_i$, according to Newton's third law. The internal forces on these particles are shown as $S_{ij}$ and $S_{ji}$ in Fig. 10.4. The first subscript identifies the particle on which a force acts, while the second subscript identifies the particle exerting this force. We can then say that

$$S_{ij} = -S_{ji} \tag{10.10}$$

Any virtual motion we give to any pair of particles must maintain a constant distance between the particles. This requirement stems from the rigid-body condition and will be true if:

1. Both particles are given the same displacement $\delta R$.
2. The particles are rotated $\delta\phi$ relative to each other.[6]

We now consider the general case where both motions are present: that is, both $m_i$ and $m_j$ are given a virtual displacement $\delta R$, and furthermore, $m_j$ is

---

[5]The sufficiency requirement again applies to an initially stationary particle.

[6]The virtual displacements $\delta r_i$ of each of the two particles must then be the result of the superposition of $\delta R$ and $\delta\phi$.

rotated through some angle $\delta\phi$ about $m_i$ (Fig. 10.4). The work done during the rotation must be zero, since $\boldsymbol{S}_{ji}$ is at right angles to the motion of the mass $m_j$. Also, the work done on each particle during the equal displacement of both masses must be equal in value and opposite in sign since the forces move through equal displacements and are themselves equal and opposite. The mutual effect of all particles of the body is of the type described. Thus, we can conclude that the internal work done for a rigid body during a virtual displacement is zero. Hence, a *necessary* condition for equilibrium is

**Figure 10.4.** Two particles of a body undergoing displacement $\delta\boldsymbol{R}$ and rotation $\delta\phi$.

$$\sum_{i=1}^{n} (\boldsymbol{K}_R)_i \cdot \delta\boldsymbol{r}_i = \delta\mathcal{W}_{\text{Virt}} = 0 \qquad (10.11)$$

*Thus, the virtual work done by active forces on a rigid body having friction-less constraints during virtual displacements consistent with the constraints is zero if the body is in equilibrium.*

We can readily prove that Eq. 10.11 is a *sufficient* condition for equilibrium of an initially stationary body by reasoning in the same manner that we did in the case of the single particle. We shall *state first that Eq. 10.11 is valid for a body*. If the body is not in equilibrium, it must begin to move. Let us say that each particle $m_i$ moves a distance $d\boldsymbol{r}_i$ consistent with the constraints under the action of the forces. The work done on particle $m_i$ is

$$(\boldsymbol{K}_R)_i \cdot d\boldsymbol{r}_i + (\boldsymbol{N}_R)_i \cdot d\boldsymbol{r}_i + (\boldsymbol{S}_R)_i \cdot d\boldsymbol{r}_i > 0 \qquad (10.12)$$

But $(\boldsymbol{N}_R)_i \cdot d\boldsymbol{r}_i$ is necessarily zero because of the nature of the constraints. When we sum the terms in the equation above for all particles, $\sum_i (\boldsymbol{S}_R)_i \cdot d\boldsymbol{r}_i$ must also be zero because of the condition of rigidity of the body as previously explained. Therefore, we may state that the supposition of no equilibrium leads to the following inequality:

$$\sum_{i=1}^{n} (\boldsymbol{K}_R)_i \cdot d\boldsymbol{r}_i > 0 \qquad (10.13)$$

But we can conceive a virtual displacement $\delta\boldsymbol{r}_i$ *equal* to $d\boldsymbol{r}_i$ for each particle to be used in Eq. 10.11, thus bringing us to a contradiction between this equation and Eq. 10.13. Since we have taken Eq. 10.11 to apply, we conclude that the supposition of nonequilibrium which led to Eq. 10.13 must be invalid, and so the body must be in equilibrium. This logic proves the sufficiency condition for the principle of virtual work in the case of a rigid body with ideal constraints that is initially stationary at the time of application of the active forces.

Consider now *several* movable rigid bodies that are interconnected by smooth pins and ball joints or that are in direct frictionless contact with each other (Fig. 10.5). Some of these bodies are also ideally constrained by immovable rigid bodies in the manner described above. Again, we may examine the system of particles $m_i$ making up the various rigid bodies. The only new kind of force to be considered is a force at the connecting point between bodies. The force on one such particle on body $A$ will be equal and opposite to the force on the corresponding particle in body $B$ at the contact point; and so on. Since such

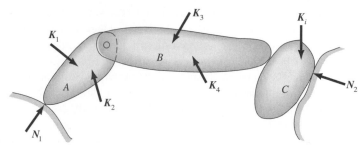

**Figure 10.5.** System of ideally constrained rigid bodies acted on by forces $K_i$.

pairs of contiguous particles have the same virtual displacement, clearly the virtual work at all connecting points between bodies is zero for any virtual displacement of the system consistent with the constraints. Hence, using the same reasoning as before, we can say *for a system of initially stationary rigid bodies, the necessary and sufficient condition of equilibrium is that the virtual work of the active forces be zero for all possible virtual displacements consistent with the constraints.* We may then use the following equation instead of equilibrium:

$$\sum_i^n (K_R)_i \cdot \delta r_i = \delta \mathcal{W}_{\text{Virt}} = 0 \qquad (10.14)$$

where $(K_R)_i$ are the active forces on the system of rigid bodies and $\delta r_i$ are the movements of the point of application of these forces during a virtual displacement of the system consistent with the constraints.

# 10.4   Degrees of Freedom and the Solution of Problems

We have developed equations sufficient for equilibrium of initially stationary systems of bodies by using the concept of virtual work for virtual displacements consistent with the constraints. These equations do not involve reactions or connecting forces, and when these forces are not of interest, the method is quite useful. Thus, we may solve for as many unknown *active* forces as there are *independent* equations stemming from virtual displacements. Then our prime interest is to know how many independent equations can be written for a system stemming from virtual displacements.

For this purpose, we define *the number of degrees of freedom of a system as the number of generalized coordinates[7] which is required to fully specify the configuration of the system.* Thus, for the pendulum in Fig. 10.6, which is restricted to move in a plane, one *independent* coordinate $\theta$ locates the pendulum. Hence, this system has but one degree of freedom. We may ask: Can't we specify $x$ and $y$ of the bob, and thus aren't there two degrees of

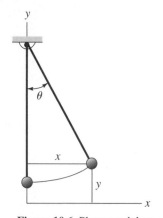

**Figure 10.6.** Plane pendulum.

---

[7]*Generalized coordinates* are any set of *independent variables* that can fully specify the configuration of a system. Generalized coordinates can include any of the usual coordinates, such as cartesian coordinates or cylindrical coordinates, but need not. We shall only consider those cases where the usual coordinates serve as the generalized coordinates.

freedom? The answer is no, because when we specify $x$ or $y$, the other coordinate is *determined* since the pendulum support, being inextensible, must sweep out a known circle as shown in the diagram. In Fig. 10.7, the piston and crank arrangement, the four-bar linkage,[8] and the balance require only one coordinate and thus have but one degree of freedom. On the other hand, the double pendulum has two degrees of freedom and a particle in space has three degrees of freedom. The number of degrees of freedom may usually be readily determined by inspection.

Since each degree of freedom represents an independent coordinate, we can, for an *n*-degree-of-freedom system, institute *n* unique virtual displacements by varying each coordinate separately. This procedure will then give *n* independent equations of equilibrium from which *n* unknowns related to the active forces can be determined. We shall examine several problems to illustrate the method of virtual work and its advantages.

Before considering the examples, we wish to point out that a torque $M$ undergoing a virtual displacement $\delta\phi$ in radians does an amount of virtual work $\delta\mathcal{W}_k$ equal to

$$\delta\mathcal{W}_k = M \cdot \delta\phi \qquad (10.15)$$

The proof of this is asked for in Problem 10.30.

SINGLE DEGREE-OF-FREEDOM SYSTEMS

Piston and crank arrangement

Four-bar linkage

Balance

MULTI DEGREE-OF-FREEDOM SYSTEMS

$\theta_1$

$\theta_2$

Double pendulum

$\theta_1$

$\theta_2$

**Figure 10.7.** Various systems illustrating degrees of freedom.

[8]The fourth bar is the base.

# Example 10.1

A device for compressing metal scrap (a compactor) is shown in Fig. 10.8. A horizontal force $P$ is exerted on joint $B$. The piston at $C$ then compresses the scrap material. For a given force $P$ and a given angle $\theta$, what is the force $F$ developed on the scrap by the piston $C$? Neglect the friction between the piston and the cylinder wall, and consider the pin joints to be ideal.

We see by inspection that one coordinate $\theta$ describes the configuration of the system. The device therefore has one degree of freedom. We shall neglect the weight of the members, and so only two active forces are present, $P$ and $F$. By assuming a virtual displacement $\delta\theta$, we will involve in the principle of virtual work only those quantities that are of interest to us, $P$, $F$, and $\theta$.[9] Let us then compute the virtual work of the active forces.

**_Force P._** The virtual displacement $\delta\theta$ is such that force $P$ has a motion in the horizontal direction of ($l\,\delta\theta\cos\theta$) as can readily be deduced from Fig. 10.9 by elementary trigonometric considerations. There is yet another way of deducing this horizontal motion, which, sometimes, is more desirable. Using an $xy$ coordinate system at $A$ as shown in Figs. 10.8 and 10.9, we can say for the position of joint $B$:

$$y_B = l\sin\theta \qquad\qquad\text{(a)}$$

**Figure 10.8.** Compacting device.

[9]If we had used a free-body approach, we would have had to bring in force components at $A$ and at $C$, and we would have had to dismember the system. To appreciate the method of virtual work even for this simple problem, we urge you to at least set up the problem by the use of free-body diagrams.

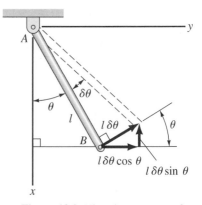

**Figure 10.9.** Virtual movement of leg $AB$.

## ■ Example 10.1 (Continued)

Now take the differential of both sides of the equation to get

$$dy_B = l \cos \theta \, d\theta \tag{b}$$

A differential of a quantity $A$, namely $dA$, is very similar to a variation of the quantity, $\delta A$. The former might actually take place in a process; the latter takes place in the mind of the engineer. Nevertheless, the relation between differential quantities should be the same as the relation between varied quantities. Accordingly, from Eq. (b), we can say:

$$\delta y_B = l \cos \theta \, \delta\theta \tag{c}$$

Note that the same horizontal movement of $B$ for $\delta\theta$ is thus computed as at the outset using trigonometry.

For the variation $\delta\theta$ chosen, the force $P$ acts in the opposite direction of $\delta y_B$, and so the virtual work done by force $P$ is negative. Thus, we have

$$\delta(\mathcal{W}_{\text{Virt}})_P = -Pl \cos \theta \, \delta\theta \tag{d}$$

**Force F.**   We can use the differential approach to get the virtual displacement of piston $C$. That is,

$$x_C = l \cos \theta + l \cos \theta = 2l \cos \theta$$
$$dx_C = -2l \sin \theta \, d\theta$$

Therefore,

$$\delta x_C = -2l \sin \theta \, \delta\theta \tag{e}$$

Since the force $F$ is in the same direction as $\delta x_C$, we should have a positive result for the work done by force $F$. Accordingly, we have

$$\delta(\mathcal{W}_{\text{Virt}})_F = F(2l \sin \theta) \, \delta\theta \tag{f}$$

We may now employ the principle of virtual work, which is sufficient here for ensuring equilibrium. Thus, we can say that

$$-Pl \cos \theta \, \delta\theta + F(2l \sin \theta) \, \delta\theta = 0 \tag{g}$$

Canceling $l \, \delta\theta$ and solving for $F$, we get

$$F = \frac{P}{2 \tan \theta}$$

For any given values of $P$ and $\theta$, we now know the amount of compressive force that the compactor can develop.

# Example 10.2

A hydraulic-lift platform for loading trucks is shown in Fig. 10.10(a). Only one side of the system is shown; the other side is identical. If the diameter of the piston in the hydraulic ram is 4 in., what pressure $p$ is needed to support a load $W$ of 5,000 lb when $\theta = 60°$? The following additional data apply:

$$l = 24 \text{ in.}$$
$$d = 60 \text{ in.}$$
$$e = 10 \text{ in.}$$

Pin $A$ is at the center of the rod.

We have here a system with one degree of freedom characterized by the angle $\theta$. The active forces that do work during a virtual displacement $\delta\theta$ are the weight $W$ and the force from the hydraulic ram. Accordingly, the virtual movements of both the platform $E$ and joint $A$ of the hydraulic ram must be found. Using reference $xy$:

$$y_E = 2l \sin \theta$$

Therefore,

$$\delta y_E = 2l \cos \theta \, \delta\theta \qquad (a)$$

For the ram force, we want the movement of pin $A$ in the direction of the axis of the pump, namely $\delta\eta$ where $\eta$ is shown in Fig. 10.10(a). Observing Fig. 10.10(b) we can say for $\eta$:

$$\eta^2 = \overline{AC}^2 + \overline{CB}^2$$
$$= [l \sin \theta - e]^2 + (d - l \cos \theta)^2 \qquad (b)$$

Hence, we have

$$2\eta \, \delta\eta = 2(l \sin \theta - e)(l \cos \theta)\delta\theta + 2(d - l \cos \theta)(l \sin \theta) \, \delta\theta \qquad (c)$$

Solving for $\delta\eta$, we get

$$\delta\eta = \frac{l}{\eta}[(l \sin \theta - e) \cos \theta + (d - l \cos \theta) \sin \theta] \, \delta\theta$$
$$= \frac{l}{\eta}(l \sin \theta \cos \theta - e \cos \theta + d \sin \theta - l \sin \theta \cos \theta) \, \delta\theta \qquad (d)$$
$$= \frac{l}{\eta}(d \sin \theta - e \cos \theta) \, \delta\theta$$

The principle of virtual work is now applied to ensure equilibrium. Thus, considering one side of the system and using half the load, we have

$$-\frac{W}{2}(\delta y_E) + \left[ p \frac{\pi(4^2)}{4} \right] \delta\eta = 0$$

# Example 10.2 (Continued)

Hence,

$$-(2,500)(2l \cos \theta \, \delta\theta) + p(4\pi)\left[\frac{l}{\eta}(d \sin \theta - e \cos \theta)\right]\delta\theta = 0 \quad \text{(e)}$$

(a)

(b)

**Figure 10.10.** Pneumatic loading platform.

The value of $\eta$ at the configuration of interest may be determined from Eq. (b). Thus,

$$\eta^2 = \left[(24)(.866) - 10\right]^2 + \left[60 - (24)(.5)\right]^2$$

Therefore,

$$\eta = 49.2 \text{ in.}$$

Now canceling $\delta\theta$ and substituting known data into Eq. (e), we may then determine $p$ for equilibrium:

$$-(2,500)(2)(24)(.5) + p(4\pi)\left\{\frac{24}{49.2}\left[(60)(.866) - (10)(.5)\right]\right\} = 0$$

Therefore,

$$p = 208 \text{ psi} \quad \text{(f)}$$

In a few of the homework problems, you have to use simple kinematics of a cylinder rolling without slipping (see Fig. 10.11). You will recall from physics that the cylinder is actually rotating about the point of contact $A$. If the cylinder rotates an angle $\delta\theta$ then $\delta C = -r\,\delta\theta$. We shall consider kinematics of rigid bodies in detail later in the text.

**Figure 10.11.** Cylinder rolling without slipping.

In concluding this section, we wish to point out that the method of virtual work is actually *not* restricted to ideal systems. Furthermore, it is permissible to give virtual displacements that *violate* one or more constraints. We then proceed by considering those friction forces that perform virtual work as active forces. And, where a constraint is violated, we consider the corresponding constraining force or torque to be active. We point out that the method of virtual work generally offers no advantage in situations where there is friction and where constraints are violated. Furthermore, the extensions of virtual work to other useful theories are primarily restricted to ideal systems. Accordingly we shall consider only ideal systems and shall take virtual displacements that do not violate constraints.

## 10.5   Looking Ahead: Deformable Solids

Let us now reconsider the virtual displacement concept. Recall that *dr* represents *a small change in the spatial coordinates of a particle arbitrary up to the point of not violating constraints.* The change in the spatial coordinates is in no way linked to a change in time. (Such a link would normally exist through Newton's laws when we are working in the field of mechanics.) This has been shown in Fig. 10.12, where we have a virtual displacement $\delta r$ from point $P(x, y, z)$ to point $P'$, which has coordinates $(x + \delta x)$, $(y + \delta y)$, and $(z + \delta z)$. We call the position vector to point $P'$, the *varied position* vector $\tilde{r}$. We can then say that

$$\delta r = \tilde{r} - r \qquad (10.16)$$

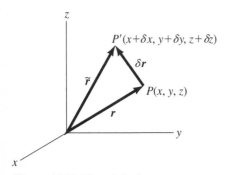

**Figure 10.12.** Virtual displacement vector.

We may at this stage consider $\delta$ above to be an *operator* acting on $r$ so as to generate the difference vector between the varied position vector $\tilde{r}$ and the position vector $r$ itself .

We may also introduce the concept of the *varied function* $\tilde{G}$ such that given a function $G(x, y, z)$ we may form

$$\tilde{G} = G(x + \delta x, y + \delta y, z + \delta z,) \qquad (10.17)$$

where $\delta x$, $\delta y$, and $\delta z$ are components of $\delta r$[10]. We now define the *variation* of $G$, denoted as $\delta G$, as

$$\delta G = \tilde{G} - G \qquad (10.18)$$

In a deformable body we can give the movement of each point in the body when deformation occurs by using the *displacement field* $\boldsymbol{u}(x, y, z)$. Specifically, when coordinates of a specific point in the underformed geometry are substituted into the particular vector function $\boldsymbol{u}(x, y, z)$ depicting a specific deformation, we get the displacement of that point resulting from this deformation.

We now extend the concept of a virtual displacement of a point to that of a virtual displacement *field*, which is a single-valued, continuous vector field representing a *hypothetical deformable body movement consistent with the constraints present.* We shall restrict ourselves here to virtual displacement fields, which result only in infinitesimal deformation.[11] We have shown the gross exaggeration of a virtual displacement field in Fig. 10.13, wherein you will notice that the constraints have not been violated. It should now be clear that we can conveniently set forth a virtual displacement field by employing the so-called variational operator $\delta$. Thus $\delta\boldsymbol{u}$ may be considered as a virtual displacement field from a given configuration to a varied configuration; the constraints present being taken into account by imposing proper conditions on the variation.

The *virtual work* concept can now be extended to the case of a deformable body. We compute the work of the *external* forces during a virtual displacement of the body with the proviso that these external forces be maintained constant. With a total body force distribution $\boldsymbol{B}(x, y, z)$ and a total traction force distribution $\boldsymbol{T}(x, y, z)$, we can then give the virtual work, denoted as $\delta W_{\text{Virt}}$, as follows:

$$\delta W_{\text{Virt}} = \iiint_V \boldsymbol{B} \cdot \delta\boldsymbol{u} \, dv + \oiint_S \boldsymbol{T} \cdot \delta\boldsymbol{u} \, dA \qquad (10.19)$$

**Figure 10.13.** Virtual displacement field consistent with constraints.

For rigid bodies, *the virtual work had to be zero for equilibrium. For deformable bodies this is no longer true.* Instead, for equilibrium the external virtual work $\delta W_{\text{Virt}}$ given above must be equal to *internal* virtual work, which

---

[10]The changes in the coordinates *x, y,* and *z* are not linked to time through the basic laws of physics as would be the case if we were considering *G* to represent some physical quantity in some real process.

[11]One need not so restrict oneself. That is, we can work with virtual displacement fields for *finite* deformation and formulate a principle of virtual work. This would take us beyond the scope of this book, however.

must be zero for rigid bodies but which is not necessarily zero for deformable solids. In your solid mechanics course you will learn that the internal work for a deformable body is given by $\iiint_V \sum_i \sum_j \tau_{ij}\delta\epsilon_{ij}\,dv$, where the indices $i$ and $j$ range over $x$, $y$, and $z$ forming terms in the integrand such as $\tau_{xx}\delta\epsilon_{xx}$, $\tau_{xy}\delta\epsilon_{xy}$, etc., (nine expressions). The satisfaction of the resulting formulation is a necessary and sufficient condition for equilibrium and can be used in place of the familiar equations of equilibrium.[12] Why would one want to do this? Actually, as we pointed in the Looking Ahead section in the chapter on structural mechanics, we can readily solve certain types of problems using virtual work, and the theorems derived from virtual work, whereas the approach for these problems using the equilibrium equations is extremely cumbersome. One important case is the solution of indeterminate truss problems. You will come to these problems later in your studies of structures and in your studies of machine design.

Virtual work and two other theorems derivable from it are called *energy displacement methods* because of the use of the virtual displacement.[13] There is an equally useful set of three formulations analogous to the three energy displacement methods and they are called *energy force methods,* wherein we hypothetically vary the forces instead of hypothetically varying the deformation.

Before moving on, the author would like to share a philosophical thought with you. In science we often physically disturb certain surroundings in the laboratory and carefully observe resulting behavior to learn to understand natural phenomena. We perhaps unwittingly mimic this approach here in the study of mechanics. That is, we have instituted mathematical "disturbances" and evaluated the results in order to understand certain vital analytically useful consequences. Thus, we instituted the mathematical "disturbance" of the virtual displacement field to arrive at extremely useful conclusions which form the basis of a considerable amount of structural mechanics. Also, we pointed out that we can institute varied force fields as our mathematical disturbances. Again, vital and useful conclusions follow.

---

[12]Thus, the virtual work equation is $\iiint_V \boldsymbol{B} \cdot \delta\boldsymbol{u}\,dv + \oiint_S \boldsymbol{T} \cdot \delta\boldsymbol{u}\,dA = \iiint_V \sum_i \sum_j \tau_{ij}\,\delta\epsilon_{ij}\,dv$

[13]The second energy displacement method is called the method of *total potential energy.* It was this principal that was presented in the "Looking Ahead" Section 6.5 for determining the pin deflections of simple trusses. The special case of this principle for conservative force fields acting on *particles* and *rigid bodies* is developed in Part B of this chapter. The third energy displacement method derivable from the second is the *First Castigliano Theorem.*

For a thorough development of these six principles with many applications and which is within reach of students who have absorbed the key contents of the statics portion of this text, see Shames, I.H., *Introduction to Solid Mechanics,* second ed., Prentice-Hall, Inc., Englewood Cliffs, N.J., Chapters 18 and 19. A good grasp of these six principles is vital for more advanced work in solid and structural mechanics, not to speak of machine design.

# PROBLEMS

**10.1.** How many degrees of freedom do the following systems possess? What coordinates can be used to locate the system?

(a) A rigid body not constrained in space.

(b) A rigid body constrained to move along a plane surface.

(c) The board *AB* in the diagram (a).

(d) The spherical bodies shown in diagram (b) may slide along shaft *C–C*, which in turn rotates about axis *E–E*. Shaft *C–C* may also slide along *E–E*. The spindle *E–E* is on a rotating platform. Give the number of degrees of freedom and coordinates for a sphere, shaft *C–C*, and spindle *E–E*.

(a)

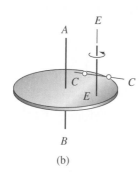

(b)

**Figure P.10.1.**

**10.2.** A parking-lot gate arm weighs 150 N. Because of the taper, the weight can be regarded as concentrated at a point 1.25 m from the pivot point. What is the solenoid force to lift the gate? What is the solenoid force if a 300-N counterweight is placed .25 m to left of the pivot point?

**Figure P.10.2.**

**10.3.** What is the longest portion of pipe weighing 400 lb/ft that can be lifted without tipping the 12,000-lb tractor?

**Figure P.10.3.**

**10.4.** If $W_1 = 100$ N and $W_2 = 150$ N, find the angle $\theta$ for equilibrium.

**Figure P.10.4.**

**10.5.** The triple pulley sheave and the double pulley sheave weigh 150 N and 100 N, respectively. What rope force is necessary to lift a 3,500-N engine?

**Figure P.10.5.**

**10.6.** What weight $W$ can be lifted with the A-frame hoist in the position shown if the cable tension is $T$?

Tension $T$

$W$

3 m

4 m

4 m

**Figure P.10.6.**

**10.7.** A small hoist has a lifting capacity of 20 kN. What is the maximum possible cable tension load?

2 m

2.5 m

.6 m

**Figure P.10.7.**

**10.8.** If $W = 1,000$ N and $P = 300$ N, find the angle $\theta$ for equilibrium.

$P$

$\frac{h}{2}$

$\frac{h}{2}$

$\theta$

$a$

$a$

$W$

**Figure P.10.8.**

**10.9.** What is the tension in the cables of a 10-ft-wide 12-ft-long 6000-lb drawbridge when the bridge is first raised? When the bridge is at 45°?

12'

12'

10'

**Figure P.10.9.**

**10.10.** Assuming frictionless contacts, determine the magnitude of $P$ for equilibrium.

$\frac{l}{2}$

$\frac{l}{2}$

$W$

$\theta$

$P$

**Figure P.10.10.**

**10.11.** A rock crusher is shown in action. If $p_1 = 50$ psig and $p_2 = 100$ psig, what is the force on the rock at the configuration shown? The diameter of the pistons is 4 in.

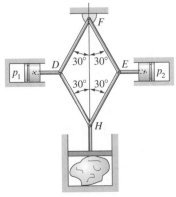

$F$

$p_1$

$D$

30° 30°

30° 30°

$E$

$p_2$

$H$

**Figure P.10.11.**

428

**10.12.** A 20-lb-ft torque is applied to a scissor jack. If friction is disregarded throughout, what weight can be maintained in equilibrium? Take the pitch of the screw threads to be .3 in. in opposite senses. All links are of equal length, 1 ft.

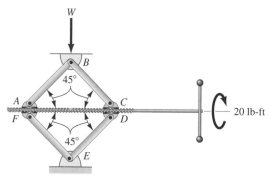

**Figure P.10.12.**

**10.13.** The 5,000-lb van of an airline food catering truck rises straight up until its floor is level with the airplane floor. What is the ram force in that position?

**Figure P.10.13.**

**10.14.** What are the cable tensions when the arms of the power shovel are in the position shown? Arm $AC$ weighs 13 kN, arm $DF$ weights 11 kN, and the shovel plus the payload weigh 9 kN.

**Figure P.10.14.**

**10.15.** A hydraulically actuated gate in a 2-m-square water-carrying tunnel under a dam is held in place with a vertical beam $AC$. What is the force in the hydraulic ram if the specific weight of water is 9818 N/m³?

**Figure P.10.15.**

**10.16.** Find the angle $\beta$ for equilibrium in terms of the parameters given in the diagram. Neglect friction and the weight of the beam.

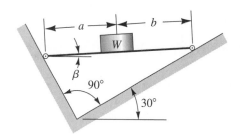

**Figure P.10.16.**

**10.17.** Do Problem 5.54 by the method of virtual work.

**10.18.** Do Problem 5.55 by the method of virtual work.

**10.19.** What is the relation among $P$, $Q$, and $\theta$ for equilibrium?

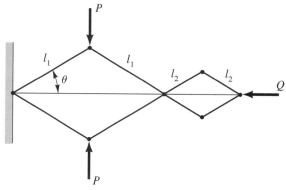

**Figure P.10.19.**

429

**10.20.** A paper collator is shown with the weight $Q$ of the collated papers equal to .2 N. The collator rests on a smooth surface and, accordingly, can slide on this surface with no resistance. What force $P$ is needed to keep the system in equilibrium for the position shown?

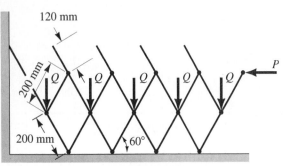

**Figure P.10.20.**

**10.21.** A stepped cylinder of weight 500 lb is connected to vehicle $A$ weighing 300 lb and to sheave $B$ weighing 50 lb. Sheave $B$ supports a weight $C$. What is the value of the weight of $C$ for equilibrium? Neglect friction.

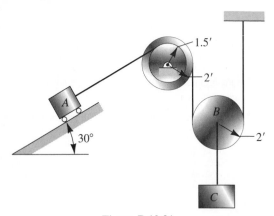

**Figure P.10.21.**

**10.22.** Do the first part of Problem 5.70 by the method of virtual work.

**10.23.** Compute the weight $W$ that can be lifted by the *differential pulley* system for an applied force $F$. Neglect the weight of the lower pulley.

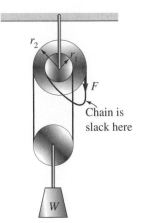

**Figure P.10.23.**

**10.24.** The pressure $p$ driving a piston of diameter 100 mm is 1 N/mm². At the configuration shown, what weight $W$ will the system hold if we neglect friction?

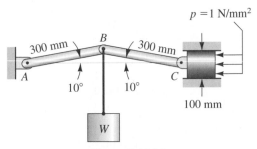

**Figure P.10.24.**

**10.25.** Blocks $A$ and $B$ weigh 200 N and 150 N, respectively. They are connected at their base by a light cord. At what position $\theta$ is there equilibrium if we disregard friction?

**Figure P.10.25.**

**10.26.** If *A* weighs 500 N, and if *B* weighs 100 N, determine the weight of *C* for equilibrium.

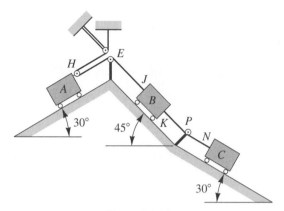

**Figure P.10.26.**

**10.27.** An embossing device imprints an image at *D* on metal stock. If a force *F* of 200 N is exerted by the operator, what is the force at *D* on the stock? The lengths of *AB* and *BC* are each 150 mm.

**Figure P.10.27.**

**10.28.** A support system holds a 500-N load. Without the load, $\theta = 45°$ and the spring is not compressed. If *K* for the spring is 10,000 N/m, how far down *d* will the 500-N load depress the upper platform if the load is applied slowly and carefully? Neglect all other weights. $DB = BE = AB = CB = 400$ mm. (*Note:* The force from the spring is *K* times its contraction.)

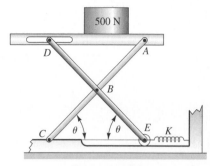

**Figure P.10.28.**

**10.29.** Rod *ABC* is connected through a pin and slot to a sleeve which slides on a vertical rod. Before the weight *W* of 100 N is applied at *C*, the rod is inclined at an angle of 45°. If *K* of the spring is 8,000 N/m, what is the angle $\theta$ for equilibrium? The length of *AB* is 300 mm and the length of *BC* is 200 mm when $\theta = 45°$. Neglect friction and all weights other than *W*. (*Note:* The force from the spring is *K* times its contraction.)

**Figure P.10.29.**

**10.30.** Show that the virtual work of a couple moment *M* for a rotation $\delta\phi$ is given as

$$\delta W = M \cdot \delta\phi$$

[*Hint:* Decompose *M* into components normal to and collinear with $\delta\phi$.]

**Figure P.10.30.**

431

# Part B:    Method of Total Potential Energy

## 10.6    Conservative Systems

We shall restrict ourselves in this section to certain types of active forces. This restriction will permit us to arrive at some additional very useful relations.

Consider first a body acted on only by gravity force $W$ as an active force and moving along a frictionless path from position 1 to position 2, as shown in Fig. 10.14. The work done by gravity, $\mathcal{W}_{1\text{-}2}$, is then

$$\mathcal{W}_{1-2} = \int_1^2 \boldsymbol{F} \cdot d\boldsymbol{r} = \int_1^2 (-W\boldsymbol{j}) \cdot d\boldsymbol{r} = -W\int_1^2 dy$$
$$= -W(y_2 - y_1) = W(y_1 - y_2) \qquad (10.20)$$

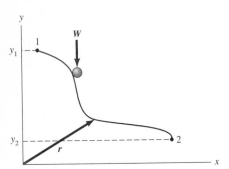

**Figure 10.14.** Particle moving along frictionless path

Note that the work done *does not depend* on the path, but depends only on the positions of the *end points* of the path. *Force fields which are functions of position and whose work like gravity is independent of the path are called conservative force fields.* In general, we can say for a conservative force field $F(x, y, z)$ that, along a path between positions 1 and 2, the work is analogous to that in Eq. 10.20. That is,

$$\mathcal{W}_{1-2} = \int_1^2 \boldsymbol{F} \cdot d\boldsymbol{r} = V_1(x, y, z) - V_2(x, y, z) \qquad (10.21)$$

where $V$, a scalar function evaluated at the end points, is called the *potential energy function*.[14] We may rewrite the equation above as follows:

$$-\int_1^2 \boldsymbol{F} \cdot d\boldsymbol{r} = V_2(x, y, z) - V_1(x, y, z) = \Delta V \qquad (10.22)$$

From the result above we can say that *the change in potential energy $\Delta V$, (equal to the final $V$ minus the initial $V$), associated with a force field is the negative of the work done by this force field in going from position 1 to position 2 along any path.* For any *closed* path, the work done by a conservative force field $F$ is then

$$\oint \boldsymbol{F} \cdot d\boldsymbol{r} = 0 \qquad (10.23)$$

How is the potential energy function $V$ related to $F$? To answer this query, consider that an arbitrary infinitesimal path segment $d\boldsymbol{r}$ starts from point 1. We can then give Eq. 10.22 as

$$\boldsymbol{F} \cdot d\boldsymbol{r} = -dV \qquad (10.24)$$

$$\therefore F_x dx + F_y dy + F_z dz = -\left(\frac{\partial V}{\partial x} dx + \frac{\partial V}{\partial y} dy + \frac{\partial V}{\partial z} dz\right) \qquad (10.25)$$

We can conclude from the equation above that since $d\boldsymbol{r}$ is arbitrary

---

[14]The context of any discussion should make clear whether $V$ refers to potential energy, or to speed, or to volume. $V$ is also denoted here as *P.E.*

$$F_x = -\frac{\partial V}{\partial x}$$

$$F_y = -\frac{\partial V}{\partial y} \qquad (10.26)$$

$$F_z = -\frac{\partial V}{\partial z}$$

Or, in other words,

$$F = -\left(\frac{\partial V}{\partial x}\,i + \frac{\partial V}{\partial y}\,j + \frac{\partial V}{\partial z}\,k\right)$$

$$= -\left(\frac{\partial}{\partial x}\,i + \frac{\partial}{\partial y}\,j + \frac{\partial}{\partial z}\,k\right)V \qquad (10.27)$$

$$= -\mathbf{grad}\ V = -\nabla V$$

The operator we have introduced is called the *gradient* operator and is given as follows for rectangular coordinates:

$$\mathbf{grad}\ \equiv \nabla \equiv \frac{\partial}{\partial x}\,i + \frac{\partial}{\partial y}\,j + \frac{\partial}{\partial z}\,k \qquad (10.28)$$

We can now say, as an alternative definition, that *a conservative force field must be a function of position and expressible as the gradient of a scalar field function.* The inverse to this statement is also valid. *That is, if a force field is a function of position and the gradient of a scalar field, it must then be a conservative force field.* Such are the following two force fields.

**Constant Force Field.**    If the force field is constant at all positions, it can always be expressed as the gradient of a scalar function of the form $V = -(ax + by + cz)$, where $a$, $b$, and $c$ are constants. The constant force field, then, is $F = ai + bj + ck$.

In limited changes of position near the Earth's surface (a common situation), we can consider the gravitational force on a particle of mass, $m$, as a constant force field given by $-mgk$. Thus, the constants for the general force field given above are $a = b = 0$ and $c = -mg$. Clearly, $V \equiv P.E. = mgz$ for this case.

**Force Proportional to Linear Displacements.**    Consider a body limited by constraints to move along a straight line. Along this line a force is developed directly proportional to the displacement of the body from some point on the line. If this line is the $x$ axis, we give this force as

$$F = -Kxi \qquad (10.29)$$

where $x$ is the displacement from the point. The constant $K$ is a positive number, so that, with the minus sign in this equation, a positive displacement $x$ from the origin means that the force is negative and is then directed back to the origin. A displacement in the negative direction from the origin (negative $x$) means that the force is positive and is directed again toward the origin. Thus, the force given above is a *restoring* force about the origin. An example of this force is that resulting from the extension of a linear spring (Fig. 10.15). The force that the spring exerts will be directly proportional to the amount of elongation or compression in the $x$ direction beyond the unextended configuration. This movement is measured from the origin of the $x$ axis. The constant $K$ in this situation is called the *spring constant*. The change in potential energy due to the displacements from the origin to some position $x$, therefore, is

$$P.E. = \frac{Kx^2}{2} \qquad (10.30)$$

since $-\nabla \left( \dfrac{Kx^2}{2} \right) = -Kx\boldsymbol{i}$.

The *change* in potential energy has been defined as the *negative* of the work done by a conservative force as we go from one position to another. Clearly, the potential energy change is then *directly equal* to the work done by the *reaction* to the conservative force during this displacement. In the case of the spring, the reaction force would be the force *from* the surroundings acting *on* the spring at point A (Fig. 10.15). During extension or compression of the spring from the undeformed position, this force (from the surroundings) clearly must do a positive amount of work. This work must, as noted above, equal the potential energy change. We now note that we can consider this work (or in other words the change in potential energy) to be a measure of the energy *stored* in the spring. That is, when allowed to return to its original position, the spring will do this amount of positive work *on* the surroundings at A, provided that the return motion is slow enough to prevent oscillations, etc.

**Figure 10.15.** Linear spring

# 10.7    Condition of Equilibrium for a Conservative System

Let us now consider a system of rigid bodies that is ideally constrained and acted on by conservative active forces. For a virtual displacement from a configuration of equilibrium, the virtual work done by the active forces, which are maintained constant during the virtual displacement, must be zero. We shall now show that the condition of equilibrium can be stated in yet another way for this system.

Specifically, suppose that we have $n$ conservative forces acting on the system of bodies. The increment of work for a real infinitesimal movement of the system can be given as follows:

$$
\begin{aligned}
d\mathcal{W} &= \sum_{p=1}^{n} \boldsymbol{F}_p \cdot d\boldsymbol{r}_p \\
&= \sum_{p=1}^{n} \left[ -\left( \frac{\partial V_p}{\partial x_p}\boldsymbol{i} + \frac{\partial V_p}{\partial y_p}\boldsymbol{j} + \frac{\partial V_p}{\partial z_p}\boldsymbol{k} \right) \right] \cdot \left( dx_p \boldsymbol{i} + dy_p \boldsymbol{j} + dz_p \boldsymbol{k} \right) \\
&= \sum_{p=1}^{n} \left[ -\left( \frac{\partial V_p}{\partial x_p} dx_p + \frac{\partial V_p}{\partial y_p} dy_p + \frac{\partial V_p}{\partial z_p} dz_p \right) \right] \\
&= -\sum_{p=1}^{n} dV_p = -d\left( \sum_{p=1}^{n} V_p \right) = -dV
\end{aligned}
$$

where $V$ without subscripts refers to *total* potential energy. By treating $\delta\boldsymbol{r}_p$ like $d\boldsymbol{r}_p$ in the equations above, we can express the virtual work $\delta\mathcal{W}_{\text{Virt}}$ as

$$\delta \mathcal{W}_{\text{Virt}} = \sum_{p=1}^{n} \boldsymbol{F}_p \cdot \delta \boldsymbol{r}_p$$

$$= \sum_{p=1}^{n} \left[ -\left( \frac{\partial V_p}{\partial x_p} \boldsymbol{i} + \frac{\partial V_p}{\partial y_p} \boldsymbol{j} + \frac{\partial V_p}{\partial z_p} \boldsymbol{k} \right) \right] \cdot \left( \delta x_p \boldsymbol{i} + \delta y_p \boldsymbol{j} + \delta z_p \boldsymbol{k} \right)$$

$$= \sum_{p=1}^{n} \left[ -\left( \frac{\partial V_p}{\partial x_p} \delta x_p + \frac{\partial V_p}{\partial y_p} \delta y_p + \frac{\partial V_p}{\partial z_p} \delta z_p \right) \right]$$

$$= -\sum_{p=1}^{n} \delta V_p = -\delta \left( \sum_{p=1}^{n} V_p \right) = -\delta V$$

But we know that for equilibrium $\delta \mathcal{W}_{\text{Virt}} = 0$, and so we can similarly say for equilibrium:

$$\delta V = 0 \tag{10.31}$$

Mathematically, this means that *the potential energy has a stationary or an extremum value at a configuration of equilibrium*, or, putting it another way, *the variation of V is zero at a configuration of equilibrium*.[15] Thus, we have another criterion which we may use to solve problems of equilibrium for conservative force systems with ideal constraints.

For solving problems, determine the potential energy using a set of independent coordinates. Then, take the variation, $\delta$, of the potential energy. For example, suppose that $V$ is a function of independent variables $q_1, q_2, ..., q_n$, thereby having $n$ degrees of freedom. The variation of $V$ then becomes

$$\delta V = \frac{\partial V}{\partial q_1} \delta q_1 + \frac{\partial V}{\partial q_2} \delta q_2 + \cdots + \frac{\partial V}{\partial q_n} \delta q_n \tag{10.32}$$

For equilibrium, we set this variation equal to zero according to Eq. 10.31. For the right side of the equation above to be zero, the coefficient of each $\delta q_i$ must be zero, since the $\delta q_i$ are independent of each other. Thus,

$$\frac{\partial V}{\partial q_1} = 0$$
$$\frac{\partial V}{\partial q_2} = 0$$
$$\vdots$$
$$\frac{\partial V}{\partial q_n} = 0 \tag{10.33}$$

We now have $n$ independent equations, which we can now solve for $n$ unknowns. This method of approach is illustrated in the following examples.

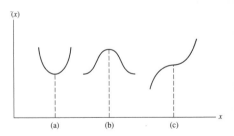

**Figure 10.16.** Stationary or extremum points.

[15]To further understand this, consider $V$ as a function of only one variable, $x$. A *stationary* value (or, as we may say, an extremum) might be a local minimum (*a* in Fig. 10.16), a local maximum (*b* in the figure), or an inflection point (*c* in the figure). Note for these points that for a differential movement, $\delta x$, there is zero first-order change in $V$ (i.e., $\delta V = 0$).

## ■ Example 10.3

A block weighing $W$ lb is placed slowly on a spring having a spring constant of $K$ lb/ft (see Fig. 10.17). Calculate how much the spring is compressed at the equilibrium configuration.

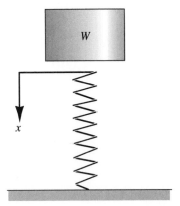

**Figure 10.17.** Mass placed on a linear spring.

This is a simple problem and could be solved by using the definition of the spring constant, but we shall take advantage of the simplicity to illustrate the preceding comments. Note that only conservative forces act on the block, namely the weight and the spring force. Using the unextended top position of the spring as the datum for gravitational potential energy and measuring $x$ from this position we have, for the potential energy of the system:

$$V = -Wx + \tfrac{1}{2}Kx^2$$

Consequently, for equilibrium, we have since there is only one degree of freedom

$$\frac{dV}{dx} = -W + Kx = 0$$

Solving for $x$, we have

$$x = \frac{W}{K}$$

# Example 10.4

A mechanism shown in Fig. 10.18 consists of two weights $W$, four pinned linkage rods of length $a$, and a spring $K$ connecting the linkage rods and which rides along a stationary vertical rod. The spring is unextended when $\theta = 45°$. If friction and the weights of the linkage rods are negligible, what are the equilibrium configurations for the system of linkage rods and weights?

  Only conservative forces can perform work on the system, and so we may use the stationary potential-energy criterion for equilibrium. We shall compute the potential energy as a function of $\theta$ (clearly, there is but one degree of freedom) using the configuration $\theta = 45°$ as the source of datum levels for the various energies. Observing Fig. 10.19, we can say that

$$V = -2Wd + \tfrac{1}{2} K(2d)^2 \tag{a}$$

As for the distance $d$, we can say (see Fig. 10.19)

$$d = a\cos 45° - a\cos\theta \tag{b}$$

Hence, we have, for Eq. (a),

$$V = -2Wa(\cos 45° - \cos\theta) + \tfrac{1}{2}K4a^2(\cos 45° - \cos\theta)^2$$

For equilibrium, we require that

$$\frac{dV}{d\theta} = 0 = -2Wa\sin\theta + 4Ka^2(\cos 45° - \cos\theta)(\sin\theta) \tag{c}$$

We can then say

$$\sin\theta\left[-W - 2Ka\left(\cos\theta - \frac{1}{\sqrt{2}}\right)\right] = 0 \tag{d}$$

We have here two possibilities for satisfying the equation. First, $\sin\theta = 0$ is a solution, so we may say that $\theta_1 = 0$ (this may not be mechanically possible) is a configuration of equilibrium. Clearly, another solution can be reached by setting the bracketed terms equal to zero:

$$-W - 2Ka\left(\cos\theta - \frac{1}{\sqrt{2}}\right) = 0$$

Therefore,

$$\cos\theta = \frac{1}{\sqrt{2}} - \frac{W}{2Ka} \tag{e}$$

The solutions for $\theta$ then are

$$\theta_1 = 0$$
$$\theta_2 = \cos^{-1}\left(\frac{1}{\sqrt{2}} - \frac{W}{2Ka}\right) \tag{f}$$

We have here two possible equilibrium configurations.

**Figure 10.18.** A mechanism.

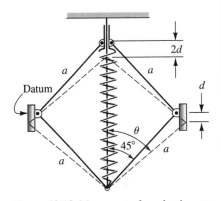

**Figure 10.19.** Movement of mechanism as determined by $\theta$.

# PROBLEMS

**10.31.** A 50-kg block is placed carefully on a spring. The spring is nonlinear. The force to deflect the spring a distance $x$ mm is proportional to the square of $x$. Also, we know that 5 N deflects the spring 1 mm. By the method of stationary potential energy, what will be the compression of the spring? Check the result using a simple calculation based on the behavior of the spring.

**Figure P.10.31.**

**10.32.** A cylinder of radius 2 ft has wrapped around it a light, inextensible cord which is tied to a 100-lb block $B$ on a 30° inclined surface. The cylinder $A$ is connected to a *torsional spring*. This spring requires a torque of 1,000 ft-lb/rad of rotation and it is linear and, of course, restoring. If $B$ is connected to $A$ when the torsional spring is unstrained, and if $B$ is allowed to move slowly down the incline, what distance $d$ do you allow it to move to reach an equilibrium configuration? Use the method of stationary potential energy and then check the result by more elementary reasoning.

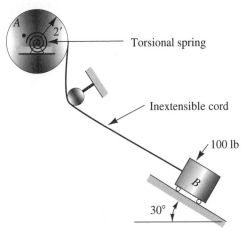

**Figure P.10.32.**

**10.33.** Find the equilibrium configurations for the system. The bars are indentical and each has a weight $W$, a length of 3 m, and a mass of 25 kg. The spring is unstretched when the bars are horizontal and has a spring constant of 1,500 N/m.

**Figure P.10.33.**

**10.34.** The springs of the mechanism are unstretched when $\theta = \theta_0$. Show that $\theta = 25.90°$ when the weight $W$ is added. Take $W = 500$ N, $a = .3$ m, $K_1 = 1$ N/mm, $K_2 = 2$ N/mm, and $\theta_0 = 45°$. Neglect the weight of the members.

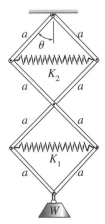

**Figure P.10.34.**

**10.35.** At what elevation $h$ must body $A$ be for equilibrium? Neglect friction. [*Hint:* What is the differential relation between $\theta$ and $l$ defining the positions of the blocks along the surface? Integrate to get the relation itself.]

**Figure P.10.35.**

**10.36.** Show that the position of equilibrium is $\theta = 77.3°$ for the 20-kg rod $AB$. Neglect friction.

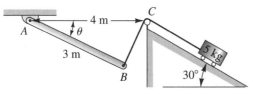

**Figure P.10.36.**

**10.37.** A beam $BC$ of length 15 ft and weight 500 lb is placed against a spring (which has a spring constant of 10 lb/in.) and smooth walls and allowed to come to rest. If the end of the spring is 5 ft away from the vertical wall when it is not compressed, show by energy methods that the amount that the spring will be compressed is .889 ft.

**Figure P.10.37.**

**10.38.** Light rods $AB$ and $BC$ support a 500-N load. End $A$ of rod $AB$ is pinned, whereas end $C$ is on a roller. A spring having a spring constant of 1,000 N/m is connected to $A$ and $C$. The spring is unstretched when $\theta = 45°$. Show that the force in the spring is 1,066 N when the 500-N load is being supported.

**Figure P.10.38.**

**10.39.** Work Problem 10.28 using the method of total potential energy.

**10.40.** Work Problem 10.29 using the method of total potential energy.

**10.41.** Do Problem 10.25 by the method of total potential energy. [*Hint:* Consider a length of cord on a circular surface. Use the top part of the surface as a datum.]

**10.42.** If member $AB$ is 10 ft long and member $BC$ is 13 ft long, show that the angle $\theta$ corresponding to equilibrium is 34.5° if the spring constant $K$ is 10 lb/in. Neglect the weight of the members and friction everywhere. Take $\theta = 30°$ for the configuration where the spring is unstretched.

**Figure P.10.42.**

**10.43.** A combination of spring and torsion-bar suspension is shown. The spring has a spring constant of 150 N/mm. The torsion bar is shown on end at $A$ and has a torsional resistance to rotation of rod $AB$ of 5,000 N-m/rad. If the vertical load is zero, the vertical spring is of length 450 mm, and rod $AB$ is horizontal. What is the angle $\alpha$ when the suspension supports a weight of 5 kN? Rod $AB$ is 400 mm in length.

**Figure P.10.43.**

**10.44.** Light rods $AB$ and $CB$ are pinned together at $B$ and pass through frictionless bearings $D$ and $E$. These bearings are connected to the ground by ball-and-socket connections and are free to rotate about these joints. Springs, each having a spring constant $K = 800$ N/m, restrain the rods as shown. The springs are unstretched when $\theta = 45°$. Show that the deflection of $B$ is .440 m when a 500-N load is attached slowly to pin $B$. The rods are each 1 m in length, and each unstretched spring is .250 m in length. Neglect the weight of the rods. Rods are welded to small plates at A and C

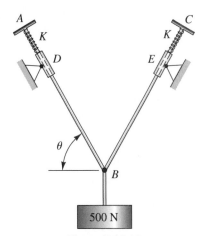

**Figure P.10.44.**

**10.45.** Do Problem 10.26 by the method of total potential energy. [*Hint:* Use $E$ as a datum and get lengths $EJ$, $KP$, and $PN$ in terms of length $HE$, including unknown constants.]

**10.46.** An elastic band is originally 1 m long. Applying a tension force of 30 N, the band will stretch .8 m in length. What deflection $a$ does a 10-N load induce on the band when the load is applied slowly at the center of the band? Consider the force vs. elongation of the band to be linear like a spring. [*Hint:* If you consider half of the band, you double the "spring constant."]

**Figure P.10.46.**

**10.47.** In Problem 10.46, the band is first stretched and then tied while stretched to supports A and B so that there is an initial tension in the band of 15 N. What is then the deflection $a$ caused by the 10-N load?

**10.48.** A rubber band of length .7 m is stretched to connect to points A and B. A tension force of 40 N is thereby developed in the band. A 20-N weight is then attached to the band at C. Find the distance $a$ that point C moves downward if the 20-N weight is constrained to move vertically downward along a frictionless rod. [*Hint:* If you consider part of the band, the "spring constant" for it will be greater than that of the whole band.]

**Figure P.10.48.**

**10.49.** The spring connecting bodies A and B has a spring constant $K$ of 3 N/mm. The unstretched length of the spring is 450 mm. If body A weighs 60 N and body B weighs 90 N, what is the stretched length of the spring for equilibrium? [*Hint:* V will be a function of two variables.]

**Figure P.10.49**

# 10.8   Stability

Consider a cylinder resting on various surfaces (Fig. 10.20). If we neglect friction, the only active force is that of gravity. Thus, we have here conservative systems for which Eq. 10.31 is valid. The only virtual displacement for which contact with the surfaces is maintained is along the path. In each case, $dy/dx$ is zero. Thus, for an infinitesimal virtual displacement, the first-order change in elevation is zero. Hence, the change in potential energy is zero for the first-order considerations. The bodies, therefore, are in *equilibrium*, according to the previous section. However, distinct physical differences exist between the states of equilibrium of the four cases.

**Case A.**   The equilibrium here is said to be *stable* in that an actual displacement from this configuration is such that the forces tend to return the body to its equilibrium configuration. Notice that the potential energy is at a *minimum* for this condition.

**Case B.**   The equilibrium here is said to be *unstable* in that an actual displacement from the configuration is such that the forces aid in increasing the departure from the equilibrium configuration. The potential energy is at a *maximum* for this condition.

**Case C.**   The equilibrium here is said to be *neutral*. Any displacement means that another equilibrium configuration is established. The potential energy is a constant for all possible positions of the body.

**Case D.**   This equilibrium state is considered *unstable* since any displacement to the left of the equilibrium configuration will result in an increasing departure from this position.

How can we tell whether a system is stable or unstable at its equilibrium configuration other than by physical inspection, as was done above? Consider again a simple situation where the potential energy is a function of only one space coordinate $x$. That is, $V = V(x)$. We can expand the potential energy in the form of a Maclaurin series about the position of equilibrium.[16] Thus,

$$V = V_{eq} + \left(\frac{dV}{dx}\right)_{eq} x + \frac{1}{2!}\left(\frac{d^2V}{d^2x}\right)_{eq} x^2 + \cdots \qquad (10.34)$$

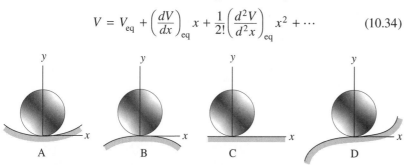

**Figure 10.20.** Different equilibrium configurations.

[16]Note that in a Maclaurin series the coefficients of the independent variable $x$ are evaluated at $x = 0$, which for us is the equilibrium position. We denote this position with the subscript eq.

We know from Eq. 10.33 applied to one variable that at the equilibrium configuration $(dV/dx)_{eq} = 0$. Hence, we can restate the equation above:

$$V - V_{eq} = \Delta V = \frac{1}{2!}\left(\frac{d^2 V}{dx^2}\right)_{eq} x^2 + \frac{1}{3!}\left(\frac{d^3 V}{dx^3}\right)_{eq} x^3 + \cdots \qquad (10.35)$$

For small enough $x$, say $x_0$, the sign of $\Delta V$ will be determined by the sign of the first term in the series, $(1/2!)(d^2 V/dx^2)_{eq} x^2$.[17] For this reason this term is called the dominant term in the series. Hence, the sign of $(d^2 V/dx^2)_{eq}$ is vital in determining the sign of $\Delta V$ for small enough $x$. If $(d^2 V/dx^2)_{eq}$ is positive, then $\Delta V$ is positive for any value of $x$ smaller than $x_0$. This means that $V$ is a local minimum at the equilibrium configuration as can be deduced from Fig. 10.20a, and we have *stable equilibrium*.[18] If $(d^2 V/dx^2)_{eq}$ is negative, then $V$ is a local maximum at the equilibrium configuration and from Fig. 10.20 we have *unstable equilibrium*. Finally, if $(d^2 V/dx^2)_{eq}$ is zero, we must investigate the next higher-order derivative in the expansion, and so forth.

For cases where the potential energy is known in terms of several variables, the determination of the kind of equilibrium for the system is correspondingly more complex. For example, if the function $V$ is known in terms of $x$ and $y$, we have from the calculus of several variables the following.

For minimum potential energy and therefore for stability:

$$\frac{\partial V}{\partial x} = \frac{\partial V}{\partial y} = 0 \qquad (10.36a)$$

$$\left(\frac{\partial^2 V}{\partial x \partial y}\right)^2 - \frac{\partial^2 V}{\partial x^2}\frac{\partial^2 V}{\partial y^2} < 0 \qquad (10.36b)$$

$$\frac{\partial^2 V}{\partial x^2} + \frac{\partial^2 V}{\partial y^2} > 0 \qquad (10.36c)$$

For maximum potential energy and therefore for instability:

$$\frac{\partial V}{\partial x} = \frac{\partial V}{\partial y} = 0 \qquad (10.37a)$$

$$\left(\frac{\partial^2 V}{\partial x \partial y}\right)^2 - \frac{\partial^2 V}{\partial x^2}\frac{\partial^2 V}{\partial y^2} < 0 \qquad (10.37b)$$

$$\frac{\partial^2 V}{\partial x^2} + \frac{\partial^2 V}{\partial y^2} < 0 \qquad (10.37c)$$

The criteria become increasingly more complex for three or more independent variables.

---

[17] As $x$ gets smaller than unity, $x^2$ will become increasingly larger than $x^3$ and powers of $x$ higher than 3. Hence, depending on the values of derivatives of $V$ at equilibrium, there will be a value of $x$—say $x_0$—for which the first term in the series will be larger than the sum of all other terms for values of $x < x_0$.

[18] That is, if the body is displaced a distance $x < x_0$, the body will return to equilibrium on release.

### ■ Example 10.5

A thick plate whose bottom edge is that of a circular arc of radius $R$ is shown in Fig. 10.21. The center of gravity of the plate is a distance $h$ above the ground when the plate is in the vertical position as shown in the diagram. What relation must be satisfied by $h$ and $R$ for stable equilibrium?

The plate has one degree of freedom under the action of gravity and we can use the angle $\theta$ (Fig. 10.22) as the independent coordinate. We can express the potential energy $V$ of the system relative to the ground as a function of $\theta$ in the following manner (see Fig. 10.23):

$$V = W[R - (R - h)\cos\theta] \qquad (a)$$

where $W$ is the weight of the plate. Clearly, $\theta = 0$ is a position of equilibrium since

$$\left(\frac{dV}{d\theta}\right)_{\theta=0} = \left[W(R - h)\sin\theta\right]_{\theta=0} = 0 \qquad (b)$$

Now consider $d^2V/d\theta^2$ at $\theta = 0$. We have

$$\left(\frac{d^2V}{d\theta^2}\right)_{\theta=0} = W(R - h) \qquad (c)$$

Clearly, when $R > h$, $(d^2V/d\theta^2)_{\theta = 0}$ is positive, and so this is the desired requirement for stable equilibrium. Thus for stable equilibrium, $R > h$.

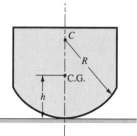

**Figure 10.21.** Plate with circular bottom edge.

**Figure 10.22.** One degree of freedom.

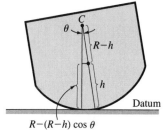

**Figure 10.23.** Position of C.G.

## 10.9   Looking Ahead: More on Total Potential Energy

When we have conservative forces acting on particles and rigid bodies, we found earlier that for establishing necessary and sufficient conditions of equilibrium we could extremize the potential energy $V$ associated with the forces. That is, we could set

$$\delta V = 0$$

to satisfy equilibrium. Recall that this result was derived from the method of virtual work. We have a similar formulation for the case of an elastic (not necessarily linearly elastic) body whereby we can guarantee equilibrium. This more-general principle is derivable from the more-general virtual work principle mentioned earlier in section 10.5. Here, we must extremize an expression more complicated (as you might expect) than $V$. This expression is denoted as $\pi$ with no relation to the number 3.1416. ... The expression $\pi$ is what we call a *functional*, wherein for the substitution of each function, such as $y(x)$, into the functional a number is established. A simple example of a functional $I$ is as follows:

$$I = \int_{x_1}^{x_2} F\left(x, y, \frac{dy}{dx}\right) dx$$

where $F$ is a function of $x$ (the so-called independent variable), $y$, and $dy/dx$. Substitution of a function $y(x)$ into $F$ followed by an integration between the fixed limits, yields for this function $y(x)$ a number for $I$. Functionals pervade the field of mechanics and most other analytic fields of knowledge. A vital step is to find the function $y(x)$ that will extremize $I$. This function then becomes known as the *extremal function*. The calculus for doing this is called the *calculus of variations*. The particular functional for the method of total potential energy is given as

$$\pi = -\iiint_V \boldsymbol{B} \cdot \boldsymbol{u} \, dv - \oiint_S \boldsymbol{T} \cdot \boldsymbol{u} \, dA + U$$

The function to be adjusted to extremize $\pi$ is $\boldsymbol{u}(x, y, z)$ taking the place of $y(x)$ in the preceding functional, and now the independent variables are $x$, $y$, and $z$ in place of just $x$ in the preceding functional. The expression $U$ is the energy of deformation that you will study later in your solids course and presented earlier in Section 10.5 as $\iiint_V \sum_i \sum_j \tau_{ij} \delta\epsilon_{ij} \, dv$. The principle of total potential energy for the case of elastic bodies is written in the following deceivingly simple looking formulation:

$$\delta\pi = 0$$

What is most intriguing about this innocent looking equation is that it is often considered to be the **most powerful equation in solid mechanics**!

We have touched on a broad area of study, namely variational methods, parts of which you will encounter in many of your studies. For example, just the method of total potential energy has the following major uses:

1. It plays an important role in *optimization theory*.
2. It can be used profitably to derive the *proper equations* and the *boundary conditions* for many areas of vital importance such as plate theory, elastic stability theory, dynamics of plates and beams, torsion theory, etc.
3. From it we can develop a number of vital approximation methods. The most prominent of these methods is the method of *finite elements*.

Clearly this is an impressive list.[19]

---

[19]You can study in some detail the contents of the two Looking Ahead sections of this chapter in I.H. Shames, *Introduction to Solid Mechanics*, 2nd ed., 1989, Prentice-Hall, Inc., Englewood Cliffs, N.J. This material should be within the reach of bright second-semester sophomores and certainly of juniors. The author has taught this material for many years to second-semester sophomores out of the aforementioned book. See Chapters 18 and 19.

**10.50.** A rod $AB$ is connected to the ground by a frictionless ball-and-socket connection at $A$. The rod is free to rest on the inside edge of a horizontal plate as shown in the diagram. The square $abcd$ has its center directly over $A$. The curve $efg$ is a semicircle. Without resorting to mathematical calculations, identify positions on this inside edge where equilibrium is possible for the rod $AB$. Describe the nature of the equilibrium and supply supporting arguments. Assume the edge of plate is frictionless.

**Figure P.10.50.**

**10.51.** In Problem 10.50, show mathematically that position $h$ is a position of unstable equilibrium for the rod.

**10.52.** Rod $AB$ is supported by a frictionless ball-and-socket joint at $A$ and leans against the inside edge of a horizontal plate. What is the nature of the equilibrium position $a$ for the rod? Assume that the edge of the plate is frictionless.

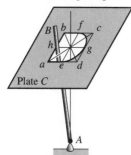

**Figure P.10.52.**

**10.53.** Consider that the potential energy of a system is given by the formulation: $V = 8x^3 + 6x^2 - 7x$. What are the equilibrium positions? Indicate whether these positions are stable or not.

**10.54.** A section of a cylinder is free to roll on a horizontal surface. If $\gamma$ of the triangular portion of the cylinder is 180 lb/ft³ and that of the semicircular portion of the cylinder is 100 lb/ft³, is the configuration shown in the diagram in stable equilibrium?

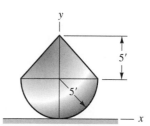

**Figure P.10.54.**

**10.55.** A system of springs and rigid bodies $AB$ and $BC$ is acted on by a weight $W$ through a pin connection at $A$. If $K$ is 50 N/mm, what is the range of the value of $W$ so that the system has an unstable equilibrium configuration when the rods $AB$ and $BC$ are collinear? Neglect the weight of the rods.

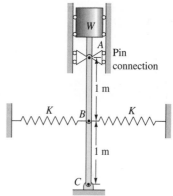

**Figure P.10.55.**

**10.56.** A weight $W$ is welded to a light rod $AB$. At $B$ there is a torsional spring for which it takes 500 ft-lb to rotate 1 rad. The torsional spring is linear and restoring and is, for rotation, the analog of the ordinary linear spring for extension or contraction. If the torsional spring is unstrained when the rod is vertical, what is the largest value of $W$ for which we have stable equilibrium in the vertical direction?

**Figure P.10.56.**

**10.57.** A light rod $AB$ is pinned to a block of weight $W$ at $A$. Also at $A$ are two identical springs $K$. Show that, for $W$ less than $2Kl$, we have stable equilibrium in the vertical position and, for $W > 2Kl$, we have unstable equilibrium. The value $W = 2Kl$ is called a *critical load* for reasons that are explained in Problem 10.58.

**Figure P.10.57.**

**10.58.** In Problem 10.57, apply a small transverse force $F$ to body $A$ as shown. Compute the horizontal deflection $\delta$ of point $A$ for a position of equilibrium by using ordinary statics as developed in earlier chapters. Now show that when $W = 2Kl$ (i.e., the critical weight), the deflection $\delta$ mathematically blows up to infinity. This shows that, even if $W < 2Kl$ and we have stable equilibrium with $F = 0$, we get increasingly very large deflections as the weight $W$ approaches its critical value and a side load $F$, however small, is introduced. The study of stability of equilibrium configuration therefore is an important area of study in mechanics. Most of you will encounter this topic in your strength of materials course.

**Figure P.10.58.**

**10.59.** Cylinders $A$ and $B$ have semicircular cross-sections. Cylinder $A$ supports a rectangular solid shown as $C$. If $\rho_A = 1,600$ kg/m$^3$ and $\rho_c = 800$ kg/m$^3$, ascertain whether the arrangement shown is in stable equilibrium. [*Hint:* Make use of point $O$ in computing $V$.]

**Figure P.10.59.**

## 10.10 Closure

In this chapter, we have taken an approach that differs radically from the approach used earlier in the text. In earlier chapters, we isolated a body for the purpose of writing equilibrium equations using all the forces acting on the body. This is the approach we often call *vectorial mechanics*. In this chapter, we have mathematically compared the equilibrium configuration with admissible neighboring configurations. We concluded that the equilibrium configuration was one from which there is zero virtual work under a virtual displacement. Or, equivalently for conservative active forces, the equilibrium configuration was the configuration having stationary (actually minimum) potential energy when compared to admissible configurations in the neighborhood. We call such an approach *variational mechanics*. The variational mechanics point of view is no doubt strange to you at this stage of study and far more subtle and mathematical than the vectorial mechanics approach.

Shifts like the one from the more physically acceptable vectorial mechanics to the more abstract variational mechanics take place in other engineering sciences. Variational methods and techniques are used in the study of plates and shells, elasticity, quantum mechanics, orbital mechanics, statistical thermodynamics, and electromagnetic theory. The variational methods and viewpoints thus are important and even vital in more advanced studies in the engineering sciences, physics, and applied mathematics.

# PROBLEMS

**10.60.** At what position must the operator of the counterweight crane locate the 50-kN counterweight when he lifts the 10-kN load of steel?

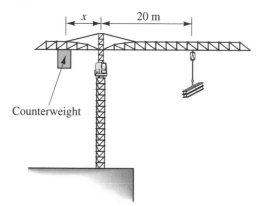

Counterweight

**Figure P.10.60.**

**10.61.** What is the relation between $P$ and $Q$ for equilibrium?

**Figure P.10.61.**

**10.62.** A 50 lb-ft torque is applied to a press. The pitch of the screw is .5 in. If there is no friction on the screw, and if the base of the screw can rotate frictionlessly in a base plate $A$, what is the force $P$ imposed by the base plate on body $B$?

50 lb-ft

**Figure P.10.62.**

**10.63.** The spring is unstretched when $\theta = 30°$. At any position of the pendulum, the spring remains horizontal. If the spring constant is 50 lb/in., at what position will the system be in equilibrium?

100 lb

**Figure P.10.63.**

**10.64.** If the springs are unstretched when $\theta = \theta_0$, find the angle $\theta$ when the weight $W$ is placed on the system. Use the method of stationary potential energy.

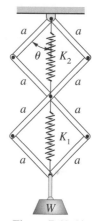

**Figure P.10.64.**

**10.65.** A mass $M$ of 20 kg slides with no friction along a vertical rod. Two internal springs $K_1$ of spring constant 2 N/mm and an external spring $K_2$ of spring constant 3 N/mm restrain the weight $W$. If all springs are unstrained at $\theta = 30°$, show that the equilibrium configuration corresponds to $\theta = 27.8°$.

**Figure P.10.65.**

**10.66.** When rod $AB$ is in the vertical position, the spring attached to the wheel by a flexible cord is unstretched. Determine all the possible angles $\theta$ for equilibrium. Show which are stable and which are not stable. The spring has a spring constant of 8 lb/in.

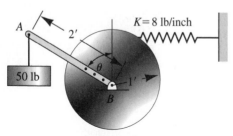

**Figure P.10.66.**

**10.67.** Two identical rods are pinned together at $B$ and are pinned at $A$ and $C$. At $B$ there is a torsional spring requiring 500 N-m/rad of rotation. What is the maximum weight $W$ that each rod can have for a case of stable equilibrium when the rods are collinear?

**Figure P.10.67.**

**10.68.** A rectangular solid body of height $h$ rests on a cylinder with a semicircular section. Set up criteria for stable and unstable equilibrium in terms of $h$ and $R$ for the position shown.

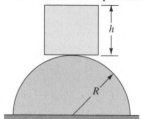

**Figure P.10.68.**

# APPENDIX I

# Integration Formulas

**1.** $\displaystyle\int \frac{x\,dx}{a+bx} = \frac{1}{b^2}\left[a + bx - a\ln(a+bx)\right]$

**2.** $\displaystyle\int \frac{dx}{a^2-x^2} = \frac{1}{2a}\ln\left(\frac{a+x}{a-x}\right)$

**3.** $\displaystyle\int \sqrt{x^2 \pm a^2}\,dx = \tfrac{1}{2}\left[x\sqrt{x^2\pm a^2} \pm a^2\ln\left(x+\sqrt{x^2\pm a^2}\right)\right]$

**4.** $\displaystyle\int \sqrt{a^2-x^2}\,dx = \tfrac{1}{2}\left(x\sqrt{a^2-x^2} + a^2\sin^{-1}\frac{x}{a}\right)$

**5.** $\displaystyle\int x\sqrt{a^2-x^2}\,dx = -\tfrac{1}{3}\sqrt{(a^2-x^2)^3}$

**6.** $\displaystyle\int x\sqrt{a+bx}\,dx = -\frac{2(2a-3bx)\sqrt{(a+bx)^3}}{15b^2}$

**7.** $\displaystyle\int x^2\sqrt{a^2-x^2}\,dx = -\frac{x}{4}\sqrt{(a^2-x^2)^3} + \frac{a^2}{8}\left(x\sqrt{a^2-x^2} + a^2\sin^{-1}\frac{x}{a}\right)$

**8.** $\displaystyle\int x^2\sqrt{a^2\pm x^2}\,dx = \frac{x}{4}\sqrt{(x^2\pm a^2)^3} \mp \frac{a^2}{8}x\sqrt{x^2\pm a^2} - \frac{a^4}{8}\ln\left(x+\sqrt{x^2\pm a^2}\right)$

**9.** $\displaystyle\int \frac{dx}{\sqrt{a^2-x^2}} = \sin^{-1}\frac{x}{a}$

**10.** $\displaystyle\int \frac{dx}{\sqrt{x^2+a^2}} = \ln\left(x+\sqrt{x^2+a^2}\right) = \sinh^{-1}\frac{x}{a}$

**11.** $\displaystyle\int x^m e^{ax}\,dx = \frac{x^m e^{ax}}{a} - \frac{m}{a}\int x^{m-1}e^{ax}\,dx$

**12.** $\displaystyle\int x^m \ln x\,dx = x^{m+1}\left(\frac{\ln x}{m+1} - \frac{1}{(m+1)^2}\right)$

**13.** $\displaystyle\int \sin^2\theta\,d\theta = \tfrac{1}{2}\theta - \tfrac{1}{4}\sin 2\theta$

**14.** $\int \cos^2 \theta \, d\theta = \frac{1}{2}\theta + \frac{1}{4}\sin 2\theta$

**15.** $\int \sin^3 \theta \, d\theta = -\frac{1}{3}\cos \theta(\sin^2 \theta + 2)$

**16.** $\int \cos^m \theta \sin \theta \, d\theta = -\dfrac{\cos^{m+1} \theta}{m + 1}$

**17.** $\int \sin^m \theta \cos \theta \, d\theta = \dfrac{\sin^{m+1} \theta}{m + 1}$

**18.** $\int \sin^m \theta \, d\theta = -\dfrac{\sin^{m-1} \theta \cos \theta}{m} + \dfrac{m - 1}{m}\int \sin^{m-2} \theta \, d\theta$

**19.** $\int \theta^2 \sin \theta \, d\theta = 2\theta \sin \theta - (\theta^2 - 2)\cos \theta$

**20.** $\int \theta^2 \cos \theta \, d\theta = 2\theta \cos \theta + (\theta^2 - 2)\sin \theta$

**21.** $\int \theta \sin^2 \theta \, d\theta = \frac{1}{4}\left[\sin \theta(\sin \theta - 2\theta \cos \theta) + \theta^2\right]$

**22.** $\int \sin m\theta \cos m\theta \, d\theta = -\dfrac{1}{4m}\cos 2m\theta$

**23.** $\int \dfrac{d\theta}{(a + b \cos \theta)^2}$

$$= \frac{1}{(a^2 - b^2)}\left(\frac{-b \sin \theta}{a + b \cos \theta} + \frac{2a}{\sqrt{a^2 - b^2}}\tan^{-1}\frac{\sqrt{a^2 - b^2}\,\tan\frac{\theta}{2}}{a + b}\right)$$

**24.** $\int \theta \sin \theta \, d\theta = \sin \theta - \theta \cos \theta$

**25.** $\int \theta \cos \theta \, d\theta = \cos \theta + \theta \sin \theta$

# APPENDIX II

# Computation of Principal Moments of Inertia

We now turn to the problem of computing the principal moments of inertia and the directions of the principal axes for the case where we do not have planes of symmetry. It is unfortunate that a careful study of this important calculation is beyond the level of this text. However, we shall present enough material to permit the computation of the principal moments of inertia and the directions of their respective axes.

The procedure that we shall outline is that of extremizing the mass moment of inertia at a point where the inertia-tensor components are known for a reference $xyz$. This will be done by varying the direction cosines $l$, $m$, and $n$ of an axis $k$ so as to extremize $I_{kk}$ as given by Eq. 9.13. We accordingly set the differential of $I_{kk}$ equal to zero as follows:

$$
\begin{aligned}
dI_{kk} = \ & 2lI_{xx}\,dl + 2mI_{yy}\,dm + 2nI_{zz}\,dn \\
& -2lI_{xy}\,dm - 2mI_{xy}\,dl - 2lI_{xz}\,dn \\
& -2nI_{xz}\,dl - 2mI_{yz}\,dn - 2nI_{yz}\,dm = 0
\end{aligned}
\tag{II.1}
$$

Collecting terms and canceling the factor 2, we get

$$
(lI_{xx} - mI_{xy} - nI_{zz})dl + (-lI_{xy} + mI_{yy} - nI_{yz})dm \\
+ (-lI_{xz} - mI_{yz} + nI_{zz})dn = 0 \tag{II.2}
$$

If the differentials $dl$, $dm$, and $dn$ were *independent* we could set their respective coefficients equal to zero to satisfy the equation. However, they are not independent because the equation

$$
l^2 + m^2 + n^2 = 1 \tag{II.3}
$$

must at all times be satisfied. Accordingly, the differentials of the direction cosines must be related as follows[1]:

$$
l\,dl + m\,dm + n\,dn = 0 \tag{II.4}
$$

[1]We are thus extremizing $I_{kk}$ in the presence of a constraining equation.

We can of course consider any two differentials as independent. The third is then established in accordance with the equation above.

We shall now introduce the *Lagrange multiplier* $\lambda$ to facilitate the extremizing process. This constant is an arbitrary constant at this stage of the calculation. Multiplying Eq. II.4 by $\lambda$ and subtracting Eq. II.4 from Eq. II.2 we get when collecting terms:

$$\left[(I_{xx} - \lambda)l - I_{xy}m - I_{xz}n\right]dl + \left[-I_{xy}l + (I_{yy} - \lambda)m - I_{yz}n\right]dm$$
$$+ \left[-I_{xz}l - I_{yz}m + (I_{zz} - \lambda)n\right]dn = 0 \qquad (II.5)$$

Let us next consider that $m$ and $n$ are independent variables and consider the value of $\lambda$ so chosen that the coefficient of $dl$ is zero. That is,

$$(I_{xx} - \lambda)l - I_{xy}m - I_{xz}n = 0 \qquad (II.6)$$

With the first term Eq. II.5 disposed of in this way, we are left with differentials $dm$ and $dn$, which are independent. Accordingly, we can set their respective coefficients equal to zero in order to satisfy the equation. Hence, we have in addition to Eq. II.6 the following equations:

$$-I_{xy}l + (I_{yy} - \lambda)m - I_{yz}n = 0$$
$$-I_{xz}l - I_{yz}m + (I_{zz} - \lambda)n = 0 \qquad (II.7)$$

A necessary condition for the solution of a set of direction cosines $l$, $m$, and $n$, from Eqs. II.6 and II.7, which does not violate Eq. II.3[2] is that the determinant of the coefficients of these variables be zero. Thus:

$$\begin{vmatrix} (I_{xx} - \lambda) & -I_{xy} & -I_{xz} \\ -I_{xy} & (I_{yy} - \lambda) & -I_{yz} \\ -I_{xz} & -I_{yz} & (I_{zz} - \lambda) \end{vmatrix} = 0 \qquad (II.8)$$

This results in a cubic equation for which we can show there are three real roots for $\lambda$. Substituting these roots into any two of Eqs. II.6 and II.7 plus Eq. II.3, we can determine three direction cosines for each root. These are the direction cosines for the principal axes measured relative to *xyz*. We could get the principal moments of inertia next by substituting a set of these direction cosines into Eq. 9.13 and solving for $I_{kk}$. However, that is not necessary, since it can be shown that the three Lagrange multipliers *are* the principal moments of inertia.

---

[2]This precludes the possibility of a trivial solution $l = m = n = 0$.

# Computer Projects: Statics

## Preface

In this appendix, you will find 12 projects specially designed for use on the computer. We have been careful to limit the programming needs for the problems not to go beyond the usual freshman course in Fortran programming. The programs conform completely to the ANSI standard for Fortran 77 and will run on any machine having an ANSI standard compiler.

At SUNY, Buffalo and at G.W., we assign one or two projects per semester **on top** of our regular program. No regular class time is used to teach programming or to go over solutions. We give our students literature for logging in, editing, filing, etc. peculiar to the machine we will work on; the students are supposed to learn to use the machine with the help of our Computing Center "consultants."

I would like to thank former graduate students, Dr. Sun Lei Chang, Dr. Anoop Dhingra and Dr. John Hu; they did the original programming under my direction. Ms. Chaishia Tracy Chan also deserves thanks for her help in rewriting the programs to make them totally standard conforming.

# Contents

| Problem | File Name |
| --- | --- |
| **Statics projects 1–12** | |
| 1. Moment Problem | MECH1.FOR |
| 2. Torque Problem | MECH2.FOR |
| 3. Equivalent Problem | MECH3.FOR |
| 4. Friction Problem | MECH4.FOR |
| 5. Centroidal Coordinates of Block | PROB5.FOR |
| 6. Centroidal Coordinates of Plate | CENTER.FOR |
| 7. Beam on Nonlinear Springs | BEAM.FOR |
| 8. Principal Axis Problem | PRINCIPL.FOR |
| 9. Area Second Moments and Products of Area | CEN.FOR |
| 10. Maximum Second Moment of Area | P883.FOR |
| 11. Moments and Products of Inertia | BLOCK.FOR |
| 12. Area Moments and Products of Area for Complex Area | TWOF.FOR |
| **Dynamics projects 13–29** | |
| 13. Projectile Problem With Friction | MECH5.FOR |
| 14. Spring Impact Problem | CLOSE.FOR |
| 15. Impact Problem (Stationary Particles) | MOM1.FOR |
| 16. Impact Problem (Moving Particles) | MOM.FOR |
| 17. Torpedo Impact (Constant Speed) | TORPEDO.FOR |
| 18. Torpedo Impact (Accelerating Freighter) | TORP.FOR |
| 19. Conservation of Energy Problem | HARM.FOR |
| 20. Conveyor Problem | EX144.FOR |
| 21. Piston Device Problem (Constant Speed of Piston) | PISTON.FOR |
| 22. Piston Device Problem (Accelerating Piston) | PISV.FOR |
| 23. Light Repair Problem | COPIT.FOR |
| *24. Four-Bar Linkage Problem With Constant Disc Speed. Find $\omega_{BC}$ | FBARS.FOR |
| *25. Four-Bar Linkage Problem With Constant Disc Speed. Find $\dot{\omega}_{BC}$ | FBAR.FOR |
| *26. Four-Bar Linkage Problem With Accelerating Disc. Find $\omega_{BC}$ | FBARVS.FOR |
| *27. Four-Bar Linkage Problem With Accelerating Disc. Find $\dot{\omega}_{BC}$ | FBARV.FOR |
| 28. Astro-Centrifuge Problem | CENT.FOR |
| *29. Constant $g$ Astro-centrifuge Problem | PLOT.FOR |

*Note that the **Newton Raphson** method can be profitably employed in Projects 24, 25, 26, 27, and 29. In each of the programs of these projects, will be found a subroutine for this well-known method. More detail is readily available in most mathematics textbooks.

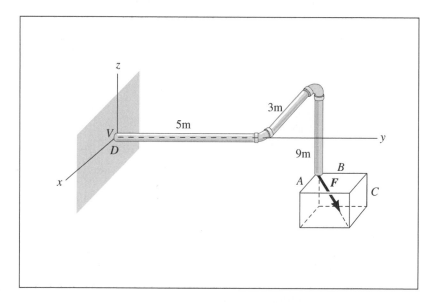

For all possible following sets of values of *A*, *B*, and *C*:

$$A \; (1., 2., 3.)$$
$$B \; (1., 2., 3.)$$
$$C \; (1., 2.5, 4.)$$

find $M_x$, $M_y$, and $M_z$ at *D*.

(There are 27 different settings for the direction of force $F = 1,000$ N.)

Print out results as shown.

| * Project 1 * | | | | | |
|---|---|---|---|---|---|
| Force = 1,000.0 N | | | | | |
| A | B | C | $M_x$ | $M_y$ | $M_z$ |
| 1.0 | 1.0 | 1.0 | 0.2309E + 04 | −0.6928E + 04 | −0.4619E + 04 |
| | | | etc. | | |

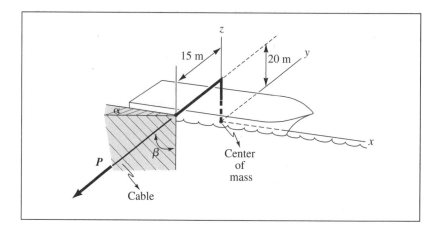

$$\begin{cases} \alpha = \text{angle of vertical plane} \\ \beta = \text{angle of cable from vertical axis} \end{cases}$$

Find the moment tending to tip the boat about its *x axis* from the cable shown, for the following conditions:

$P$ = 100 N, 200 N, 300 N, 400 N
for each $\alpha$ (0°, 10°, 20°, 30°, 40°, 50°)
and for each $\beta$ (0°, 10°, 20°, 30°, 40°, 50°, 60°, 70°, 80°, 90°)

Print out data as shown.

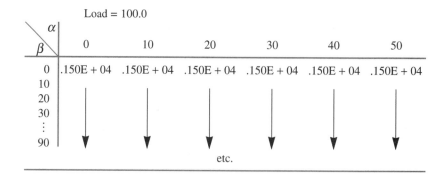

Load = 100.0

| $\beta$ \ $\alpha$ | 0 | 10 | 20 | 30 | 40 | 50 |
|---|---|---|---|---|---|---|
| 0 | .150E + 04 | .150E + 04 | .150E + 04 | .150E + 04 | .150E + 04 | .150E + 04 |
| 10 | | | | | | |
| 20 | | | | | | |
| 30 | | | | | | |
| ⋮ | | | | | | |
| 90 | | | | | | |

etc.

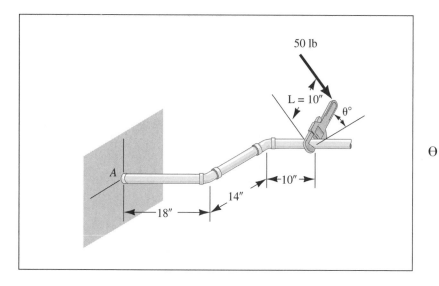

For setting of $\theta$, going from zero to 330° in steps of 30°, compute *components* of *equivalent force system* at "A" for the 50-lb force for two distances of $L$ being 0 in and 10 in.

Print out data as shown.

**Note:**  I →î, J → ĵ, K → k̂.

| L θ | 0 in | 10 in |
|---|---|---|
| 0° | F =     0.00I +     0.00J + −50.00K<br>C = −75.00I + −100.00J +     0.00K | F =     0.00I +     0.00J + −50.00K<br>C = −116.67I + −100.00J +     0.00K |
| 30° | F =<br>C = | F =<br>C = |
| 60° | | |
| ⋮ | | |
| 330° | etc. | |

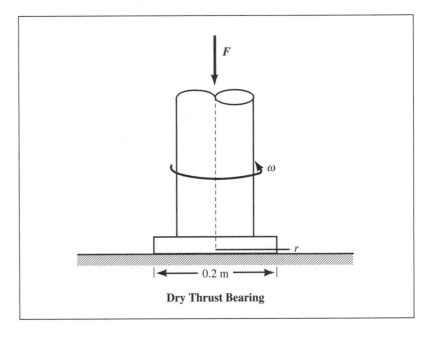

**Dry Thrust Bearing**

$F = 100, 200, 300$ N

Pressure $= P = K[\log(2 + \sqrt{r})](\cos 2\pi r)$ Pa

($K$ is a constant you must determine)

$\mu$ varies with $r$

for

| | |
|---|---|
| $r = 0.005$ m | $\mu = 0.40$ |
| $r = 0.015$ m | $\mu = 0.25$ |
| $r = 0.025$ m | $\mu = 0.15$ |
| $r = 0.035$ m | $\mu = 0.10$ |
| $r = 0.045$ m | $\mu = 0.07$ |
| $r = 0.055$ m | $\mu = 0.05$ |
| $r = 0.065$ m | $\mu = 0.04$ |
| $r = 0.075$ m | $\mu = 0.03$ |
| $r = 0.085$ m | $\mu = 0.02$ |
| $r = 0.095$ m | $\mu = 0.01$ |

Find Resisting Torque due to *Coulomb friction*.

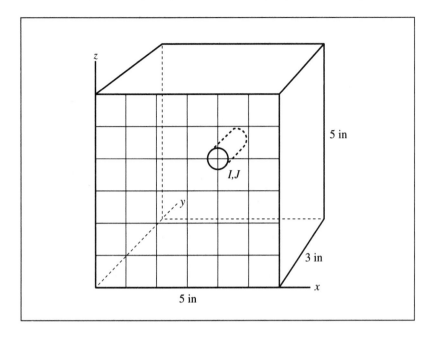

**Find Centroidal Coordinates.** There are 25 *equally spaced* holes in the block each with its *own diameter* and *length* that will be asked for *interactively* during the problem. Please *echo* the data you use.

$D(I, J)$    Diameter of each hole

$L(I, J)$    Length of each hole

$$\begin{cases} D(I, J) + D(I + 1, J) \leq \tfrac{5}{3}'' \\ D(I, J) + D(I, J + 1) \leq \tfrac{5}{3}'' \end{cases}$$

Indicate cases where

$$D(I, J) + D(I + 1, J) > \tfrac{5}{3}'' \text{ or } D(I, J) + D(I, J + 1) > \tfrac{5}{3}''$$

For *your problem*

$$D(I, J) = \begin{bmatrix} .1 & .2 & .3 & .2 & .1 \\ .2 & .1 & .3 & .1 & .2 \\ .3 & .4 & .5 & .4 & .3 \\ .4 & .5 & .4 & .5 & .4 \\ .5 & .3 & .1 & .3 & .5 \end{bmatrix} \text{in.} \qquad L(I, J) = \begin{bmatrix} .1 & .2 & .3 & .2 & .1 \\ .2 & .1 & .3 & .1 & .2 \\ .3 & .4 & .5 & .4 & .3 \\ .4 & .5 & .4 & .5 & .4 \\ .5 & .3 & .1 & .3 & .5 \end{bmatrix} \text{in.}$$

Print out data as shown:

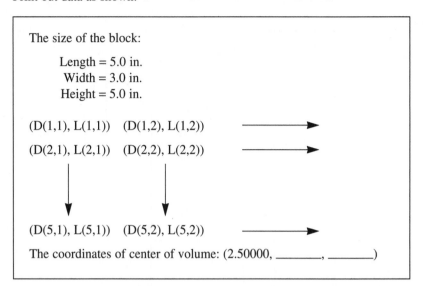

The size of the block:

      Length = 5.0 in.
      Width = 3.0 in.
      Height = 5.0 in.

(D(1,1), L(1,1))   (D(1,2), L(1,2))

(D(2,1), L(2,1))   (D(2,2), L(2,2))

(D(5,1), L(5,1))   (D(5,2), L(5,2))

The coordinates of center of volume: (2.50000, _____, _____)

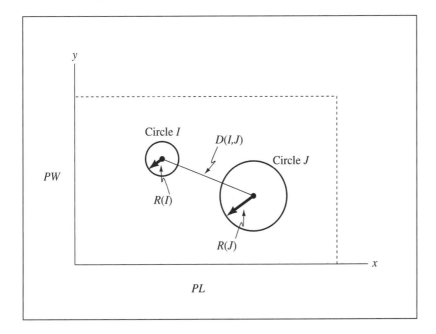

There are up to 100 circular holes randomly distributed in a rectangular plate, each with its own radius and *x, y* coordinates that will be asked for interactively during the program. The program will check if *overlaps* occur between holes or hole and boundary. If overlaps occur then the program will print a message indicating kind of overlap. Otherwise, calculate the *centroidal coordinates* of the plate. The plate has *length PL* and *width PW* that will be asked for interactively.

**Procedure**

a) Input data; Echo
b) Check *overlap* with *boundary*. (Print out which holes have overlap with boundaries.)

$X(I) + R(I) > X_{max}$    $X(I) - R(I) < 0$
$Y(I) + R(I) > Y_{max}$    $Y(I) - R(I) < 0$

c) Check *overlay* between *circles*. (Print out which holes overlap with each other.)

$D(I,J) < R(I) + R(J)$

d) Calculate area and first moments about X and Y axes.
e) Calculate the centroidal coordinates.
f) Print out in the following format:

Length = _____    Width = _____

| Hole | XCOOR | YCOOR | RADIUS |
|------|-------|-------|--------|
| 1 | | | |
| 2 | ↓ | ↓ | ↓ |
| ⋮ | | | |

Area of plate is _____.

The centroidal coordinates are

XCOOR = _____

YCOOR = _____

Use your program for the following cases:

**Case 1**

PL = 50.000
PW = 30.000

| X(I) | Y(I) | R(I) |
|------|------|------|
| 13.00 | 21.00 | 4.00 |
| 45.00 | 24.00 | 2.00 |
| 38.00 | 25.00 | 4.00 |
| 5.00 | 23.00 | 3.00 |
| 25.00 | 2.00 | 1.00 |

**Case 2**

PL = 28.00
PW = 14.00

| X(I) | Y(I) | R(I) |
|------|------|------|
| 5.0 | 12.0 | 3.0 |
| 18.5 | 2.0 | 1.0 |
| 25.0 | 13.0 | 0.5 |
| 16.0 | 10.0 | 2.0 |
| 22.0 | 12.0 | 3.0 |
| 18.0 | 3.0 | 2.0 |

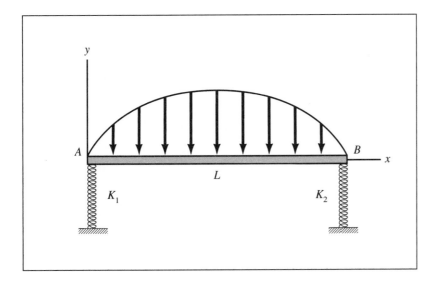

A nonuniform force distribution acts on a rigid beam *AB* supported by two different nonlinear springs. Compute the slope of beam *AB* for the following conditions.

Force:
$$-\left[10\sin\left(2\pi\frac{X}{L}\right)+15\log_{10}\left(\frac{X}{5}+1\right)\right]e^{-X}\ \frac{\text{N}}{\text{m}}$$

Spring 1:   $K_1 = |Y|^{3/2}\times 5\times 10^4\ \dfrac{\text{N}}{\text{mm}}$

Spring 2:   $K_2 = |Y|^{4/3}\times 10^4\ \dfrac{\text{N}}{\text{mm}}$

The length of the beam *L* will be inputted interactively.

Work out for $L = 2.0$ m and $L = 2.5$ m

In your output give:

> Length of beam
> $Y_1$ and $Y_2$ in mm
> Slope of AB in radians and in degrees.

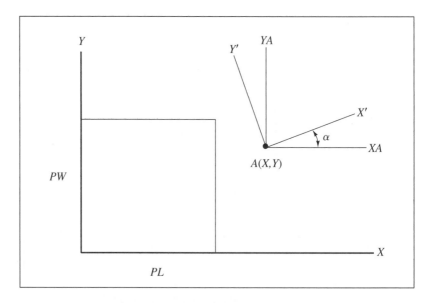

For *any set* of values given interactively.

Length: *PL*
Width: *PW*
Position: *A(X,Y)*
Angle: $\alpha$

Find at *A* and for the following data

$I_{X'X'}, I_{Y'Y'}, I_{X'Y'}$:     Second moments of area (cross-hatched).
$I'_1, I'_2$:               Principal second moments of area.

**Note:** Point *A* can be inside area.

Work out for the *following data*:

$$PL = 18.00 \text{ mm}$$
$$PW = 12.00 \text{ mm}$$
$$\alpha = 22.5°$$
$$(X,Y)_A = 15.00 \text{ mm}, -8.00 \text{ mm}$$

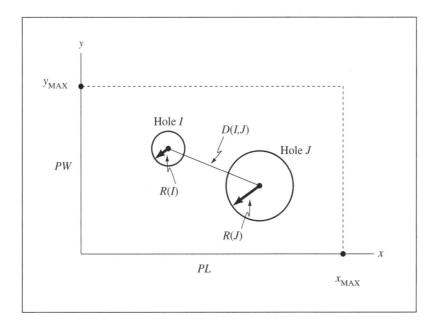

There are up to 100 circular holes randomly distributed in a rectangular plate. Each hose has its own radius and $x, y$ coordinates that will be asked for interactively during the program. The program will check if *overlaps* occur between *holes* or *hole* and *boundary*. If overlaps occur then the program will print a message. Otherwise, the program calculates $I_{XX}$, $I_{YY}$, $I_{XY}$ for area of the plate surface.

## Procedure

a) Input. Specify PL, PW, X(I), Y(I), R(I), and N (the number of holes).
b) Check overlap with boundary

For $1 \leq I \leq N$

If $\begin{cases} X(I) + R(I) > PL \\ X(I) - R(I) < 0 \\ Y(I) + R(I) > PW \\ Y(I) - R(I) < 0 \end{cases}$

*Then* print out: "Circle I has overlap with boundary."

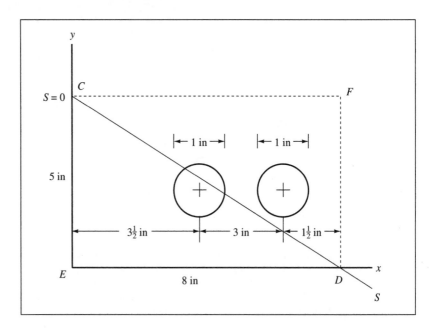

At what point $S$ along $CD$ do you get the maximum second moment of area for area $CEDF$ having 2 holes. What is its value? (Even if you know intuitively where this point is, show this via the computer.)

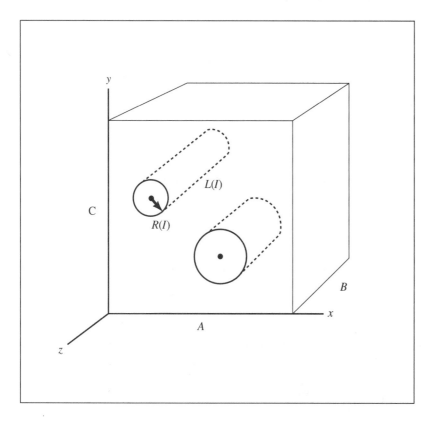

User selects dimensions *A*, *B*, *C* for a block with a random system of holes having

$$X(I)$$
$$Y(I)$$
$$R(I)$$
$$L(I)$$

Check for overlap of holes with boundary. If so print out which holes overlap with boundary. Check for overlap of holes with each other. If so print out which holes overlap with each other. Compute $I_{xx}$, $I_{yy}$, and $I_{xy}$ for unit density of block. Run program for the following cases.

**Case 1**

A = 50        B = 60        C = 70

| Hole | X(I) | Y(I) | L(I) | R(I) |
|------|------|------|------|------|
| 1 | 12.00 | 18.00 | 29.00 | 5.50 |
| 2 | 27.00 | 32.00 | 3.00 | 5.10 |
| 3 | 35.00 | 41.00 | 41.00 | 1.70 |
| 4 | 20.00 | 20.00 | 23.00 | 0.10 |

**Case 2**

A = 20        B = 30        C = 40

| Hole | X(I) | Y(I) | L(I) | R(I) |
|------|------|------|------|------|
| 1 | 4.00 | 4.00 | 13.00 | 2.00 |
| 2 | 4.00 | 15.00 | 23.00 | 2.50 |
| 3 | 14.00 | 3.00 | 24.00 | 1.20 |
| 4 | 13.00 | 19.00 | 1.00 | 1.10 |
| 5 | 19.00 | 25.00 | 28.00 | 3.00 |
| 6 | 19.00 | 26.00 | 1.00 | 0.10 |
| 7 | 10.00 | 15.00 | 30.00 | 0.30 |
| 8 | 2.00 | 15.00 | 1.00 | 0.10 |

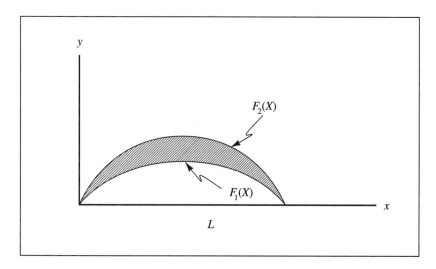

Write a program for computing the second moments and products of area $I_{xx}$, $I_{yy}$, $I_{xy}$ for the cross-hatched area. User inputs $L$.

The functions describing the curves are:

$$F_1(X) = X\left(\frac{X - L}{2}\right)\left(\sinh\frac{X}{L}\right)e^{-X/L}$$

$$F_2(X) = X\left(\frac{L - X}{L^2}\right)\cosh\left(\frac{L - X}{L}\right)$$

Run problem for $L = 1$, $L = 50$.

# ANSWERS

**2.2.** $F = 38.5$ N @ $66.6°$ from $x$ axis.

**2.4.** $L = 2.75$ km.

**2.6.** $B = 17.32$ N $\quad \alpha = 60°$

**2.8.** $F_A = 100$ N @ $-120°$ with horizontal.
$F_B = 76.53$ N @ $-67.5°$ with horizontal.
$F_C = 76.53$ N @ $-22.5°$ with horizontal.

**2.10.** $F = 846$ N @ $17.68°$ with horizontal.

**2.12.** $T_{AC} = 767.2$ N $\quad \alpha = 36.8°$

**2.14.** $F = 1,206$ N.

**2.16.** $F = 137.5$ N @ $43.34°$ from $x$ direction.

**2.18.** $F = 242$ lb @ $3.07°$ from $x$ direction.

**2.22.** $F_{slot} = 36.4$ lbf; $F_{vertical} = 81.5$ lbf.

**2.24.** $F_{BC} = .630F \quad F_{vert} = .590F$.

**2.26.** $2,690$ lb $803.8$ lb.

**2.28.** $F_{AC} = 707$ N $\quad \alpha = 90°$.

**2.30.** $F = 215.9i + 1968j + 3151k$ N.

**2.32.** $F = 37.4$ lb $\quad l = .267 \quad m = .535 \quad n = -.802$

**2.34.** $F_1 + F_2 = 918.6i - 1581j - 835.8k$ N.

**2.36.** $F_x = 400$ N $\quad F_y = -1007$ N.

**2.40.** $F = 25.7i + 24.7j + 16k$ lb.

**2.42.** $A = \pm 5\sqrt{2}\,i \mp 5\sqrt{2}\,k$.

**2.44.** $f = .465i + .814j + .349k$.
$F = 46.5i + 81.4j + 34.9k$ N.

**2.46.** $-164 \quad -.465 \quad -10.5$.

**2.48.** $D = 10i - .769j - 3.77k$.

**2.52** $.418$ ft.

**2.54.** $37.05$ N.

**2.56.** $A = 2.5$ N $\quad \alpha = 45.7°$

**2.58.** $-75$ ft$^2$ $\quad 95.94°$.

**2.60.** $-28.83$ N.

**2.62.** $47.51°$.

**2.64.** $(18i + 20j - 42k) \quad 47$.

**2.68.** $.804i + .465j + .372k$

**2.70.** $640$ m$^2$.

**2.74.** $-29.6$.

**2.76.** (a) $-43i + 49j + 2k$.
(b) $-136$.
(c) $-136$.

**2.78.** $L_{DT} = 373$ miles $\quad L_{TC} = 689$ miles
$162$ miles longer.

**2.80.** $F_x = 57.1$ N $\quad F_y = 342.8$ N $\quad F_z = 971.4$ N.

**2.82.** $F = 231.3$ N.

**2.84.** $32.4$ ft-lb.

**2.86.** $112.4°$ $\quad -190.72$ lb.

**2.88.** $35.7$ kN.

**2.90.** $19.87$ m.

**2.92.** $F = -80.1 \times 10^{-13}k$ N.

**2.94.** $251$ ft$^2$.

**2.96.** $18.21$m $\quad (l, m, n) = (.5145, .6860, .5145)$.

**2.98.** $T_{AB} = 500$ N $\quad T_{AC} = 866$ N $\quad \alpha = 30°$.

**2.100** $15.90°$.

**3.2.** $4i - 16j - 3k$ ft.

**3.4.** $6i + 7.16j + 7.598k$ m.

**3.6.** $-1067.4$ N-m $\quad 1959.3$ N-m.

**3.8.** $\pm 2\sqrt{2z}\,j + zk$.

**3.10.** $-18,026$ N-m $\quad -6,824.8$ N-m.

**3.12.** $13.8$ m.

**3.14.** $-257.5k$ N-m.

**3.16.** $8944$ ft-lb $\quad 4472$ ft-lb $\quad 0$ ft-lb.

**3.18.** $180i - 50k$ kN-m.
$30i + 75j - 50k$ kN-m.
$-1,551.8i + 75j + 389.4k$ kN-m.

**3.20.** $-84i + 94j - 46k$ N-m.

**3.22.** $M_A = \frac{10}{\sqrt{3}}ak - \frac{10}{\sqrt{3}}aj$ lb-ft.
$M_D = 0$ lb-ft.
$M_I = -\frac{10}{\sqrt{3}}ai + \frac{10}{\sqrt{3}}aj$ lb-ft
etc.

**3.24.** $(7.277F_{BA} + 8.575F_{CD} - 10,000)i +$
$(2.911F_{BA} - 5.145F_{CD})j +$
$(-2.911F_{BA} + 5.145F_{CD})k$ ft-lb.

**3.26.** $1.3296F$ ft-lb.

**3.28.** $22.295$ N-m $\quad M_D = -30i + 100j - 36k$ N-m.

**3.30.** $-146.2$ lb $\quad 77.00$ lb-ft.

**3.32.** $5,769$ lb-ft.

**3.34.** $-113.5$ kN-m $\quad -117.5$ kN-m.

**3.38.** $300$ lb-ft $\quad 600$ lb-ft $\quad 1,200$ lb-ft.

**3.40.** $175$ N.

**3.42.** $-7,270k$ lb-ft $\quad 10,405k$ lb-ft.

**3.44.** $M_A = M_P = -261i - 261j$ N-m.

**3.46.** $420$ ft-lb.

**3.48.** $C = 100i + 50j - 224k$ lb-ft.

**3.50.** $-1,857$ N-m.

**3.52.** $M_P = 48i - 36j - 225.6k$ N-m.
$M_E = -146.9$ N-m.

**3.54.** $M = 3,635$ lb-in @ $68.2°$ to horizontal.

**3.56.** $C = 35.35i + 22.36j + 80.07k$ N-m.

**3.58.** $120$ lb-ft $\quad 160$ lb-ft $\quad 60$ lb.

**3.60.** $1,750$ N-m.

**3.62.** $408.4$ N-m.

**3.64.** $1,179$ ft-lb.

**3.66.** $-452k$ lb-ft $\quad -1,202k$ lb-ft.

| | |
|---|---|
| **3.68.** | $10,600i + 4,500j - 5,000k$ m. |
| | $12,554$ m. |
| **3.70.** | $-27.2$ N-m. |
| **3.72.** | $277.6$ N-m. |
| | |
| **4.2.** | $F = 15$ kN    $C = 30$ kN-m C.C.W. |
| **4.4.** | $F = 150$ N    $C = 187.5$ N-m. |
| **4.6.** | $M_A = 8k$ kN-m    $F_A = F_B = -10j$kN. |
| | $M_B = 22.72k$ kN-m. |
| **4.8.** | Move 200 lb force 5 ft to left. |
| **4.10.** | $F_A = -200j - 150k$ N. |
| | $C_A = 34.0i + 26.0k$ N-m. |
| **4.12.** | $F = 20i - 60j + 30k$ N. |
| | $C_A = 900i + 680j + 760k$ N-m. |
| **4.14.** | $12.375$ m from origin. |
| **4.16.** | $F = -44.567i - 33.425j - 22.283k$ N. |
| | $C = -31.007i - 160.46j + 302.71k$ kN-m. |
| **4.18.** | $F_R = -204i + 604j + 408k$ lb. |
| | $C_A = 8,160i + 2,032j + 3,264k$ ft-lb. |
| | $C_B = 2,032j - 816k$ ft-lb. |
| **4.20.** | $F_R = 400i - 1,900j + 600k$ N. |
| | $C_R = 24,200i + 2,000j - 3,300k$ N-m. |
| **4.22.** | $F_R = -50k$ kN    $C_R = 7.5i + 50j$ kN-m. |
| **4.24.** | $F_1 = 26.9$ lb    $F_2 = 17.6$ lb    $F_3 = 35.5$ lb. |
| **4.26.** | $F = 2,100i - 500j$ lb    $\bar{x} = -.05$ ft. |
| **4.28.** | $F_R = 400i - 1600j$ N |
| | $C_R = -1,800i - 2,400j - 12,600k$ N-m. |
| **4.30.** | $F_R = 39.44i + 74.7j$ N. |
| | $\bar{x} = .669$ m. |
| **4.32.** | $F_R = -100k$ N    $\bar{x} = 2.5$ m    $\bar{y} = 2.2$ m. |
| | $F_R = 0$    $C_R = 280i + 450j$ N-m. |
| **4.34.** | $F_R = 0$    $C_k = 860i + 900j$ N-cm. |
| **4.36.** | $F_R = 353.5i - 653.5j$ N    $\bar{x} = 9.2739$ m. |
| **4.38.** | $F_R = 43,000$ lb    $\bar{x} = 23.209$ ft. |
| **4.42.** | $F = 3,164k$ lb. |
| **4.44.** | $\bar{x} = 10.37$ in.    $\bar{y} = \bar{z} = 0$. |
| **4.48.** | $\bar{x} = .289$    $\bar{y} = .400$. |
| **4.50.** | $10.84$ m. |
| **4.52.** | $\bar{x} = 1.292$ ft    $\bar{y} = 7.63$ ft. |
| **4.54.** | $\bar{x} = .210$ ft    $\bar{y} = .0756$ ft. |
| **4.56.** | $\bar{y} = 3.808$ m. |
| **4.58.** | $F = -37.5k$ kN    $\bar{x} = .844$ m    $\bar{y} = 1.067$ m |
| | (origin at front left lower corner of the load). |
| | original weight 90 kN. |
| | lost weight = 52.5 kN. |
| **4.60.** | $F_R = -111,780k$ lb    $\bar{x} = -3.78$ ft |
| | (in front of trailer C.G.). |
| | $F_R = -86,386k$ lb    $\bar{x} = -7.329$ ft. |

| | |
|---|---|
| **4.62.** | $W = .02739\gamma.$    $x_c = .2042$ m    $y_c = .1$ m |
| | $z_c = -.1968$ m. |
| **4.64.** | $F_R = 2.315 \times 10^6$ lb    $\bar{x} = 20$ ft    $\bar{y} = 12.44$ ft. |
| **4.66.** | $F_R = 262$ N. |
| **4.68.** | $4,025$ N-m    $596.3$ N-m. |
| **4.70.** | $p_a = 7.8644 \times 10^4$ Pa. |
| **4.72.** | $F = 61,740$ lb    $10.386$ ft. |
| **4.74.** | $F = 8761$ lb    $2.615$ ft. |
| **4.76.** | $F = 1.697 \times 10^{10}$ N    $24.04$ m from base along in- |
| | clined surface. |
| **4.78.** | $F = 666 \times 10^8$ N. |
| | @ $\frac{40}{3}$ m from base. |
| **4.80.** | $\bar{x} = 21.2$ ft    $\bar{y} = 4.67$ ft. |
| **4.82.** | $x_c = .07441$ m. |
| **4.84.** | $F = -400j$ kN    $M_{\text{outside lane}} = -900k$ kN-m |
| | $M_{\text{inside lane}} = -300k$ kN-m. |
| **4.86.** | $1,415$ kN    $1.915$ m from front of load. |
| **4.88.** | $F_R = 4.698i - 10.09j$ kN. |
| | $M_R = -3i + 3.2316k$ kN-m. |
| **4.90.** | $r' = 17.57x^2 - 7.027x + .7026$. |
| **4.92.** | $l = 3.51$ m. |
| **4.94.** | $F = 2,200$ lb    $\bar{x} = 4.136$ ft |
| **4.96.** | $-217.2$ N-m. |
| **4.98.** | $F_R = 500i - 1,220j$ N |
| | $C_R = 240i - 2,190k$ N-m. |
| **4.100.** | $F_R = 58,900$ lb    $\bar{x} = 3.919$ from bottom left corner. |
| **4.102.** | $\bar{x} = 1.20$ ft    $\bar{y} = 1.125$ ft    $\bar{z} = \frac{t}{2}$ ft. |
| **4.104.** | $\bar{y} = 1.299$ ft. |
| **4.106.** | $-1,550$ lb    $\bar{x} = 20.04$ ft. |
| **4.108.** | $\bar{x} = 10.94$ ft    $\bar{y} = 12.67$ ft. |
| **4.110.** | $485.397$ kN @ $1.636$ m below top of gate. |
| | $132.381$ kN @ 1 m below top of gate. |
| **4.112.** | $\bar{x} = 4.98$ m    $\bar{y} = 2.65$ m. |
| | |
| **5.18.** | $T_{BC} = 26.79$ lb    $T_{BA} = 37.9$ lb. |
| **5.20.** | $T_{AB} = 111.8$ lb    $T_{BD} = 109.9$ lb    $T_{BC} = 63.7$ lb. |
| **5.22.** | $27$ lb    $36.35$ lb    $4.653$ in. |
| **5.24.** | $671$ N,    $447$ N,    $1,000$ N. |
| **5.26.** | $409.8$ N-m. |
| **5.28.** | $A_x = 4,233$ N,    $A_y = 3,140$ N,    $B_x = 4,233$ N, |
| | $B_y = -1,840$ N. |
| **5.30.** | $A_x = 2,000$ N,    $A_y = 6,970$ N,    $M_A = 62,360$ N-m |
| | $B = 530.4$ N. |
| **5.32.** | $A_x = 8.00$ kN    $A_y = -12.50$ kN    $B_x = -8.00$ kN |
| | $B_y = 22.5$ kN. |
| **5.34.** | $F_{BD} = 462$ lb    $F_{BC} = 215$ lb    $F_{AB} = 375$ lb. |
| **5.36.** | $T_A = 125.0$ N    $f = 125$ N. |
| **5.38.** | $T = 117.4$ N    $\alpha = 79.8°$ |
| **5.40.** | $A_x = 433$ N,    $A_y = 790$ N    $M_A = 1,167$ N-m. |

**5.42.** $A_x = 200$ N, $A_y = 3,464$ N $M_A = 1,192$ N-m.

**5.44.** $A_x = 9,736$ N $A_y = -2,670$ N $B_x = -9,736$ N
$B_y = 6,670$ N.

**5.46.** $\alpha = 17.46°$.

**5.48.** 58.7 ft-lb.

**5.50.** $\beta = \tan^{-1}\left[\dfrac{\dfrac{W_2}{\tan \alpha_1} - \dfrac{W_1}{\tan \alpha_2}}{W_1 + W_2}\right]$.

**5.52.** $A_x = 408$ N $A_y = -258$N $G_x = -749.5$ N
$G_y = 583$ N.

**5.54.** $T = 20$ N-m.

**5.56.** $T_{AB} = 40.5$ kN $T_{BC} = 39.3$ kN.

**5.58.** 72 lb 447 lb.

**5.60.** 4 m.

**5.62.** $F = \dfrac{W}{2}\dfrac{r_1 - r_2}{r_2}$

**5.64.** $R_x = 0$ $R_y = 58,900$ lb $M_B = 142,500$ ft-lb.

**5.66.** $R_x = 0$, $R_y = 4,953$ lb $M = 61,082.5$ ft-lb.

**5.68.** $R_C = 1,006$ lb $M_C = 10,108$ ft-lb.
$R_D = 294$ lb $M_D = 2,770$ ft-lb.

**5.70.** 270 lb 253 lb.

**5.72.** $W_{max} = 2.5$ ton $B_y = 10.8$ ton $A_y = 1.7$ ton
$C_y = 1.7$ ton $M_c = 34$ ft-ton $D_y = 1.7$ ton
$M_D = 85$ ft-ton.

**5.74.** 17.55 ton (2 supports) 24.9 ton $M = 70$ ton-ft.

**5.76.** $F_1 = 293$ N $F_2 = 52.37$ N $F_3 = 254.3$ N.

**5.78.** $A = -500j + 1,500k$ lb
$M_A = 26,260i - 9,100k$ ft-lb
$B = -500j$ lb
$M_B = -9,100k$ ft-lb.

**5.80.** $F = 57.14$ lb $A_y = 160.1$ lb
$B_y = 93.0$ lb $B_z = 22.6$ lb $A_z = 18.08$ lb.

**5.82.** $A_x = 0$ $A_y = 50$ N $A_z = 100$ N
$M_x = -200$ N-m.
$M_y = -220$ N-m $M_z = 110$ N-m.

**5.84.** $B_y = 3,000$ lb $B_z = 26,500$ lb $C_z = 17,000$ lb.

**5.86.** 3 ft $A_z = 50$ N $B_z = 50$ N.

**5.88.** $P = 33.7$ lb $A_z = 25$ lb $A_y = -7.50$ lb.

**5.92.** 11,394 lb.

**5.94.** 65,300 lb.

**5.96.** $F_{EC} = 4.36$ kN $A_x = .800$ kN $A_y = 196.7$ kN.

**5.98.** $B_x = 165.9$ N $B_y = 206.3$ N $F_{CB} = 208.8$ N.

**5.100.** $C_x = 576$ lb $C_y = 1,326$ lb $M_c = 8009.6$ N-m.

**5.102.** $T_1 = 209$ N $T_2 = 202$ N $F = 286$ N.

**5.104.** $A = 10,640$ lb $C = 6,520$ lb $D = 6,520$ lb
$G = H = 7,500$ lb $F = 10,640$ lb.

**5.106.** 1154 N

**5.108.** $9.418 \times 10^6$ Pa.

**5.110.** $\tau = (6.833 - 6.50z - \frac{z^3}{3}) \times 10^4$ Pa.

**5.112.** $\tau_{zz} = 281.7$ psi

**5.114.** $G_x = 750$ lb $G_y = 650$ lb.

**5.116.** $F_x = A_x = 806$ lb $F_y = A_y = 500$ lb.

**5.118.** $R_1 = 41$ kN $R_2 = 88$ kN $R_3 = 141$ kN.

**5.120.** $F_{CD} = 1,694$ lb $F_{EF} = 4,931$ lb.

**5.122.** $A_x = 13.15$ lb $A_y = 142.6$ lb.

**5.124.** $T_E = 1,688$ N $T_D = 2,840$ N.

**5.126.** $T = 30.67$ lb.

**5.128.** $C_x = -322.3$ lb $C_y = -161.3$ lb $C_z = -583.6$ lb
$D_x = -322.7$ lb $D_z = -398.9$ lb.

**5.130.** $T = 2,121$ lb $H = 3,000$ lb $V = 3,000$ lb
$T = 1,148$ lb $H = 2,121$ lb $V = 5,121$ lb.

**5.132.** $T_{DB} = 353$ lb $T_{AC} = 289$ lb $T_{EC} = 394$ lb.

**5.134.** $T_R = 1.928$ N $T_P = 709$ N $T_A = 2,129$ N
$T_D = 1,012$ N $l = .0828$ $m = .728$ $n = .618$.

**5.136.** $F_{AB} = 8,000$ lb $E_x = 5,000$ lb $E_y = 6,928$ lb
$M_E = 5,000$ ft-lb.

**5.138.** $A_x = -3,984$ N $A_y = -7,800$ N $B_x = 3,984$ N
$B_y = 4,800$ N.

**5.140.** $A_x = -925$ N $A_y = 150$ N.

**5.142.** 63.75 kN-m.

**5.144.** $F = 996.7$ N.

**5.146.** $A_y = B_y = 18,170$ N $A_x = 63,660$ N.

**5.148.** $C_x = 600$ lb $C_y = 0$ $C_z = -320$ lb
$F_D = 1,033$ lb $F_A = 312.5$ lb.

**5.150.** $p = 208$ psi gage.

**5.152.** $A = 433i - 533j - 19.00k$ lb $E = 240$ lb.
$H = -433i + 433j - 520k$ lb $K = 0$.

**5.154.** $A_x = -28.13$ lb $A_y = -98$ lb $B_x = 70$ lb
$B_y = 0$.

**5.156.** $f = 15.08$ kN.

**5.158.** $A_x = 0$ $A_y = 1,938$ lb $C_y = B_y = 3,062$ lb
$D_y = 1,938$ lb. Ram force $= 9,798$ lb.

**5.160.** $T_{EF} = 1,174$ N.

**5.162.** $A_x = 1.399 \times 10^4$ N $A_y = 5.549 \times 10^3$ N.
$M_A = 1.355 \times 10^4$ N-m $B = 2,887$ N.

**6.2.** $EF = 2,000$ lb T $AF = 4,808$ lb T
$BC = 2,670$ lb. C etc.

**6.4.** $DE = 18,720$ lb C $DH = 8,825$ T
$HG = 18,720$ lb T etc.

**6.6.** $AB = DE = 26.8$ kN C $GC = CF = 3.39$ kN C
$AG = FE = 19$ kN T $BC = CD = 20$ kN C.
$BG = DF = 3.14$ kN T $GF = 21.36$ kN C.

**6.8.** $AB = 707$ lb C
$AC = CE = 500$ lb T
$EB = 0$
$BD = 500$ lb C
$DF = 707$ lb C.

**6.10.** $AB = 6,950$ lb C
$AC = 4,914$ lb T
$BC = 3,207$ lb T
$CD = 0$.

**6.12.** $AB = 353.5$ kN C
$AL = 250$ kN T
$LK = 250$ kN T
$LB = 100$ kN T
$BC = 316.2$ kN T
$BK = 70.8$ kN T.

**6.14.** $JI = 25$ kN T $\quad BH = 8.64$ kN C
$AJ = 31.25$ kN C $\quad BC = 50$ kN C
$IH = 56.25$ kN T $\quad CD = 50$ kN C
$BJ = 6.25$ kN T $\quad HC = 0$
etc.

**6.16.** $CD = 583$ lb C
$AC = BC = AD = 0$
$BD = 1,158$ lb T
others are zero.

**6.18.** $AC = 14.14$ kN T
$CD = 14$ kN T
$DE = 2$ kN C
$AD = 24.5$ kN C
others are zero.

**6.20.** $F_{AB} = 24,246$ lb C, $F_{BC} = 32,709.5$ lb T
$F_{DC} = 1,350$ lb C, $F_{DF} = 22,709.5$ lb C.

**6.22.** $DC = 1,202$ lb T $\quad DE = 0$.

**6.24.** $GF = 21.3$ kN T.

**6.26.** $BF = 0 \quad AB = 258$ kN T.

**6.28.** $BC = 115.5$ kN T $\quad BK = 57.7$ kN C
$DE = 77.0$ kN T $\quad DI = 38.50$ kN T
$EF = 38.50$ kN T.

**6.30.** $DG = 90$ kN C $\quad DF = 56.6$ kN C
$AB = 127.3$ kN T $\quad AC = 90$ kN C
$CB = 50$ kN C $\quad CD = 90$ kN C.

**6.32.** $FH = HE = 4.72$ K T $\quad FE = 2.5$ K C $\quad FC = 0$ K.

**6.34.** $FI = .558$ K C $\quad EF = .5$ K C $\quad DH = 1.118$ K T.

**6.36.** $F_{LC} = 41.63$ K C $\quad F_{KL} = 162.5$ K T
$F_{HC} = 83.38$ K T $\quad F_{FG} = 50.03$ K C.

**6.38.** $M = -500x + 2,500$ ft-lb
$M = -15x^2 - 200x + 1,000$ ft-lb.

**6.40.** $V = -737.5$ lb $\quad M = 3,690$ ft-lb
$V = 262.5$ lb $\quad M = 6,850$ ft-lb
$V = 262.5$ lb $\quad M = 1,312.5$ ft-lb.

**6.42.** $\underline{0 < s < 14.14} \quad V = -37.1$ lb $\quad H = -37.1$ lb
$M = 37.1s$ ft-lb
$\underline{14.14 < s \leq 29.14} \quad V = -52.5$ lb $\quad H = 0$
$M = 52.5s - 217$ ft-lb.
etc.

**6.44.** $\underline{0 \leq x < 10} \quad H = 0 \quad V_x = 0 \quad V_y = -20s$
$M_y = 0 \quad M_z = 0 \quad M_x = -10s^2$
$\underline{10 < s < 20} \quad H = 0 \quad V_y = 200$ N $\quad V_z = 0$
$M_x = 1,000$ N-m $\quad M_z = 200(s - 10) \quad M_y = 0$
$\underline{20 < s < 30} \quad H = 200$ N $\quad V_z = -1,000$ N
$V_x = 0 \quad M_y = 0 \quad M_x = -1,000s + 21,000$ N-m
$M_z = 2,000$ N-m.

**6.46.** $\underline{0 < \theta < \frac{\pi}{4}} \quad V = -70.7 \sin \theta$ N
$H = -70.7 \cos \theta$ N
$M = 424 - 424 \cos \theta$ N-m
$\underline{\frac{\pi}{4} < \theta < \frac{\pi}{2}} \quad V = 29.3 \sin \theta$ N
$H = 29.3 \cos \theta$ N
$M = 176 \cos \theta$ N-m.

**6.48.** $\underline{\text{Section } AB} \quad H = 0 \quad V_x = 0 \quad V_z = 11.36s - 262$ lb
$M_y = 410$ ft-lb $\quad M_z = 0$
$M_x = -5.68s^2 + 262s - 2,194$ ft-lb
$\underline{\text{Section } BC} \quad V = 130.02 + 5.64s$ lb
$H = 227.54 - 9.87s$ lb
$M_{x'} \equiv (\text{axial torque}) = -70.47$ ft-lb
$M_{z'} \equiv (\text{bending torque}) = -123.33$ ft-lb
$\underline{\text{Section } CD} \quad V_z = -262 + 11.36s$ lb
$M_y \equiv (\text{axial torque}) = -.477$ ft-lb
$M_x = -3,021 + 262s - 5.58s^2$ ft-lb.

**6.50.** $\underline{0 < x < 15}$
$V = M = 0$
$\underline{5 < x \leq 10}$
$V = 500$ lb
$M = -500(x - 5)$ ft-lb
$\underline{10 \leq x < 25}$
$M = -500(x - 5) - 30(x - 10)^2/2$.

**6.52.** $\underline{0 < x < 10}$
$V = -737.5$ lb
$M = 737.5x$ ft-lb
$\underline{10 < x < 25}$
$V = 262.5$ lb
$M = 737.5x - 1,000(x - 10)$
$\underline{25 < x < 10}$
$V = 262.5$ lb
$M = -262.5x + 10,500$ ft-lb.

**6.54.** $\underline{0 < x \leq 5}$
$V = -1.333$ N
$M = 1.333x$ N-m
$\underline{5 \leq x \leq 15}$
$V = -20x^2 + 600x - 3,833$ N
$M = 1,333x - 200(x - 5)^2 + \frac{20}{3}(x - 9)^3$ N-m
$\underline{15 \leq x < 25}$
$V = 667$ N
$M = 667x + 16,660$ N-m.

**6.56.** $-500$ N-m.

**6.58.** $M_{\max} = -5,334$ N-m.

**6.60.** $M_{max} = -6,500$ ft-lb.

**6.62.** $\underline{0 < x < 3}$
$V = -60 + 5x^2$ N
$M = 60x - \frac{5}{3}x^3$ N-m
$\underline{3 < x < 6}$
$V = -60 + 5x^2$ N
$M = 60x - \frac{5}{3}x^3 - 5,000$ N-m
$\underline{6 < x < 11}$
$V = -1,000$ N
$M = 1,000x - 16,000$ N-m
$M_{max} = 5,000$ N-m.

**6.64.** $\underline{0 < x < 5}$
$V = .596x^{3/2}$ kN
$M = -.238x^{5/2}$ kN-m
$\underline{5 < x < 10}$
$V = 6.66 - 3(x - 5) + \frac{3}{10}(x - 5)^2$ kN
$M = -6.66x + 19.99 + \frac{3}{2}(x - 5)^2 - \frac{1}{10}(x - 5)^3$
$M_{max} = -22.6$ kN-m

**6.66.** length = 230 ft  $T_{max} = 565.8$ lb.

**6.68.** $T_{max} = 17,088$ lb  $L = 81.5$ ft.

**6.70.** $T_{max} = 35,200$ N.

**6.72.** $T_{max} = 109$ lb  h = 118.1 ft.

**6.74.** elevation @ $B = 17.84$ m
61.68 m dragging.

**6.76.** $A_x = 675$ N  $B_x = 675$ N  $A_y = 650$ N
$B_y = 500$ N
$T_{AC} = 937.1$ N  $\alpha = 43.92°$.

**6.78.** $BC = 45.3$ kN C
$DC = 32$ kN C
$DE = 32$ kN C
$BA = 45.3$ kN T
$DB = BE = AE = 0$.

**6.80.** $AE = 1,250$ lb T
$AD = 1,250$ lb T
$ED = 1,030$ lb C
$CE = 1,060$ lb C
$CA = 1,060$ lb C
$CD = 1,500$ lb T.

**6.82.** $\underline{0 \leq x < 3}$
$V = 10x$ N
$M = -\frac{10x^2}{2}$ N-m
$\underline{3 < x < 9}$
$V = 10x - 19.44$ N
$M = -10\frac{x^2}{2} + 19.44(x - 3)$ N-m
$\underline{9 \leq x < 12}$
$V = 70.6$ N
$M = -90(x - 4.5) + 19.44(x - 3) + 500$ N-m.

**6.84.** $\underline{0 \leq x < 10}$
$V = M = 0$
$\underline{10 < x < 20}$
$V = -650$ lb
$M = 650(x - 10)$ ft-lb
$\underline{20 < x < 40}$
$V = 350$ lb
$M = 650(x - 10) - 1,000(x - 20)$ ft-lb
$\underline{40 < x < 65}$
$V = 0$
$M = -500$ ft-lb.

**6.86.** $y = h(1 - \cos\frac{\pi x}{l})$.

**6.88.** $DE = 10,928$ N T  $CE = 8,000$ N C
$DC = 0$  $BD = 10,928$ N T  $BA = 4,000$ N T.

**6.90.** $\underline{3 < x \leq 6}$
$V = 1.110$ kN
$M = -1.110x + 50$ kN-m

$\underline{6 \leq x \leq 14}$
$V = 1.110 + \frac{(x - 6)^2}{20}$ kN

$M = 50 - 1.110x - \frac{(x - 6)^3}{60}$ kN-m

$\underline{14 \leq x < 20}$
$V = 4.31$ kN
$M = 86.27 - 4.31x$ kN-m.

**7.4.** 600 lb.

**7.6.** 151.0 lb  139.0 lb.

**7.8.** $\mu_s = 0.322$.

**7.10.** $F = 52.76$ N.

**7.14.** $(\mu_s)_{max} = 0.249$.

**7.16.** at start $F = 4,142$ lb
for movement $F = 4,414$.

**7.18.** $P = 733.6$ N.

**7.20.** $f = 115.5$ N
$\mu_{min} = .578$.

**7.24.** $C$ moves 1.5 m to left.

**7.26.** $T = 123.8$ ft-lb.

**7.28.** 750 N (clockwise)
50 N (counterclockwise).

**7.30.** $F = 196.7$ N.

**7.34.** 32.68 lb $< W_1 <$ 67.32 lb.
$W_1 = 50$ lb (no friction).

**7.36.** $\mu_{min} = 0.225$.

**7.38.** $\alpha_{min} = 28.0°$.

**7.40.** 4.76 ft.

**7.40.** 1.691 m.

**7.44.** 251.5 lb.

**7.46.** $W_2 = 7.245$ N  $\mu_{min} = 0.0279$.

**7.48.** $T = \dfrac{P\mu}{3}\left[\dfrac{D_2^3 - D_1^3}{D_2^2 - D_1^2}\right].$

**7.50.** $T = \dfrac{PD\mu_d}{4}.$

**7.52.** $T = 320$ N.

**7.54.** $T = 188.6$ N-m.

**7.56.** 832 N.

**7.58.** 267 lb.

**7.60.** 136.8 N.

**7.62.** $\mu_s = 0.260.$

**7.64.** $\mu_s = 0.048.$

**7.66.** $P = 202$ N    $\mu_5 = 0.556.$

**7.68.** $x = 277$ mm.

**7.70.** $F_x = 256$ N    $F_y = 150.4$ N.

**7.72.** $F_{min} = 2,210$ N    $\mu_s = 0.4.$

**7.78.** 133.6 in.-lb    155.6 in.-lb.

**7.80.** $W = 6,100$ lb.

**7.84.** $0.2387°.$

**7.86.** $\mu_s = .00308.$

**7.88.** $F = 128.2$ N.

**7.90.** $\theta = 30.96°$    $T = 292$ lb.

**7.92.** 7.11 N.

**7.94.** 3,333 N.

**7.96.** $P = 13.22$ lb C.W.    $P = 163.2$ lb C.C.W.

**7.98.** 40.7 N-m.

**7.100.** 94.7 mm.

**7.102.** 35.55° above horizontal.

**7.106.** $\theta = 41.9°.$

**7.108.** 256 N.

**7.110.** $1.002 \times 10^4$ N.

**7.112.** $W_2 = 429.18$ N.

**7.114.** $W_c = 810$ N.

**7.116.** $T = 877$ N-m.

**8.4.** $M_x = 125$ ft$^3$    $M_y = 200$ ft$^3$.

**8.8.** $\bar{x} = 14.30$ ft    $\bar{y} = 5.74$ ft.

**8.12.** $x_c = 0.1691$ m    $y_c = 0.363$ m.

**8.14.** $x_c = 2.02$ mm    $y_c = 83.6$ mm.

**8.16.** $x_c = 0.1217$ m    $y_c = 0.4478$ m.

**8.18.** $x_c = 7.23$ in.    $y_c = 2.55$ in.    $M_{z'} = 226.1$ in.$^3$

**8.20.** $x_c = y_c = 242.22$ mm.

**8.22.** $x_c = 1.7172$ m    $y_c = 0.1722$ m.

**8.24.** $x_c = 2.889$ m    $y_c = 2.667$ m.

**8.26.** $y_c = 315.7$ mm.

**8.28.** $x_c = 40.1$ mm    $y_c = 19.51$ mm.

**8.30.** $\bar{x} = 12.92$ ft.

**8.32.** $x_c = \frac{a}{2}$    $y_c = \frac{2}{3}b$    $z_c = \frac{4}{3}c.$

**8.34.** $\bar{z} = \frac{3}{8}a.$

**8.36.** $r_c = 1.179i + .955j + .284k$ m.

**8.38.** $x_c = 2.571$ m.

**8.40.** $\bar{x} = 8.22$ m    $\bar{y} = \bar{z} = 0.$

**8.42.** $x_c = y_c = 0$    $z_c = 9.78$ in.

**8.44.** center of volume $\bar{x} = -9.75$ mm
$\bar{y} = 122.55$ mm    $\bar{z} = 0$
center of mass and gravity
$\bar{x} = -32.76$ mm    $\bar{y} = 447$ mm    $\bar{z} = 0.$

**8.46.** $\bar{x} = 41.95$ mm    $\bar{y} = 493$ mm    $\bar{z} = 608$ mm.

**8.48.** $r_c = .742j - .1720k$ m.

**8.50.** $y_c = \frac{2}{3}\left(\frac{b}{\pi}\right)$

**8.52.** $A = 1,751$ in.$^2$    $V = 5,242$ in.$^3$

**8.54.** $A = .8624$ m$^2$    $V = .06329$ m$^3$.

**8.56.** $r_c = 3.39i + 3.44j - 3k$ in.

**8.58.** $I_{xx} = \dfrac{ab^3\pi}{4}.$

**8.60.** $I_{xx} = 1.113$ units$^4$    $I_{yy} = .535$ units$^4$.
$I_{xy} = .750$ units$^4$.

**8.62.** $I_{yy} = \frac{5}{2}\pi^2 - 8$ ft$^4$.

**8.66.** $I_{xx} = 666.7$ ft$^4$    $I_{yy} = 1,257$ ft$^4$.
$I_{xy} = 791.7$ ft$^4$.

**8.68.** $I_{xx} = 4,540$ ft$^4$    $I_{yy} = 11,250$ ft$^4$.

**8.70.** $I_{xx} = 6.646$ ft$^4$    $I_{yy} = 15.39$ ft$^4$.
$I_{xy} = 9.52$ ft$^4$.

**8.76.** $I_{xx} = 1,512$ ft$^4$    $I_{xy} = 559$ ft$^4$.

**8.78.** $I_{xx} = \frac{5\sqrt{3}}{16}R^4$    $I_{yy} = \sqrt{3}R^4\left(\frac{5}{16}\right)$    $I_{xy} = 0.$

**8.80.** $I_{x_c x_c} = 8.463 \times 10^7$ mm$^4$    $I_{y_c y_c} = 1.8869 \times 10^8$ mm$^4$.

**8.82.** $I_{x'x'} = 140.6$ ft$^4$    $I_{y'y'} = 439$ ft$^4$    $I_{x'y'} = 213$ ft$^4$.

**8.84.** $I_{xx} = 91.92$ ft$^4$    $I_{yy} = 1,455$ ft$^4$    $I_{xy} = -96.52$ ft$^4$.
$J_p = 1,547$ ft$^4$.

**8.88.** 109.2 mm$^4$    2,367 mm$^4$.

**8.90.** 1,299 mm$^4$    79.1 mm$^4$.

**8.94.** $I_1 = .003781$ m$^4$    $I_2 = .0002557$ m$^4$.

**8.96.** $\bar{x} = 3.16$ mm    $\bar{y} = 1.323$ mm.
$I_1 = 2.11 \times 10^4$ mm$^4$    $I_2 = 8.0675 \times 10^3$ mm$^4$.

**8.98.** $y_c = 1.671$ m.

**8.100.** $I_{xx} = 3.29 \times 10^7$ mm$^4$    $I_{x_c x_c} = 9.85 \times 10^6$ mm$^4$
$I_{yy} = 2.89 \times 10^6$ mm$^4$    $I_{y_c y_c} = 1.607 \times 10^6$ mm$^4$
$I_{xy} = 6.90 \times 10^6$ mm$^4$    $I_{x_c y_c} = 1.457c \times 10^6$ mm$^4$.

**8.102.** $x_c = \frac{5}{6}h$    $y_c = \frac{20}{21\pi}a$    $z_c = 0.$

**8.104.** 6.75° 96.75°.

**8.106.** $A = 151.24$ m$^3$    $V = 113.0$ m$^3$.

**8.108.** $x_c = 1.500$ in.    $I_{x_c x_c} = 83.250$ in$^4$.
$I_{y_c y_c} = 19.250$ in$^4$    $I_{x_c y_c} = 0.$

**8.110.** 12,708 m$^4$.

**8.112.** $M_z = 10.616s$ in$^3$
$M_x = 185.7 - 0.465y^2$
$M_x = (.1886)[316.8 + (4 + 3\eta)\sqrt{(\eta + 2)^3}.$

**9.2.** 4,333 lbm-ft$^2$  15,333 lbm-ft$^2$.

**9.4.** $I_{yy} = \frac{1}{12}M(a^2 + b^2)$  $I_{zz} = \frac{1}{12}M(b^2 + l^2)$
$I_{xx} = \frac{1}{12}M(a^2 + l^2)$.

**9.6.** $\frac{1}{2}Mr^2$.

**9.8.** $2.67 \times 10^7$ grams-mm$^2$.
$3.95 \times 10^8$ grams-mm$^2$.

**9.10.** $1.723 \times 10^6$ grams-mm$^2$.

**9.12.** 3,959 kg-mm$^2$.

**9.14.** $I_{xx} = 1.004 \times 10^5$ mm$^4$.
$I_{yy} = 1,875$ mm$^4$
$I_{xy} = 1.172 \times 10^4$ mm$^4$
$(I_{xx})_M = 205$ kg-mm$^2$
$(I_{yy})_M = 3.82$ kg-mm$^2$
$(I_{xy})_M = 23.9$ kg-mm$^2$.

**9.16.** 37.0 kg-mm$^2$  166.8 kg-mm$^2$  203.8 kg-mm$^2$.

**9.20.** $I_{y'y'} = 6.21$ slug-ft$^2$  $I_{z'z'} = I_{x'x'} = 29.0$ slug-ft$^2$
$I_{xx} = 119.1$ slug-ft$^2$  $I_{yy} = 18.63$ slug-ft$^2$
$I_{zz} = 106.6$ slug-ft$^2$.

**9.22.** $I_{xx} + I_{yy} + I_{zz} = \frac{M}{6}(a^2 + b^2 + c^2)$
$+ 2M(x^2 + y^2 + z^2)$

**9.24.** $I_{x''x''} = 3,870$ kg-m$^2$
$I_{y''y''} = 4,615$ kg-m$^2$.

**9.26.** $1.258 \times 10^{-2}$ kg-m$^2$  $1.239 \times 10^{-2}$ kg-m$^2$
$3.91 \times 10^{-3}$ kg-m$^2$.

**9.28.** 1.650 kg-m$^2$  1.438 kg-m$^2$  .513 kg-m$^2$.

**9.30.** $k = 2.54$ ft  $I = 28,400$ lbm-ft$^2$.

**9.32.** $I_{xx} = 8,093$ lbm-in.$^2$  $I_{yy} = 1,603$ lbm-in.$^2$
$I_{xy} = 1,282$ lbm-in.$^2$

**9.34.** $I_{cc} = 37.81$ kg-m$^2$.

**9.36.** 3,026 lbm-ft$^2$.

**9.38.** .524 kg-m$^2$.

**9.40.** $-.00696$ kg-m$^2$.

**9.44.** 362 kg-mm$^2$.

**9.46.** $(I_{xx})_M = 534$ kg-mm$^2$
$(I_{yy})_M = 16.59 \times 10^2$ kg-mm$^2$
$(I_{zz})_M = 21.9 \times 10^2$ kg-mm$^2$
$(I_{xy})_M = 5.68 \times 10^2$ kg-mm$^2$
$(I_{xz})_M = (I_{yz})_M = 0$.

**9.48.** .0493 kg-m$^2$  .00187 kg-m$^2$  .0503 kg-m$^2$.

**9.50.** 1,075 kg-m$^2$  92.35 kg-m$^2$.

**9.52.** $I_{yy} = 7.58\rho_0$ kg-m$^2$  $I_{yz} = -1.789\rho_0$ kg-m$^2$.

**10.2.** $S = 1,250$ N
$S = 750$ N.

**10.4.** $\theta = 19.48°$.

**10.6.** $W = .351\,T$.

**10.8.** $\theta = 8.53°$.

**10.10.** $P = (W/2) \cot \theta$.

**10.12.** 2,513 lb.

**10.14.** $T_{CE} = 7.42$ kN
$T_{DE} = 28.8$ kN.

**10.16.** $\tan \beta = \frac{1}{\sqrt{3}}[(a - 3b)/(b + a)]$.

**10.18.** $T' = 40$ N-m.

**10.20.** $P = .0545$ N.

**10.22.** $W = 270$ lb.

**10.24.** $W = 2,770$ N.

**10.26.** $C = 108.6$ N.

**10.28.** $d = 72.0$ mm.

**10.32.** $d = .2$ ft.

**10.34.** $25.9°$.

**10.36.** $77.3°$.

**10.38.** 1,066 N.

**10.40.** $\theta = 19.22°$.

**10.42.** $34.5°$.

**10.44.** .440 m.

**10.46.** $a = .358$ m.

**10.48.** $a = .1126$ m.

**10.52.** When $d > 2$, stable equilibrium.
When $d < 2$, unstable equilibrium.

**10.56.** $W_{\max} = 250$ lb.

**10.60.** $x = 4$ m.

**10.61.** $Q = 3P$.

**10.62.** $P = 7,540$ lb.

**10.63.** $\theta = 27.7°$.

**10.64.** $\cos \theta = \dfrac{\cos \theta_0 (aK_1 + aK_2) + W}{aK_1 + aK_2}$.

**10.66.** $\theta = 0$ (unstable)
$\theta = 28.1°$ (stable).

**10.67.** $W_{\max} = 1,000$ N.

**10.68.** $R > (h/2)$ for stable equilibrium.
$R < (h/2)$ for unstable equilibrium.

# Index

**A**

Active forces, 414
Addition of vectors, 24–27
Angle of repose, 292
Angle of wrap, 304
Archimedes, 3
Associative law, 24
Axial force, 247
  sign convention, 248, 249

**B**

Beam:
  definition of, 247
  differential equations of equilibrium,
    259–60
  section forces, 247–49
Belt friction, 301–04
  centrifugal effects, 316
Bending moment:
  definition of, 247
  diagrams, 261–63
  sign conventions, 249
Body forces, 117
Bound vectors 18–19
Bowden, F. P., 281
Broyt X-20 digger, 206
Bucyrus-Erie Dynahoe digger, 218

**C**

Cables, 266
  coplanar, 266–67, 270–71
Calculus of variations, 444
Catenary curve, 273
Center of gravity, 118, 343
Center of mass, 343
Center of pressure, 129–30
Center of volume, 342
Centroid:
  coordinates, 332

  for volume, 342–43
  of plane area, 332
Chains, 266
Coefficient of friction, 282
  table of values, 283
Coefficient of rolling resistance, 320
  table of values, 320
Coefficient of viscosity, 22
Commutative law, 16
  for dot product, 42
Compass settings, 28
Composite areas, 334
  for second moments and products of area,
    360–62
Composite volumes, 127, 344
Cone clutch, 326,
Conservative force fields, 433
Constraining force, 414
Continuum, 12
Control volume, 158
Coulomb friction:
  definition of, 281
  laws of, 282–83
Couple moment, 77–79
  as a free vector, 79
Couples, 77–79
  addition of, 80–82
  moment about a line, 82
  moment of, 77–79
  Cross product, 47–49

**D**

Deformable solids:
  displacement field, 425
  virtual work, 425
Degrees of freedom, 418–19
Dimensional homogeneity, 8–9
Dimensions:
  basic, 4–5

Dimensions (cont.)
  force, 9
  law of dimensional homogeneity, 8–9
  length, 4
  mass, 5
  primary, 4
  secondary, 4, 7
  time, 4–5
Directed line segment, 14
Direction cosines, 32
Displacement field, 425
Displacement vector, 15
Distributed force systems, 117–118
Distributive law:
  for cross product, 47–48
  for dot product, 42
Dot product, 41–43
Dym, C. L., 413
Dynamic coefficient of friction, 282–83

E
Einstein, Albert, 3
Ellipsoid of inertia, 407–09
Energy methods, 413, 426
Equality of vectors, 17, 18
Equations of equilibrium:
  concurrent forces, 165–66
  coplanar forces, 170
  general case, 187
  parallel forces, 183
  summary of special cases, 183
Equilibrium:
  for concurrent forces, 165–66
  for conservative forces, 432–33
  for coplanar forces, 170
  for parallel forces, 183
  general equations, 162–164
  method of total potential energy,
      434–35
  necessary and sufficient conditions,
      162
  neutral, 441
  stable, 441
  unstable, 441
Extremum points, 435

F
Finite elements, 444
Fink truss, 235
First moment:
  for composite areas, 334
  with axis of symmetry, 334
First moment of area, 331–32
First moment of mass, 343
First moment of volume, 342–43
First-order tensors, 402
Flyball governor, 208
Force polygon, 168
Forces:
  active, 414
  body, 117

conservative, 432–33
constraining, 414
distributed, 117–18
parallel, coplanar distribution, 133
parallel distribution, 129–30
surface, 117
translation of, 94–95
Frame, 155
Free body diagram, 152–54
Free vectors, 18
Friction:
  Coulomb, 281
    belt friction, 301–03
    coefficients of friction, 282
    complex contact problems, 299
    impending slippage, 282
    laws of, 282–83
    simple contact problems, 284–85
  definition, 281
  viscous, 281
Functional, 243

G
Generalized coordinates, 418
Gradient operator, 483
Gravitational attraction, 21
Gravitational constant, 21
Gravity:
  law of attraction, 21
Gusset plate, 221, 224

I
Idealizations of mechanics, 12–13
Impending slippage, 282
Inertia ellipsoid, 407–09
Inertial reference, 19
Integration formulas, XVII–XXII
Intensity of loading, 118

J
Just-rigid truss, 224

L
Law of cosines, 25
Laws of gravitational attraction, 20–21
Law of sines, 25
Length, 4
Line of action, 17
Lubrication, 281

M
MacLaurin series, 441
Mass, units of, 10
Mass moments of inertia, 379–81
  invarient of, 381
  principal axes, 389, 409–10
  relation to areas, 387
  rotation of axes, 395–98
  transfer theorem, 392–93
Mass product of inertia, 380–98
  involving plane of symmetry, 382–83

relation to areas, 386–87
rotation of axes, 395–98
transfer theorem, 393
Mechanics, laws of, 19–21
Meter, 4
Method of joints, 225–26
Method of sections, 238–39
Method of total potential energy, 434–35
Mohr's circle, 372
Moment of a force:
  about an axis, 69–73
  about a point, 62–64
Moment of area, 331–32
  for mass (see Mass moment of inertia)
Multiple integration, 120, 126

**N**
Neutral equilibrium, 441
Newton, Isaac, 3
Newton's first law, 19–20
Newton's law, 9
Newton's second law, 19–20
Newton's third law, 19
Newton's viscosity law, 22
Normal stress, 401

**O**
Optimization theory, 444
Over-rigid truss, 224

**P**
Pappus-Guldinus theorems, 347–50
Parallelogram law, 15, 21
Particle, 13, 14
Pascal unit, 131
Plane areas, radii of gyration, 355
Plane strain, 401
Plane stress, 401, 410
Plane surfaces:
  centroid, 332
  first moment, 331–34
  second moment, 355–56
Point force, 13
Polar moment of area, 369
Position vector, 61
Potential energy, 388
  linear spring, 390
Potential energy function, 432
Pratt truss, 235
Principal axes:
  for areas, 370–71
  for mass, 389, 407–9
Principal moments of inertia, for mass, 389,
  407–9
Principle of virtual work (see Virtual work)
Product of area, 335
  composite areas, 360–62
  computations of, 357–62
  rotation of axes, 366–67
Properties of plane areas, principal axes,
  370–72

**R**
Radius of gyration, 355
  for mass, 382
Random walk, 60
Reference:
  inertial, 19
  right-hand, 48
Relativistic effects, 20
Resolution of vectors, 30–32
Resultant:
  definition of, 102–03
  general, 103
  simplest:
    coplanar system, 106–07
    general case, 112
    parallel system, 110
Rigid body, 13
Rolling resistance, 319–20
Rotation, finite, 16

**S**
Scalar components, 30–33
  orthogonal, 31–32
  rectangular, 31–32
Scalar fields, 117
Scalar product, 41–43
Scalars, 15
  as zero-order tensors, 402
Scalar triple product, 51–52
Schrödinger's equation, 20
Second moment of area, 355
  composite areas, 360–62
  computations of, 357–60
  principal axes, 370–72
  rotation of axes, 366–67
Second-order tensor, 400–02
Shames, I. H., 210, 247, 372, 408, 410, 413,
  444
Shear force:
  definition of, 247
  diagrams, 261-63
  sign convention, 249
Shear stress, 401
Simple trusses, 224
Specific weight, 119, 343
Speed, 14
Spring constant, 434
Square screw thread, 317
  lead, 317
  pitch, 317
  self locking, 318
Stress, 401
Stability, 441-43
Stable equilibrium, 441
  requirements for, 441-42
Static coefficient of friction, 283
Static determinancy:
  definition of, 204–205
  external, 205
  internal, 205
Static indetermancy, definition of, 204–205

# PROPERTIES OF VARIOUS AREAS

**Rectangle**

$$A = bh$$

$$x_c = \frac{b}{2}$$

$$y_c = \frac{h}{2}$$

$$(I_{xx})_c = \frac{1}{12}\,bh^3$$

$$(I_{yy})_c = \frac{1}{12}\,hb^3$$

$$(I_{xy})_c = 0$$

$$J_c = \frac{1}{12}\,bh\,(b^2 + h^2)$$

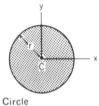
**Circle**

$$A = \pi r^2$$

$$x_c = 0$$

$$y_c = 0$$

$$(I_{xx})_c = \frac{1}{4}\,\pi r^4$$

$$(I_{yy})_c = \frac{1}{4}\,\pi r^4$$

$$J_c = \frac{1}{2}\,\pi r^4$$

**Semi circle**

$$A = \frac{\pi r^2}{2}$$

$$x_c = 0$$

$$y_c = .424r$$

$$(I_{xx})_c = .00686\,d^4$$

$$(I_{yy})_c = \frac{1}{8}\,\pi r^4$$

**Quarter circle**

$$A = \frac{\pi r^2}{4}$$

$$x_c = \frac{4r}{3\pi}$$

$$y_c = \frac{4r}{3\pi}$$

$$(I_{xx})_c = .0549\,r^4$$

$$(I_{yy})_c = .0549\,r^4$$

**Ellipse**

$$A = \pi ab$$

$$(I_{xx})_c = \frac{\pi ab^3}{4}$$

$$(I_{yy})_c = \frac{\pi ba^3}{4}$$